# Welcome to Greater Edendale

MCGILL-QUEEN'S STUDIES IN URBAN GOVERNANCE
Series editors: Kristin Good and Martin Horak

In recent years there has been an explosion of interest in local politics and the governance of cities – both in Canada and around the world. Globally, the city has become a consequential site where instances of social conflict and of cooperation play out. Urban centres are increasingly understood as vital engines of innovation and prosperity and a growing body of interdisciplinary research on urban issues suggests that high-performing cities have become crucial to the success of nations, even in the global era. Yet at the same time, local and regional governments continue to struggle for political recognition and for the policy resources needed to manage cities, to effectively govern, and to achieve sustainable growth.

The purpose of the McGill-Queen's Studies in Urban Governance series is to highlight the growing importance of municipal issues, local governance, and the need for policy reform in urban spaces. The series aims to answer the question "why do cities matter?" while exploring relationships between levels of government and examining the changing dynamics of metropolitan and community development. By taking a four-pronged approach to the study of urban governance, the series encourages debate and discussion of: (1) actors, institutions, and how cities are governed; (2) policy issues and policy reform; (3) the city as case study; and (4) urban politics and policy through a comparative framework.

With a strong focus on governance, policy, and the role of the city, this series welcomes manuscripts from a broad range of disciplines and viewpoints.

1 Local Self-Government and the Right to the City
*Warren Magnusson*

2 City-Regions in Prospect?
Exploring Points between Place and Practice
*Edited by Kevin Edson Jones, Alex Lord, and Rob Shields*

3 On Their Own
Women, Urbanization, and the Right to the City in South Africa
*Allison Goebel*

4 The Boundary Bargain
Growth, Development, and the Future of City–County Separation
*Zachary Spicer*

5 Welcome to Greater Edendale
Histories of Environment, Health, and Gender in an African City
*Marc Epprecht*

# Welcome to Greater Edendale

## Histories of Environment, Health, and Gender in an African City

Marc Epprecht

McGill-Queen's University Press
Montreal & Kingston • London • Chicago

ISBN 978-0-7735-4773-5 (cloth)
ISBN 978-0-7735-4774-2 (paper)
ISBN 978-0-7735-9965-9 (ePDF)
ISBN 978-0-7735-9966-6 (ePUB)

Legal deposit third quarter 2016
Bibliothèque nationale du Québec

Printed in the United Kingdom

This book has been published with the help of a grant from the Canadian Federation for the Humanities and Social Sciences, through the Awards to Scholarly Publications Program, using funds provided by the Social Sciences and Humanities Research Council of Canada.

McGill-Queen's University Press acknowledges the support of the Canada Council for the Arts for our publishing program. We also acknowledge the financial support of the Government of Canada through the Canada Book Fund for our publishing activities.

Library and Archives Canada Cataloguing in Publication

Epprecht, Marc, 1957–, author
Welcome to greater Edendale : histories of environment, health, and gender in an African city / Marc Epprecht.

(McGill-Queen's studies in urban governance ; 5)
Includes bibliographical references and index. Issued in print and electronic formats.
ISBN 978-0-7735-4773-5 (cloth).–ISBN 978-0-7735-4774-2 (paper).—
ISBN 978-0-7735-9965-9 (ePDF).–ISBN 978-0-7735-9966-6 (ePUB)

1. Edendale (KwaZulu-Natal, South Africa)–Social conditions.
2. Edendale (KwaZulu-Natal, South Africa)–Environmental conditions.
3. Edendale (KwaZulu-Natal, South Africa)–Economic conditions.
4. Edendale (KwaZulu-Natal, South Africa)–Politics and government.
5. Edendale (KwaZulu-Natal, South Africa)–History.
I. Title. II. Series: McGill-Queen's studies in urban governance ; 5

DT2199.E66 2016 968.4'7 C2016-903032-6 C2016-903033-4

# Contents

# Illustrations

*Uncredited items are courtesy of Marc Epprecht.*

# Abbreviations

| | |
|---|---|
| AFRA | Association for Rural Advancement |
| ANC | African National Congress |
| AP | Alan Paton Centre and Struggle Archives |
| AZAPO | Azanian People's Organization |
| AZASO | Azanian Students Organization |
| BAD | Department of Bantu Administration and Development |
| BEE | Black Economic Empowerment |
| BESG | Built Environment Support Group |
| CNC | Chief Native Commissioner |
| DA | Democratic Alliance |
| DUCT | Duzi-uMgeni Conservation Trust |
| EDPHA | Edendale and District Public Health Area |
| ELHA | Edendale Lot-Holders' Association |
| GAA | Group Areas Act |
| GEAR | Growth, Employment and Redistribution policy |
| GEDI | Greater Edendale Development Initiative |
| GEVDI | Greater Edendale, Vulindlela Development Initiative |
| GREEN | Greater Edendale Environmental Network |
| ICU | Industrial and Commercial Workers Union |
| IDP | Integrated Development Plan |
| IFP | Inkatha Freedom Party |
| IMF | International Monetary Fund |
| INR | Institute of Natural Resources |
| IPCC | Intergovernmental Panel on Climate Change |
| KTA | KwaZulu Territorial Assembly |
| KZN | KwaZulu-Natal |

| | |
|---|---|
| LHC | Local Health Commission |
| LPSA | Liberal Party of South Africa |
| NAD | Native Affairs Department |
| NAR | National Archives Repository |
| NC | Native Commissioner |
| NUAA | Natives (Urban Areas) Act of 1923 |
| *NW* | *Natal Witness* |
| PACSA | Pietermaritzburg Agency for Christian Social Awareness |
| PAR | Pietermartizburg Archives Repository |
| PHA | Public Health Area |
| RDP | Reconstruction and Development Programme |
| SANNC | South African Native National Congress |
| SC | Sinomlando Centre for Oral History and Memory Work in Africa |
| UDF | United Democratic Front |
| UKZN | University of KwaZulu-Natal |

# Acknowledgments

This book emerged from an unanticipated turn of events and many serendipitous connections. It goes back to 2003, when my partner, Allison Goebel, invited me to participate in brainstorming a new project with colleagues of hers at the University of KwaZulu-Natal (UKZN), Trevor Hill and Robert Fincham. The role offered to me as the project cohered was to provide the other investigators with the historical context related to health and the environment in the peri-urban areas of Pietermaritzburg. Though it was quite far from my most recent gender and sexuality research in Zimbabwe, it took me back to earlier work I had done on related topics in colonial Lesotho. I was able, so I was later informed, to provide some helpful input to the project.

Years went by, and short consultative meetings grew into a substantive research initiative. By 2009 I realized I had accumulated enough data and made enough fruitful contacts in Msunduzi to warrant devoting myself to a project of my own. Towards that goal, I spent the first half of 2010 on sabbatical in the city for intensive archival work. This research was further enriched by data collected by graduate students at Queen's (Ann McNeely, Colleen Sutton, Leah Savage, Ralph Callebert, Andrea Kolbe, and Mary Caesar), the University of Western Ontario (Janet Roukema and Krista McMullen), and the UKZN (Adrian Pole, Pascal Karemera, Renusha Chanda, Cyprian Alokwu, and Sithabile Hlahla); oral interviews conducted by the Sinomlando Centre for Oral History and Memory Work in Africa for the Greater Edendale Development Initiative; and aerial photographs retrieved by Doreen Grossman. The opportunity to begin writing, with the goal of producing a conceptually coherent package, was facilitated by my stay at the University of Basel in the (long) winter/spring of 2013. I thank Veit Arlt for his enormous assistance in arranging that stay. Amanda-Bea Rehman assisted with archival photos reproduced here, while Adriane Epprecht very ably assisted with data transcription as well as bibliographic and index work. Kudos are owed to Brice Gijsbertsen for his excellent production of the maps.

My biggest debts of gratitude I owe to Allison Goebel for involving me in the initial ecohealth project, always encouraging me to contribute more, and providing the many intangibles (like love and fun) that make sustained research possible. Trevor Hill introduced me to spaces and places not just in the Msunduzi municipality but in the countryside that surrounds it, while Rob Fincham introduced me to many very helpful people at the level of municipal government. Over the years, many great colleagues either shared insights from their personal engagements with the city, very constructively challenged me on my understanding of South African historiography in various workshop or conference settings, and/or encouraged me to continue my work at those times when I was having existential doubts about it, and I would like to thank Peter Alegi, Brian Bassett, Cameron Brisbane, Philippe Denis, Mark Dent, Mbongeleni Dlamini, Belinda Dodson, Julie Dyer, Myron Echenberg, Bill Freund, Mike Greatwood, Robert Haswell, Simonne Horwitz, Mark Hunter, Jeff Guy, R. Simangaliso Kumalo, Mary Lawhon, Peter Limb, Shula Marks, Nokhaya Makiwane, Nhlaka Mdunge, Chris Merrett, Robert Morrell, Garth Myers, Pieter Nel, Noor Nieftagodien, Vanessa Noble, Radikobo Ntsimane, Julie Parle, Howard Phillips, Hilton Ryder, Andrew Simpson, Prem Singh, Jabulani Sithole, Nicholas Southey, and Diane Wylie. Thank you as well to Beverly Ellis, both for her path-breaking scholarship on the area's environmental history, which got me pointed in the right direction, and for mentoring Adriane at Maritzburg Girls' High School. Staff and colleagues at the Centre for Environment, Agriculture, and Development, and in the Department of Geography at the UKZN, were unfailingly welcoming and helpful, as were those at the national and provincial archives as well as the Alan Paton Centre and Struggle Archives. I feel an especially warm sense of gratitude for the hospitality and intellectual inputs shared by Sheila Meintjes and Patrick Harries. I am moved as well by the confidence shown in me – and the professionalism demonstrated – by Jacqueline Mason, Barbara Tessman, and the rest of the McGill-Queen's University Press team.

The research was made possible by the generous funding of the Social Sciences and Humanities Research Council, the International Development Research Centre of Canada, and my employer over this long haul, Queen's University. Finally, the book's publication was made possible by the Canadian Federation for Humanities and Social Science.

# A Note on Terminology

The terms used to denote different groups of people by race in South Africa are laden with controversy and continue to upset or confound people two decades after the transition to non-racial democracy. Apartheid-era designations are still in widespread use, including by the central government for census taking and analysis. In this book I will be conforming to common parlance without in any way accepting an essential meaning to the terms. I use the lower case (black, white, coloured), except when the reference is to a geographical origin (European, Indian). *African* these days is usually taken to be synonymous with black in the sense promoted by pioneering African intellectual Magema Fuze in the 1920s – that is, a dark-skinned person with ancestors indigenous to the region sometime in the last two millennia (hence not including those people who for political reasons identify as black when in the past they would have been more commonly referred to as non-whites or other formal racial designations). Other colonial or early apartheid terms such as *Native*, *Bantu*, *tribal*, *Asiatic*, and so on are preserved in selections from documents of the time, with the hope that no offence is conveyed.

A number of other terms (notably, *middle class*, *conservative*, *progressive*, *liberal*) are deeply problematic when applied in a literal sense, and so probably warrant quotation marks when used. I have eschewed that clumsy and distracting practice, trusting that readers understand the commonly shared metaphorical meanings of these terms.

Following the style used by Carton, Laband, and Sithole (2009), among others, I apply prefixes to indicate the predominant African languages and cultures in the district (isiZulu and seSotho) and discrete groups of people among isiZulu speakers (*amakhosi*, *amakholwa*, *amarespectables*, and so forth).

Msunduze is spelled different ways in contemporary sources. I have opted for the commonest way to distinguish the river from the municipality.

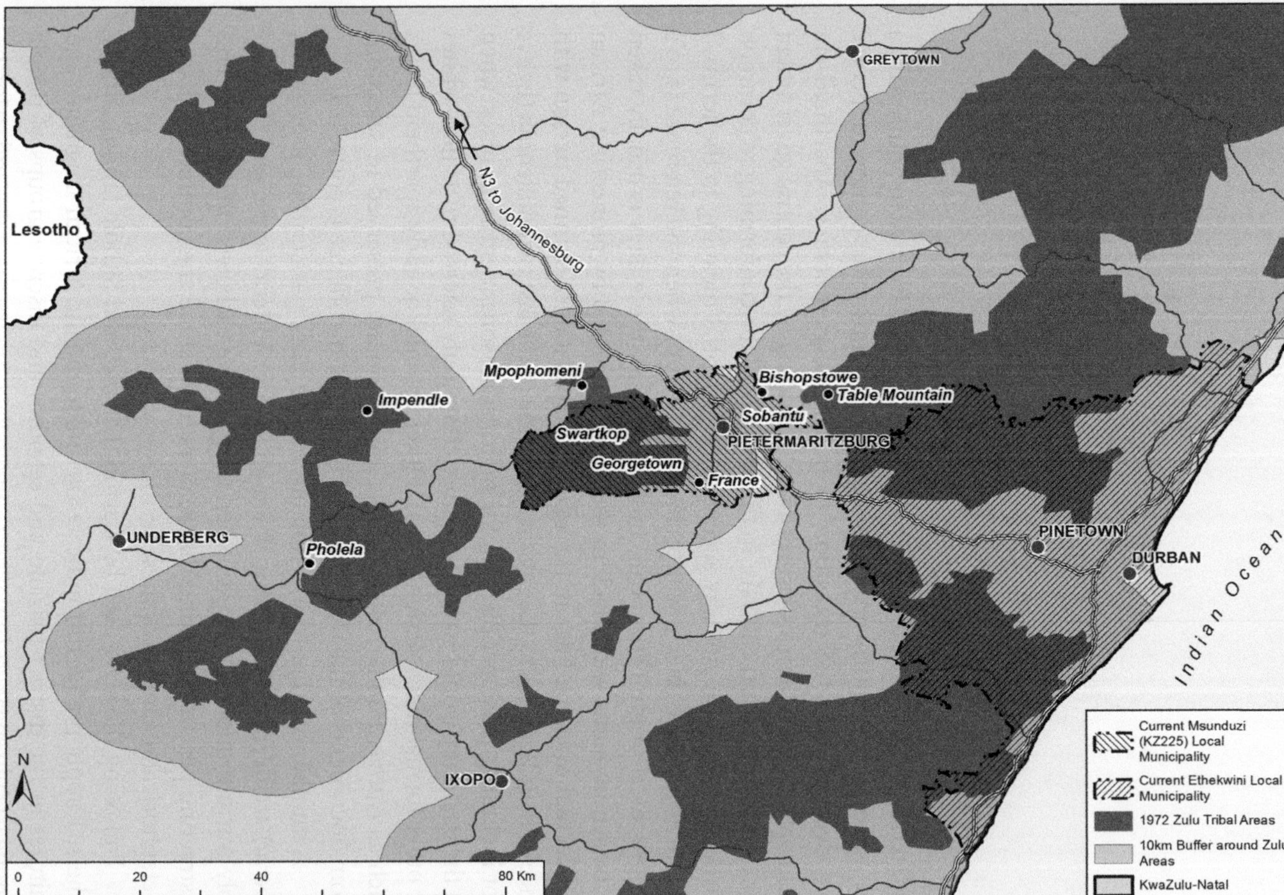


Map 0.1 Msunduzi, KwaZulu, and the “border zones”

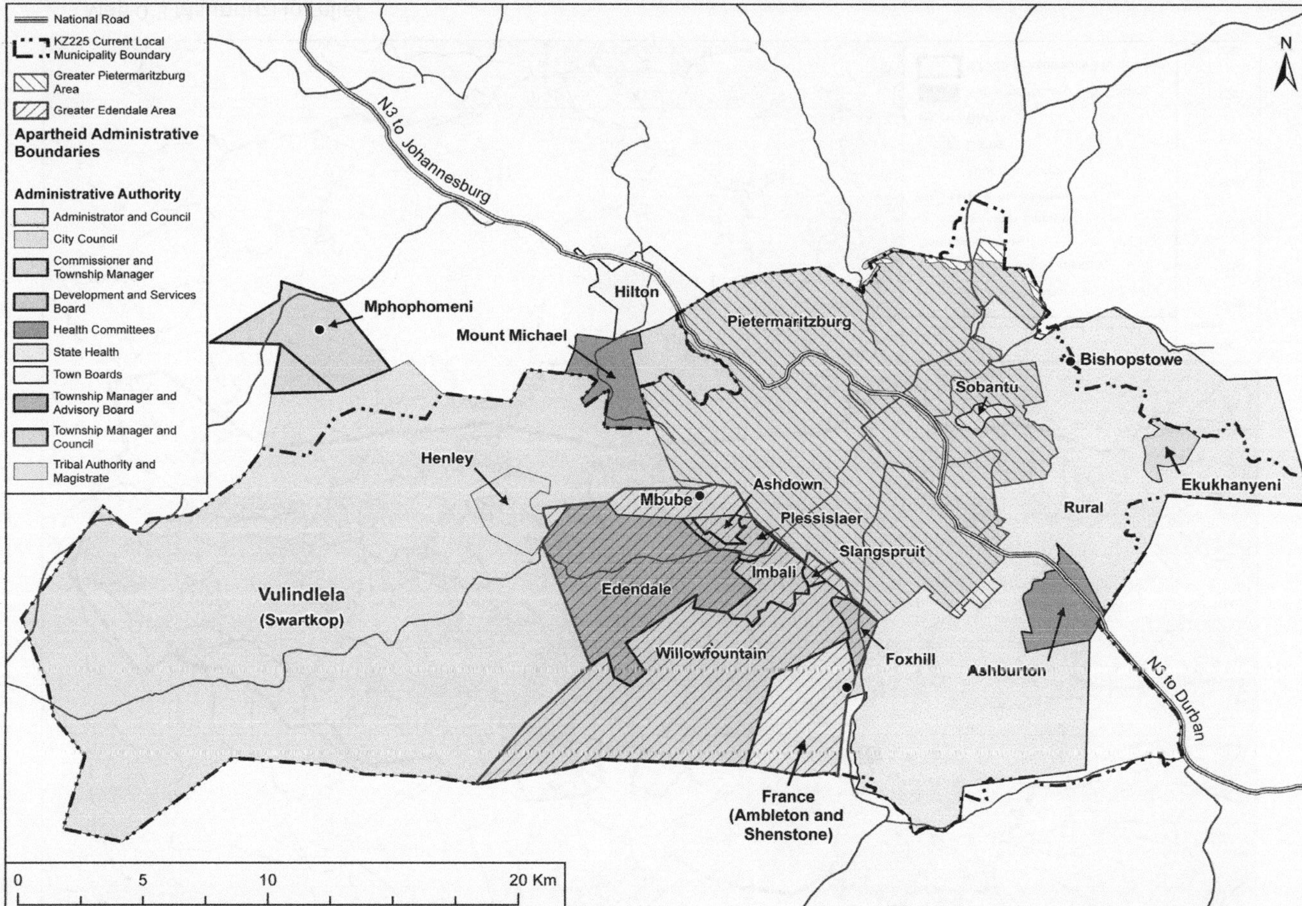


Map 0.2 Msunduzi with apartheid-era jurisdictions

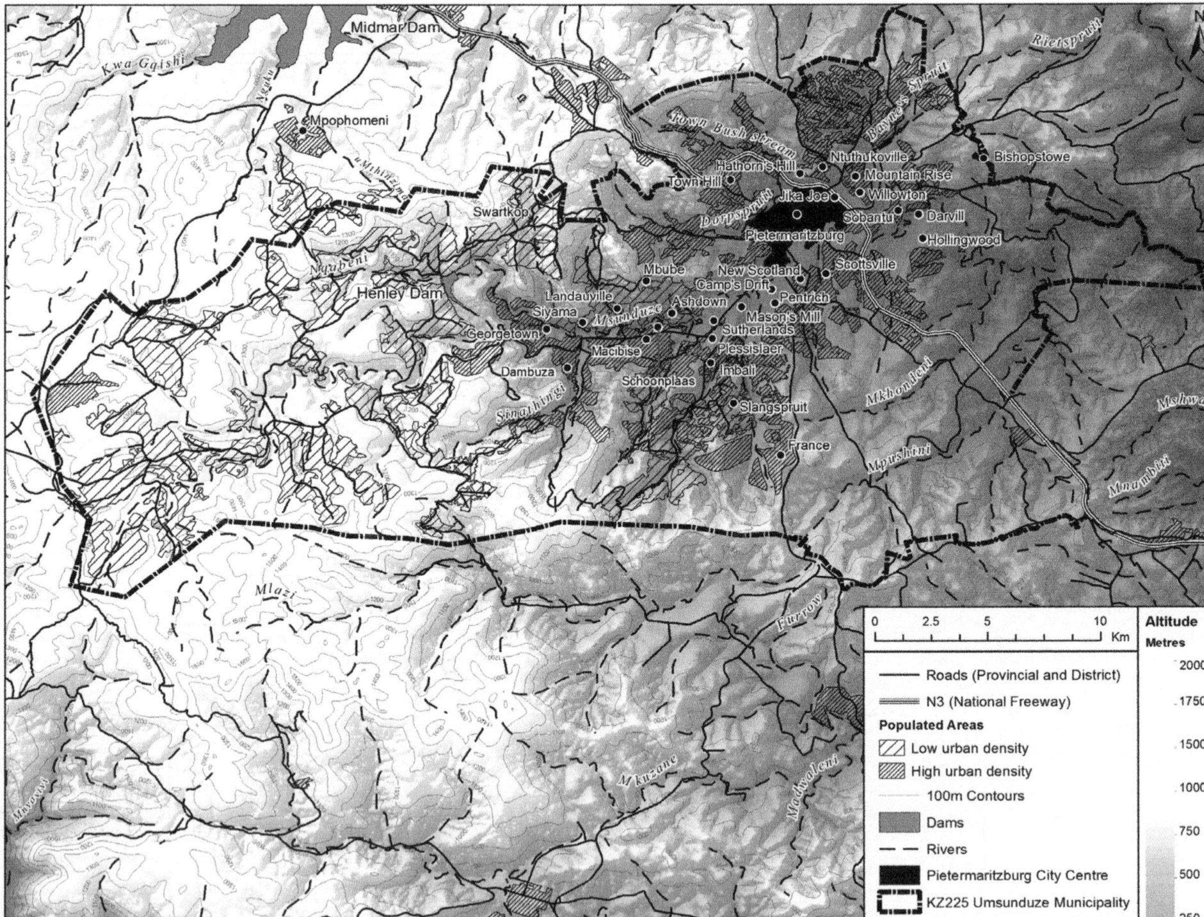


Map 0.3 Msunduzi in relief

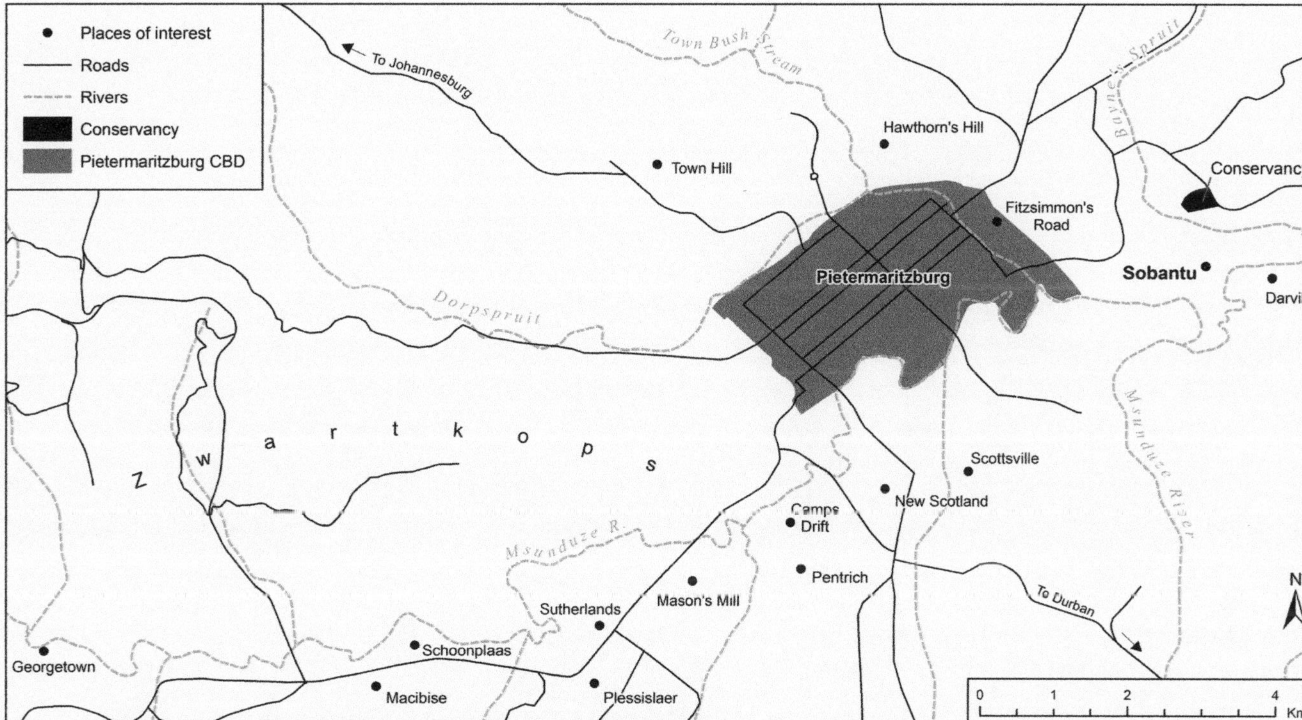


Map 0.4 Communities of the middle Msunduze River Valley

# Welcome to Greater Edendale

# ONE | Setting the Scene

To approach a lecture as complex as this one requires some deftness and guile, as if one is walking over eggs without an intention of breaking them, as if one is trying to make an omelette without breaking those eggs. At the same time, it offers the opportunity to be daring in one's exposition so as to extricate from the crevices of historical oblivion, and bring light, on the life of a person [or a place?] about whom not much was written, but from whom many a comrade, a clergy, a parent and a child, found a reference point to lead responsible, fulfilling and worthwhile lives.

*The Hon. Jeff Radebe*, "Rev. Enos Sikhakhane and the Art of Positioning Edendale on the Political and Religious Map of the World," 2009

Edendale hospital is the most disgusting place out. We've had major things, major problems there with them ... That's not industrial effluent, but that's medical waste being dumped into the sewers. And when I say medical waste it's everything. You feed a sheet down that's full of blood, and instead of sending it to the washer to be washed or incinerated, you feed it down in the sluice room until it disappears, and the nurse says now she doesn't have to take it anywhere else, and wonders why it blocks the sewer. And then when it blocks the sewer everything pours out the sewer into the river. And we've had things, horrific things, you know when I'm ... you're standing in mounds of disposable gloves just all over the place, needles, syringes.

*Mike Greatwood*, Chief Water and Sewage Engineer,
Msunduzi Municipality, 2001

That we should like to be informed under what law we are really to stand, and that we so much like to be treated as people than as dogs.

*John Kumalo*, "Meeting of Civilised Natives at the Edendale Mission Station," 1863

A sprawl of communities runs for about seven kilometres upwards along the Msunduze River valley from the southwest side of Pietermaritzburg, South Africa. Welcome to "Greater Edendale," "the Edendale complex," or, in common but technically incorrect usage, just plain Edendale. It constitutes a bit less than a quarter of the area and has a third to half of the total population of the municipality of Msunduzi, the jurisdiction created in 2000 to unify formerly segregated places under one democratic local authority. Now more than twenty years after the end of apartheid, however, de facto racial segregation remains. Virtually all of the variously estimated 200,000 to 400,000 people who currently live in Greater Edendale are black or African, in the conventional sense of those words. Even the old buffer zone between the former white group area and the nearest black township remains an undeveloped, littered strip of grass.[1]

To draw a somewhat misleading analogy that might resonate with readers, Edendale is to Pietermaritzburg as Soweto is to Johannesburg, except, as we shall see, with more hills, with a poorer and sicker population, and with a generally less developed infrastructure. It is the distressed side of Msunduzi, and at first sight it seems to conform to a common pattern found throughout the region (an impoverished black township on the margin of a historically white city and comfortable northern suburbs). In fact, as we shall see, Greater Edendale has a much longer history and more complex make-up than most predominantly black urban areas in southern Africa. Greater Edendale today includes once-prosperous freehold villages and farmhouses, colonial-era buildings, apartheid-era townships, vast tracts of post-1994 "RDP houses," a sprinkling of middle-class neighbourhoods and mansions of the nouveau riche, and dense pockets of informal housing that elide into the former "tribal" areas of Vulindlela. Some forests, pastures, and wetlands remain.

This book offers a history of that complexity, with a particular focus on how the strained relationship between Edendale, Pietermaritzburg, and various levels of government left a legacy of inequality and environmental damage that is proving extremely difficult to address. It focuses primarily on contests over how to manage evidently growing environmental and public health problems in Edendale and neighbouring peri-urban areas. I argue that repeated failures to resolve these problems had long-term negative consequences for the people of the whole of Msunduzi but most acutely for those living in Greater Edendale. These are manifest not only in severe pollution and seemingly intractable poverty, but also in gender inequalities and gender-based violence that directly fuel the HIV/AIDS pandemic. Can connections be made between the history of local environmental and health mismanagement and the fact that Greater Edendale today is among the worst-affected cities *in the world* for HIV? I am going to suggest that the answer is yes and that the inability of either the present democratically constituted local authority or economic growth to redress environmental, health, economic, and other disparities and injustices is rooted in this history. I will also show that debates in the past were often much more sophisticated than tend to be acknowledged in the simplistic, patriotic, or progress/growth narratives that predominate in current public discourse. For example, the past – even the apartheid past – includes some moderately successful achievements that were actively promoted by leading African nationalists but that tend now to be forgotten or disparaged in the historiography of racialized urbanization in South Africa. I will highlight a little-remembered experiment known as the Local Health Commission (LHC or *KwaPoyinandi*), a unique form of local authority that governed Edendale from 1942 to 1974. I will also draw out evidence of relationships between gender struggle and environmental management that are only rarely alluded to in conventional political and even social histories.

Not many South Africans know about Edendale, and, indeed, it is probably safe to say that not even many Maritzburgers have ever been there. My hope is that a greater appreciation of the complexity of debates around environment, health, gender, race, and governance with respect to this neglected community might contribute fruitfully to re-imagining ways to make Msunduzi a healthy place to live in the broadest and most sustainable sense of good health. This history may also hold potential lessons for other distressed communities in the region as they too struggle with an exhausted model of economic development.

≈

Both Edendale and Pietermaritzburg are historically significant by any measure. Pietermaritzburg (also known colloquially as Maritzburg or PMB) was the first town laid out by the Afrikaner trekkers in their short-lived republic of Natalia (1838–43). It served as their capital until the British asserted control over the district and the majority of Afrikaners upped camp for freer pastures. Thereafter it became the capital of British-ruled Natal, the main garrison from which British troops extended their conquests over the area, and the hub of a relatively diversified and prosperous "Garden Colony" economy. A long period of economic stagnation set in in the early twentieth century, however and, in the shadow of the burgeoning port of Durban eighty kilometres away, Maritzburg acquired the reputation of a "Sleepy Hollow." That era of stagnation allowed one of the continent's richest legacies of colonial architecture to survive. Notwithstanding numerous late-apartheid architectural offences, much of old Maritzburg remains distinguished by a distinctive red brick and gables style commonly described as "charming." Among hundreds of historically noteworthy buildings, monuments, and gardens are those created by one of the oldest and largest Indian communities in the country, including mosques, bungalows, and barracks-style housing for indentured labourers. Today Pietermaritzburg is the seat of government of South Africa's second-most populous province, KwaZulu-Natal.

Edendale, for its part, is anchored on one of the first and largest freehold black settlements in South Africa. It was founded as a Wesleyan mission station in 1851 on 6,172 acres of land (roughly 2,500 hectares), with a neatly planned village (Georgetown) about ten kilometres as the crow flies from central Maritzburg. Old Edendale is thus much older, more self-consciously urban, and bigger in area than more storied freehold counterparts elsewhere in the country.[2] It was also, as whites and Indians increasingly acquired property there in the early twentieth century, more racially and culturally diverse. By the mid-1950s it had a functioning multiracial local authority advisory board (with a black chairman), was the home of an internationally renowned hospital and several enduring industrial establishments, and produced a thriving cultural/music scene. Three names in particular stand out from the latter: R.R.R. (Reggie) Dhlomo, who penned South Africa's first published novel by a black author (Dhlomo 1928); Reuben Caluza, South Africa's first doctor of music, "national treasure," and pioneering African businessman (Couzens 1985; Verwey and Sonderling 1995); and Gerard Bhengu, South Africa's first black visual artist with an international reputation (Savory 1965).

In the sphere of politics, Edendale and Pietermaritzburg tend to be recalled mainly for two celebrated sojourners – Mohandas Gandhi, famously turfed off the train in the city's main station in an act of crude racism that sparked the development of his political consciousness, and Nelson Mandela, whose stirring speech at the All-In Africa Conference in 1961 was his last public appearance prior to his capture by police and subsequent decades of imprisonment. Yet numerous other figures in South Africa's political history actually came from Msunduzi and went on play important roles, for better or worse, in shaping the country's liberation movements. Already in the 1860s, men such as John Kumalo were ruffling feathers with their outspoken criticism of colonial injustices and indignities. Several of the founders of the African National Congress (ANC) and the Industrial and Commercial Workers Union (ICU) hailed from Edendale. Indeed, it was a place where ideas and personalities connected – in school, through marriage, in intellectual life, in the professions, and across racial lines. Albert Luthuli, Africa's first Nobel Peace Prize winner, notably recalled his years at Edendale College from 1915 to 1917 as formative to his commitment to non-racialism and to his belief in the value of cross-cultural learning (Luthuli 1966, 28). Josiah T. Gumede was a lawyer and leading black property owner in Maritzburg who became president of the ANC in the late 1920s and was a key figure in promoting the party's transition to a social democratic, mass nationalist movement in the 1940s. From the 1930s to the 1950s, Reggie Dhlomo, together with his even more famous brother Herbert and their childhood friends Caluza and Bhengu, emerged as leading figures in the New African movement that sought to instill pride in a progressive African culture against the dominant racist and tribalist ideologies of the times, a predecessor in spirit to Black Consciousness. In 1976, the Federal Theological Seminary of Southern Africa (Fedsem) became a national centre for black consciousness and liberation theology, with a multiracial faculty and student body based in Greater Edendale that (mostly) defied the spirit of apartheid. Not to belabour this point too much, but since Alan Paton is perhaps Pietermaritzburg's most famous son internationally, let me also note the special mention he gives to Edendale in the formation of his own political consciousness: it was the first place he ever shook hands with blacks as social equals (Paton 1980, 63).[3]

Locally, the best-remembered local politician is undoubtedly H. Selby Msimang – part of the main highway linking Edendale and Maritzburg today bears his name. Msimang will feature prominently in this book, but by way of introducing him I simply want to show how intertwined people and the politics of African nationalism coming out of Edendale could be. Msimang came from one of the founding Edendale families and, together with his brother Richard and his second partner in marriage, Noluthandu, played an important role in the early trajectories of both the ANC and ICU. This included helping to engineer the ouster of Josiah Gumede from the ANC presidency in the late 1920s and of ICU leader (and later Gumede's son-in-law) A.W.G. Champion in the 1930s. He was a key ally of Albert Luthuli in the latter's rise to the national ANC leadership in the early 1950s. Luthuli and Msimang became politically estranged, however, as Luthuli led the ANC leftwards under the influence of his close friend Moses Mabhida (the eventual general secretary of the Communist Party and a leading proponent of armed struggle against apartheid). Mabhida had been born on a farm just south of Maritzburg, the city where he grew up and cut his teeth as a trade union activist. As Mabhida's influence in the ANC rose, Msimang left that party to become one of the founders of the Liberal Party of South Africa (along with Paton and Peter Brown, also of Pietermaritzburg). After that party was effectively banned in 1968, Msimang then became a leading figure in the rise of the modern Inkatha movement. Yet Msimang remained close to many of his former ANC comrades, including his cousin through marriage, Archibald Gumede (Josiah's son). Archie Gumede was one of the Treason Trialists in 1956, along with fellow Maritzburger Dr Mohammed Chota Motala of the Natal Indian Congress.

Family ties clearly played a part in nurturing Edendale's distinctive political culture, but so too did the fact that Africans could own land there as private property. This did not always serve them well. On the contrary, private property ownership in the absence of effective local government contributed to the conditions that tended to undermine the economic health and social cohesion of the community. As early as 1880, the colonial government noted with concern the "existence of certain evils in the Edendale station … Beer drinking parties are constantly taking place. The people concerned in this practice disregard all counsel, and defy all authority. It appears that the occupants of the houses reside upon their own land, and think they can act without any restraint … The orderly and respectable people are greatly distressed at this state of things, but powerless to alter it."[4] Conditions only got worse through subsequent decades of neglect and sometimes active discrimination against the community by various levels of government. Many landowners went bankrupt and either abandoned their property or sold out to whites and Indians. By the 1930s, a village once known for its neatness within verdant hills and valleys had become notorious for slum dwellings, a large transient population, and thriving illicit sex and home-brewed liquor industries.

State initiatives to clean up the area with housing, health, and infrastructural development from the 1940s were at the forefront of progressive thinking in the country and perhaps the world at that time. One of Greater Edendale's neighbourhoods is actually named after its first medical officer of health, David Landau, the man who "probably wrote much of the

[Gluckman] Report" that in 1944 recommended a National Health Service based on a social medicine or primary health care approach (Marks 2014, 176). Such initiatives were not sustained, however, and did not successfully address the community's fundamental problems. Tightening racial bureaucracy through the 1960s then exacerbated many injustices and inequalities, which came to fuel political violence. By the time of the transition to democracy in the late 1980s through the mid-1990s, Edendale had become a focal point of Natal's civil war. South Africa's worst so-called black-on-black fighting resulted in tens of thousands of refugees flooding into Edendale and up to 4,000 dead throughout the surrounding district. Given that no criminal charges were ever laid from this violence and that refugees are now claiming rights to the lands that they occupied in those chaotic years, it is small wonder that Jeff Radebe, minister for justice and constitutional development, advised "deftness and guile" in approaching the topic of Edendale's modern history (Radebe 2009).

In much of this history, and in the sharpness of its divergence from the developmental path of predominantly white Pietermaritzburg, Edendale has a great deal in common with the history of other predominantly black urban communities in South Africa, such as Alexandra, Evaton, and Cato Manor. However, the scale of inequality was perhaps more extreme than most places in South Africa. Where Pietermaritzburg used to promote itself as the healthiest city in South Africa, if not the world – "the Mecca of those who seek healthy surroundings" (Pietermaritzburg [1904–90] 1936, 75) – sanitary conditions in Edendale were, in the words of the chief Native commissioner in 1933, "the worst he had ever seen," surpassing Johannesburg's infamous slumyards (*Natal Witness*, 28 March 1933, 1).[5] Moreover, unusually for a city with a colonial layout, such poverty was in relatively close proximity to, and located upstream from, the historically affluent centre. Msunduzi's geography thus adds to the drama. The city stretches out in the shadow of a great escarpment that rises nearly six hundred metres from dry thornveld (grassland dotted with mimosa and acacia trees, which is to say, "typical African savannah") to misty rolling hills and deep kloofs (gorges). A rise of land opposite the bottom of the escarpment through which the Msunduze River flows out from the city creates a bowl-like effect that catches smoke and dust. Under certain wind conditions and a typical wintertime temperature inversion, this renders the air in the central business district among the most polluted in the country. One can often look down from the clear skies over Edendale's pretty

1.1 Temperature inversion, winter, looking down the valley towards the city

upper reaches to see the sharp line of choking air trapped by temperature inversion in the city below.

The transition to democracy beginning in 1990 brought high hopes for addressing the legacies of inequality and injustice that burden Edendale and environs, including those downstream and downwind. Yet more than two decades on, the people of Greater Edendale continue to face daunting economic, health, service delivery, and governance challenges. Unemployment is estimated to be over 50 per cent; the prevalence of HIV/AIDS among adults is likely as high as 25 per cent and is often accompanied by virulent new strains of drug-resistant tuberculosis; criminal violence, including gender-based violence, is pervasive; and much of the infrastructure – including significant swathes of post-1994 housing – is crumbling. Adding to the woes in the past few years has been a marked level of mismanagement and corruption at the municipal level of government. In 2010, the province intervened by dismissing the entire political executive of Msunduzi, while filing criminal charges in some cases, and imposing its own emergency administrator to

1.2 A Msimang house waiting to be transformed into a museum and teahouse as a key component of the GEDI proposal for history for development

sort out the mess. Reflective of the multi-layered crisis, Msunduzi has experienced an exodus of Indians and whites that is among the most extreme of any of the major cities in the country. In the context of South Africa's inherited inequalities, this exodus translates into a serious "brain drain" and an erosion of the rate-paying population that directly affects the municipality's ability to deliver services to the historically disadvantaged parts of the city. The opening epigragh by city engineer Mike Greatwood attests to how that breakdown sometimes manifests in extremely hazardous ways.[6]

A number of important studies have examined diverse aspects of the roots of Msunduzi's contemporary crisis. Yet with a few important exceptions to be discussed below, Msunduzi's two main components – Greater Edendale and Pietermaritzburg – have tended to be neglected in the wider historiography of South Africa. Bonner and Nieftagodien (2008, 3), for one of many examples, clearly forgot about Edendale in making their claim that Alexandra was "singular and distinctive" as a freehold urban community that survived apartheid. But who can blame them? Edendale even gets short shrift in the public history of Msunduzi itself. The municipality's premier historical museum devotes a single sentence on a single panel to the place (theme: architecture), with a glaring typographical error and a highly misleading photograph. Edendale does not even appear on several of the maps posted on Msunduzi's official website for tourism, a gaffe that apartheid-era ideologues might find wryly amusing.[7] Meanwhile, the signature piece in a city proposal to develop historical tourism and community pride – the original adobe house of the Msimang family, donated to the city by their heir – has acquired an expensive protective roof but little else in the eight years I have been visiting. Steadily crumbling, it appears at the time of writing to be occupied by squatters.[8]

This book aims in part to address that gap in the historiography and in public history. However, it is not a history of Msunduzi, Pietermaritzburg, or Edendale in the conventional narrative sense. Rather, it will examine a series of case studies that focus on specific themes and controversies in the city's past, such as where and how to house Africans, how to administer a multiracial community, how to balance industrial growth with environmental and health protections, and what to do about women's particular health issues. In doing so, it will consider several questions. What light does available evidence from these local struggles shed on wider historiographic debates about the relationships between environment, health, gender, race, governance, and other pressing concerns that emerge from the history of urbanization in South Africa, and indeed throughout the region more broadly? Can these local debates help us to engage young people in imagining ways to realize aspirations for health, justice, and dignity that are not and, I will argue, *cannot* be met by the currently prevalent models of development? To put that question more pointedly, can the history of local struggles in a middling-sized city help us think of new ways to resist the juggernaut of "really existing empire" and "global coloniality" that South Africa faces today, with all the devastating impacts that current trends in global capitalism and the paradigm of unending growth imply (Ndlovu-Gatsheni 2013)?

The very rich historiography of urban South Africa engages in these complex questions and debates to an impressive extent, and I am deeply indebted to pioneers in the pertinent fields for many of the questions I will be asking. However, the scholarship at present – including if we extend our gaze to geography, anthropology, and urban planning – is very heavily focused on the country's big three cities (Johannesburg, Durban, and Cape Town) and their various satellite communities.[9] Yet it is widely predicted that "secondary cities" rather than the current existing megacities will experience the most dramatic growth as Africa rapidly urbanizes (UN-Habitat 2014, 23). Moreover, relatively sheltered from the glare of the big national media, secondary cities offer tempting sites for political patronage and corruption. Indeed, not coincidentally, some of the worst violence accompanying so-called service delivery protests in recent years in South Africa have taken place in small cities like Kokstad and Sasolburg.[10] As Sue Parnell suggested some time ago (1995), local urban histories of smaller cities and towns could thus enrich and perhaps profoundly challenge our understandings of historical change that at present derive primarily from studies of the big, most politically vocal, and best-scrutinized cities. Such is the suggestion as well of Peter Limb in his very different type of study, a "revisionist political history" that shifts the national narrative of political struggle against colonialism/apartheid through insights from local, sociocultural histories (Limb 2010, xiii). Local studies, as a type of "history from below," might also contribute to the objectives Martin Murray and Garth Myers set as a priority for understanding the pressing problems, and vitality, of contemporary cities in Africa: "what is needed is a much more nuanced and rounded view that not only acknowledges the spatial unevenness of the urbanization process but also treats African urban residents as active agents in constructing meaningful lives for themselves rather than simply passive victims of inexorable structural processes beyond their control" (2006, 3; see also Soske 2012).

Of course there *are* structural processes far beyond the control of even South Africa's central government, let alone a struggling town council, a smallish chamber of commerce, and a seriously disjointed civil society such as those found in Msunduzi today. The so-called great recession that started with bank failures in the United States in 2008 had ripple effects around the world and hit South Africa especially hard. It revealed underlying, overlapping crises of neoliberal economic policy, environmental stress, and political malaise that in South Africa are commonly subsumed within "service delivery crisis" discourse. In the past few years, a calamitous fall in global market prices for many of South Africa's key export commodities has further profoundly undercut the national government's growth, employment, and welfarist ambitions.[11]

Local historians thus take their eyes off transnational forces at their great peril. In considering many such forces, I join Jane Carruthers (2014) in seeing global climate change as among the most urgent concerns in southern Africa, one that demands transnational analysis, civil society action, and far-reaching policy responses. An average global temperature rise of two degrees by the end of this century would impose a rise of four degrees for South Africa as a whole (South Africa 2015). Climate change is expected to affect KwaZulu-Natal directly through more intense storms and an expanding malarial belt that could impose significant costs (IPCC 2014). The province is already bearing the brunt of the 2015 "epic drought" linked to that year's exaggerated el Nino effect (Essa 2015). Impacts on the rest of the country and the continent are likely to be even more severe, including desiccation of already dry areas and the inundation and salinization of densely populated coastal areas by rising sea levels. In the context of an imminent end to cheap sources of conventional energy, global predatory capitalism, emergent and drug-resistant diseases, and an anticipated demographic explosion north of the Zambezi, South Africa can hardly expect business as usual in the near future. Indeed, when even leading cheerleaders for neoliberal economics now concede that there is a problem, South Africans should be worried. The Economics Department of the Organisation for Economic Co-operation and Development anticipates a significant widening of socio-economic inequality in G-20 countries over the next four decades, plus "soaring health costs and productivity losses related to local pollution." "Risks of catastrophic environmental events will rise, and the [anticipated 100 per cent] increase in concentration of greenhouse-gases in the atmosphere up to 2060 will lock-in further, and potentially more serious, environmental damages beyond the 2060 horizon" (OECD 2014, 8). We might also well consider climate change in light of recent figures from the sober-minded demographers at the United Nations, who project that Nigeria's currently very young population may expand more than fivefold by 2100. Without some radical change of direction or cataclysmic event, they anticipate the population of Mozambique to surpass South Africa's in 2055, while dirt-poor Malawi (smaller in area than KwaZulu-Natal) could be home to over 100 million people by the end of the century.[12]

Of course, we should take these estimates with a grain of salt and consciousness of the polemical abuses to which demographic predictions easily lend themselves. But who can deny that even a fraction of such population growth or climate change is going to deliver shocks in the coming decades? My point here is simply to stress that South Africa will likely attract economic migrants on a scale that makes today's pressures seem mild. It is timely therefore to consider one city that is manifestly ill prepared to absorb such external shocks. Fresh empirical evidence from a local setting can then be weighed against broad contemporary questions of social and environmental justice, and of strategies for resistance to global coloniality that are unfolding on the streets, and in imaginations, throughout the world today.

≈

To begin understanding any local history, we need to specify the principal geographical terms and pertinent features, and the questions these have raised for historians. To begin at the biggest scale: the municipality of Msunduzi covers a total area of 634 square kilometres, very close to the area of Toronto (which has perhaps three times as many people living in it). Within Msunduzi, four zones show fairly distinctive land use patterns and biophysical environments. While there is much diversity within each zone and much blurring of lines between them, they can reasonably be generalized as urban (Pietermaritzburg proper, which includes extensive parks and "townlands" of plantation forest), tribal or communal land tenure with subsistence agriculture (Vulindlela, the largest zone by area), freehold land tenure for commercial agricultural production (Bishopstowe), and mixed, peri-urban (Greater Edendale).[13]

These zones reflect natural geographical or climatological features that acquired greater significance over time through the cumulative impacts of many political and economic decisions. Maritzburg, notably, was the first zone to be delineated, following the typical rectangular grid pattern of Boer settlements in the Cape. The settlers chose the area with an eye to its spacious, gently sloping land – not too high in elevation, not too low – and ample water supply to provide town-dwellers with the means to do market gardening. Over time it filled in with relatively dense urban settlement and industry, with suburbs like Scottsville, racially designated "villages" and townships like Sobantu and Northdale, and protected or managed parks and forests on the fringes.

In the 1840s, white settlers also staked out the farms that became the Bishopstowe zone. Along with its commercial farms and isolated freehold villages such as Hollingwood and Maryville, Bishopstowe was home to one of the oldest Christian mission stations in the region (Ekukhanyeni, created in 1856).

The next zone to be demarcated (1848) was the "native location" called Zwartkop (sometimes spelled Zwaartkops or Swartkop; presently Vulindlela). Zwartkop is the Afrikaans term for Mbube, the mountain that links the zone to Maritzburg's western border and that is the most prominent topographical feature in the whole valley. Zwartkop the "location" (latterly "reserve" or "homeland") originally consisted of three main Zulu chieftaincies on lands that were not particularly attractive for commercial enterprise (high elevation, excessive fog and cold, steep slopes). Chiefs held court there in a more or less traditional manner over communally held lands under the jurisdiction of what eventually became known as the Native Affairs Department (NAD). Politically, during the apartheid era it became an enclave within the complicated KwaZulu apparatus where ostensibly traditional cultural values and land husbandry could be preserved. That still shows in the upper reaches of the district such as the village of Elandskop, where a strong rural ambiance remains with small homesteads in the traditional kraal style. Yet despite this structured conservatism, Zwartkop came to be closely entwined economically with Edendale and Maritzburg. By the early twentieth century a railroad and highway ran through to tie it directly to the city. Unofficial private property relations steadily crept in over time over an expanding area, with rental units and some substantial private dwellings. As a result, the areas on its south and east borders are today difficult to distinguish from the general peri-urban sprawl of Greater Edendale.

Greater Edendale itself has its origins as a distinctive zone in 1851. In that year, roughly 400 Africans settled on the farm they called Edendale in reference to its idyllic features (plentiful water, rich soil, not too high elevation or extremes of climate). Those settlers eventually came to hold freehold title to about half of the farm area and so (in theory) possessed the same legal rights over their property as did any other private landowners in the colony. The other half of the farm area was designated as commonage and was managed by a trust originally consisting of three white men but later including an elected member – in the early twentieth century that was the titular African chief Stephen Mini. Georgetown, the village at the heart of Edendale, was laid out in a grid in the colonial fashion. Its inhabitants were known as *amakholwa* ("Christians" or "believers," sometimes also taken to mean "progressives") or *oNonhlevu* ("first converts," also sometimes taken to mean "chosen" or "difficult" depending on one's point of view) (Kumalo 2009, 117). The founding families of Georgetown came from a variety of ethnic groups, including Swazi, Basotho, Griqua, Hlubi, and Zulu, as well as a contingent of freed slaves from the Cape Colony. Afrikaans (Cape Dutch) was one of their original lingua francas, then English. They explicitly renounced much of their African cultural heritage and in particular the practice of polygyny.

They fought on the side of the British to defeat the Zulu kingdom, various rebellions against colonial rule, and Afrikaner attacks on Natal in 1899–1902. As spokesperson "Johannes" reportedly said in requesting to be exempt from Native Law in 1863, "We have left the race of our forefathers; we have left the black race and clung to the white. We imitate them in everything we can. We feel we are in the midst of a civilised people … we are happy" (*Natal Witness*, 27 March 1863).[14]

Over time a number of unplanned new villages sprouted around Georgetown, often named after prominent landowners (Dambuza, Msimangville, Siyamu, Caluza). Non-*amakholwa* immigrants were able to settle there as renters, and the *amakholwa* gradually became culturally "re-Africanized" or "Zulu-ized" to a significant degree. However, faith in the power of modern education and legal rights to bring social and economic advancement remained central to the community, and as a consequence the *amakholwa* continued to produce a disproportionate share of black South Africa's teachers, lawyers, clerks, nurses, and ministers as well as members of the intelligentsia well into the 1950s. The principle of private property ownership was also never breached. Edendale proper did become what was known as a "released area" in 1936, meaning that non-Africans would not be allowed to purchase any further land without express permission from the provincial administration. Existing non-African landowners, however, were not expropriated and their numbers grew steadily until the late 1950s. Even during the height of apartheid, government curtly informed Africans in Edendale that if they wanted to make it an all-black area they would have to purchase the land from remaining non-African owners using their own funds (Msimang 1975a).

Heading downwards along the valley from Vulindlela and Edendale proper, several farms owned by white settlers created a rural buffer that eventually abutted the expansive (and theoretically uninhabited) "town lands" encircling Pietermaritzburg city – Ashdown, Smeroe, Plessislaer (also commonly referred to as Sutherlands after one of the largest owners), Slangspruit, Giles, Mount Partridge, Politique, Ambleton, Shenstone, and Wilgefontein (Willowfountain). Over the years, these farms sprouted informal and technically illegal settlements, some of which were very densely packed with shanties and shebeens (notably, Macibise and Schoonplaas, which straddled the road connecting Edendale to Maritzburg near where the hospital is today). Others came to house a working or professional black middle class, Indian farmers, and renters of all races including poor whites. From 1942

1.3 A shebeen (informal boozer), in "downtown" Macibise. This is one of the oldest surviving structures from a time when the village was among the most densely populated parts of the valley (particularly on weekends)

these diverse settlements were swept up together into the Edendale and District Public Health Area (EDPHA). This new jurisdiction was roughly double the size of old Edendale and included many of the above farms. It came under the administration of a new form of local authority known as the Local Health Commission (LHC). Beginning in 1943, the LHC laid out its first formal township on Ashdown farm. This was followed over the years by smaller townships for different levels of income, with attendant leases, rents, and other regulations (Landauville, for example, for people cleared out of slum dwellings in Georgetown), plus boundary extensions to encompass ever-proliferating informal settlements. A huge expansion of formal township housing mostly on Slangspruit farm just outside the public health area came in the form of Imbali in the 1960s. Eventually amounting to several thousand rental properties under tight lease conditions, Imbali was initially administered directly by Maritzburg city as its showcase commitment to grand apartheid. Yet another farm far to the south, Ambleton, was set aside in the dying days of the apartheid regime as "Edendale East" (not contiguous to the original Edendale). This was to accommodate refugees from violence in the surrounding countryside. The area has since mostly been gobbled up in the post-1994 proliferation of "RDP housing" (free homes with title provided by the state to qualified poor through the Reconstruction and Development Programme), and is now colloquially known as "France."

In current usage, many people do not distinguish between (in fact are often completely unaware of) these historical distinctions. Many of the old names and the memory of their idiosyncratic origins have simply disappeared in isiZulu-ized place names (former predominantly Indian-populated Mount Partridge has become KwaPata, for example). When people say "Edendale" today, they thus commonly use it as shorthand for the whole southwest side of the municipality. The municipality uses the term "Greater Edendale" (or "Greater Edendale and Vulindlela" when it wants to blur that border as well) to capture the fact that it encompasses multiple communities with different origins, forms of land tenure, and modes of administration. Common features of underdevelopment, not the complexities of scarcely remembered local variation, are the driving concern of the municipal government.

The historiography of Greater Edendale does better, although frankly not that much so. The first published accounts of Georgetown came from European missionaries and settlers. An underlying tone in these accounts is one of surprise, and perhaps bemusement, that Africans were capable of adapting to Christianity and "civilized" life. To proponents of African advancement, it was a pleasant surprise or vindication. Georgetown was already a "remarkable success" in 1854, according to Bishop John Colenso (1855, 51), who visited that year. By the 1870s, Lady Barker (wife of the lietenant governor of the colony) waxed lyrical in her description of the "thoroughly nice respectable little houses of adobe brick," "luxuriant gardens," and "inexpressibly home-like and fertile" ambience of the village, which she compared favourably to the shabby appearance of Maritzburg (1877, 194–9, 59). So too F.R. Statham, who rhapsodized about Edendale and its fashionable, eloquent, industrious, talented, and loyal "honoured citizens," the *amakholwa* ([1881] 1969, 194).

Unfortunately, as Statham himself conceded, "honoured" was not necessarily the prevailing sentiment among Edendale's white neighbours. I quote J.F. Holliday's satiric account of life in Natal to remind us of the more common and enduring racist sentiment among reputable citizens of Pietermaritzburg at the time. Referring to the possibility of Georgetown becoming a properly incorporated town with its own mayor and council and the powers to regulate civic ordinances and to raise funds (which seemed imminent to Holliday in 1865), the long-standing city businessman "joked": "When that time arrives, won't the swells in pegtops, long shirt collars, and no shoes, accompanied by their lubly Dinahs, decked in all the colours of the rainbow, make a display, while the young dirty-noses make mud pies in the gutter to their heart's content. What a glorious day for Darkeydom!" ([1865] 1890, 17).

As it happened, the colonial and subsequently provincial governments never did grant Georgetown its expected town status, notwithstanding its growing size, the ardent petitions of its leading residents, and the bravery of Edendale cavalry in the defence of the empire. As Georgetown remained outside of Pietermaritzburg's borders, historians of that city saw no reason to mention, let alone to analyze, it in their writing. The bulk of the historiography of Maritzburg as late as 1981 thus reads like a boosterist account of great white men and their plucky wives who tamed the wilderness and built the city. Non-whites appear in the background, if at all. Edendale simply does not exist in key works of early professional historians such as Alan Hattersley (1938, 1951) and John Clark (1972). Ruth Gordon (1981, 133–6) does acknowledge it as a colourful extension of the city proper, but primarily as a further point of civic pride in Maritzburg's liberal tradition.

Yet Edendale as a problem that the city would need to deal with was much reported in the popular press through the 1930s. Snippets of its history of frustration and decline were aired to various commissions of enquiry at that time. The first overview of its history in published form came in an introductory section of a socio-economic and health survey conducted in the late 1940s. *Experiment at Edendale* (University of Natal 1951) sketched the basics of the community's mission station origins, its early economic prosperity, its subsequent economic decline, and the emergence of non-African communities within its borders. The report laid the blame for the frustration of *amakholwa* aspirations on the "open market" (8). Reading between the lines, however, the report implicitly linked the area's general economic decline to the collective failure of Edendale's trustees, Pietermaritzburg city council, and the colonial/provincial governments to honour their promises to the people of Edendale. Its proposed solutions were closely in line with liberal thinking at the time about poverty and environmental degradation in the communal areas (Brookes and Hurwitz 1957) and included a range of economic reforms, education, and gradual progress towards modern forms of self-governance.

Professional historians began to explore the causes of that collective failure of the liberal vision in the 1970s with reference to the wider colony/province and nation as a whole.[15] Although with very little specific attention to Edendale or Maritzburg, the radical or materialist historiographical tradition in South Africa sought to expose – and assess the relationships between – the racialist presumptions and capitalist structures established by white settler regimes that tended to undermine African development and dignity to the benefit of the white minority. The title of John Lambert's analysis of how the government of colonial Natal in the late nineteenth century increasingly targeted legislation against African producers put it bluntly: *Betrayed Trust* (1995). The government did so in order both to suppress Africans' ability to compete with white settlers in the agricultural market and to preclude effective African labour or political organization that might inhibit the flow of cheap labour to white employers.

Only with the work of Sheila Meintjes did disciplined historical researchers turn their gaze directly on Edendale. Meintjes was an activist with the Association for Rural Advancement (AFRA) in the early 1980s. AFRA was a Maritzburg-based civil society group that strove to monitor and resist the apartheid state's efforts to eradicate so-called black spots in its drive to impose racial separation throughout the country. Edendale, so close to Maritzburg and standing upon such an attractive landscape, was a prime candidate for removal to make way for white suburbs and gardens. Documenting the early history of Edendale as an economically successful village with a culturally modern and Christian elite was for Meintjes a contribution to efforts to protect Edendale's current black population from removal (Meintjes 1984, 1988a, 1988b, and personal communication). Meintjes (1990) added a further contribution to the radical historiography of the day by documenting the role that African women and gender relations played in the development of the community to 1906, an astute insight for a time when such issues were commonly disregarded in the scholarship.

Another contribution from the early "struggle" historiography was Timothy Nuttall's honours thesis (1984). A precocious piece of work that was never published, Nuttall's study makes a number of points of direct pertinence to the present one. He analyzed conflicts among Africans in Edendale around the extension of a liberal-minded developmentalist state in the years just preceding apartheid (a tale that I will be picking up in chapters 5 and 6). In the process, he drew attention to the complexity both of class relations among Africans and of the state itself – both important insights that commonly get erased in the struggle narrative that predominates so much of today's public history.

Critical studies also began to appear in the mid-1980s on the other non-white components of the wider city region, beginning with the city's first formal township, Sobantu (Peel 1987; Napier and Mtimkulu 1989). Laband and Haswell (1988) brought a range of such studies together in an accessible format on the occasion of the city's 150th anniversary. Despite its "coffee table book" appearance, *Pietermaritzburg, 1838–1988* provided a major contribution to our understanding of how even a supposedly typical apartheid city involved highly complex, often hidden struggles leading to its present form. Some of the more sordid underpinnings of Maritzburg's genteel self-image were revealed in the book, including its dependence on brutally exploited African and Indian labour, monopoly beer sales to subsidize the administration costs over an impoverished population, and grossly inequitable (by race and class) levels of taxation and services (see also Wills 1991, 1994; Seethal 1993).

Since the 1990s several dissertations and monographs have touched upon specific aspects of Msunduzi's history. These included Kentridge (1990), who documented the political violence in the 1980s; Atkins (1993) on the development of African working-class consciousness in the nineteenth century; Morrell (2001) on the making of elite white Natalian masculinity; Butler and Harley with Aitchison (1993) on Imbali; Garner (1998, 2004) on religion and development in near-to-contemporary Edendale; Pole (2002) on post-apartheid efforts to control water pollution; Merrett (2009) on sports as a factor in entrenching racial inequalities in the city; Mkhize (2012) on the "beer riots" of 1959 and (2015) on Selby Msimang; Denis, Ntsimane, and Cannell (2010) on political violence in Vulindlela; and Denis and Duncan (2011) on Fedsem, an early focal point of the gathering tension between the apartheid state, Inkatha-aligned elites, and the mass democratic movement. Julie Dyer's compendious history of health in Maritzburg (2012) is another important contribution to our understanding of the sharply contrasting lived experiences of racialized groups within the city.

Together these works add up to a profound indictment of the modernization paradigm as it unfolded particularly over the apartheid years. Yet ironically, in focusing on the exploitative structures and state violence that Maritzburg's white liberals hypocritically benefited from if not actively stoked, and on African resistance to the same, this scholarship has an unintended effect of somewhat homogenizing the struggles and communities

1.4 Topography – Maritzburg has lots of it. Here, a rainy season view from the edge of the city looking past townland plantations towards Vulindlela

involved. One could be forgiven if, for example, having read all of this literature thoroughly but never actually been to Edendale, one mentally pictured it as flat, entirely isiZulu-speaking, and mostly populated by men. Particularly noticeable in the historiography of the anti-apartheid struggle is the non-appearance of the LHC, of African liberals, and of the *amakholwa* wing of the predecessor to the Inkatha Freedom Party (IFP), all of which were at times vocal opponents of racial segregation, tribal structures, and state violence. Houston et al. (2013), for example, does not even mention the existence of the Liberal Party of South Africa (LPSA) in their overview of anti-apartheid political movements for a proposed heritage trail. Yet the LPSA had more black members in Natal than anywhere else in South Africa. Headquartered in Maritzburg, it was the party to which Msimang belonged when he was arrested by the apartheid regime, and his membership served as the pretext for compelling him to resign (or dismissing him) from the LHC advisory board. The rise and fall of Edendale's Indian communities has also not been substantively detailed, although we can note the strong nostalgic tone in the recollections of Indian women published in Diesel (2006). The biophysical environment meanwhile makes only cameo appearances in much of the scholarship, usually related to the occasions when it most dramatically misbehaved (from a human perspective), with spectacular floods, droughts, and epizootics.

A further critique of the literature to this point derives from the post-colonial turn in urban studies that Jennifer Robinson (2006) articulates so well and that clearly weighs heavily in the minds of proponents of "new urbanism," or re-thinking the fundamentals of form and functionality in Africa's burgeoning cities. The anti-apartheid scholarship is obviously a vast step forward from Pietermaritzburg's "elegiac" tradition of writing about the place (Laband 1988). Yet Laband and Haswell, as an important example, do not escape marginalizing the majority population to "the outer city," buried within the book between chapters on the royal visit of 1947, prominent white leaders, and parks, buildings, sports, and other amenities of primary interest to a presumed mostly white readership. In characterizing Maritzburg as a "typical apartheid city," Wills (1991) and Merrett (2009) fall into a similar discursive trap. That judgement asserts three things: the place is low on the interesting scale ("typical") and low on the scale of moral, political, environmental, and social intelligence ("apartheid") but nonetheless high on the modern scale ("city"). Edendale, by contrast, does not qualify for city status, or even, in much of the contemporary planning documents, as a "suburb." Rather, it ranks below these categories of urban space as "a township" (which is, of course, technically incorrect), "urban place," "gateway," "functional area," "hub," "node," or (most ambitiously) a "city within a city," a subordinate urbanism that needs to be "fixed," when seen against North American ideals of modernity. Two new malls with their vast parking lots and fast food joints directly opposite each other on the main highway leading into Edendale from town powerfully symbolize that ambition.[16]

Those ideals have no shortage of critics. Adam Ashforth (1990), James Ferguson (1990), and Jonathan Crush (1995), among numerous others, all persuasively demonstrated that development and modernization discourses in southern Africa historically helped to obscure, and hence enable, the extension of bureaucratic power over Africans in support of hegemonic capitalist interests in the region.[17] Such power camouflages itself within coolly scientific language and that can transcend political transformations. With reference to the transformation of urban planning at the national scale in the 1990s, Harrison, Todes, and Watson (2008) argue that strong legacies

1.5 Georgetown today: Shepstone Street, in the winter looking west

of modernist thinking inherited uncritically by today's planners can have frustrating results for the people living today in the planned areas. Vanessa Watson (2014) is damning in her analysis of the broader trend across the continent to construct "fantasy cities" whose slick modernism appears oblivious to African culture and the existing political economy, let alone future demographic, health, and environmental challenges. That gloss is abundantly apparent in current planning for Edendale but, as may be inferred from photo 1.5, it has a long way to go to offset existing blight, let alone achieve consumerist ideals.

The point of the critique is clear. Perceiving inequalities and alienation structured into modern urban forms and discourses will require more creative imagination and boldness than we have seen thus far. Unfortunately, there is little evidence that this is happening at the level of the National Development Plan (which assumes high levels of economic growth in conventional measurements) or on the ground at the local level. Indeed, notwithstanding some impressive initiatives and achievements that I will duly acknowledge, many profoundly alienating and unsustainable wheels are currently being reinvented, are implicit, or are mooted in the official development vision for Msunduzi: in addition to more cars, malls, and fast food outlets, there will be buffer zones, low-density gated communities, far-flung RDP "estates," sprawling export-oriented, low-wage industrial corridors, private hospitals, and high-end tourism and recreational events. A single, supposedly unifying national struggle mythology will inspire feelings of patriotic citizenship. Are these really the most appropriate and sustainable means to transform Greater Edendale into a healthy place to live?

≈

My approach to the challenges that Edendale presents to the re-making of Msunduzi is somewhat eclectic. I started my original work in the region within the theoretical framework of socialist feminism and critical men's studies (Epprecht 1993, 2000). In a place like southern Africa, such an approach almost invariably leads one to questions of health and environment, as documents pertaining to the overlapping crises of health and environment from the 1930s to the 1960s are among the few archival sources that substantively discuss African women, either as a problem to be fixed or as a potential resource to help manage the corrosive impacts of the male migrant labour system and indirect rule. A fair reading of such documents, taken together with African women's own reflections obtained through oral history, makes it hard (and I am understating) to be sympathetic to the confident assertions of colonialist and apartheid propaganda about those systems and the developmentalist project as a whole. Hence, I situated my work under the broad tent of "radical," with a critique of the androcentrism and cultural essentialisms so deeply embedded within colonial/apartheid development projects and often uncritically re-affirmed by male African nationalist leaders of the era (and since).

My understanding of the interconnectedness of gender relations, sexuality, governance, health, and environment resonated well with the framework put forward by the Ecosystems and Human Health in South Africa project to which I was invited to contribute in 2004. This project described its analysis as an "ecohealth approach," meaning a holistic and transdisciplinary conception of research that aims to inform policy interventions seeking to improve people's health and overall quality of life in the spirit of environmental justice (Forget and Lebel 2001; Goebel, Dodson, and Hill 2010). Over the years, other members of the team used, among other research methods and questions, scientific instruments to measure indoor air pollution; large household surveys to assess health problems and attitudes towards place; statistical analysis; in-depth interviews with women on wide-ranging topics around gender, health, and local environment; and key informant interviews with policy makers (Goebel 2007 and 2015; Goebel, Hill, et al. 2010). While I tagged along for much of this research, my main task was to troll through the archives in search of evidence that could shed light on decision-making processes in the past and that would thus help the other researchers properly contextualize their findings. By "properly," I had in mind the following: awareness of the history of struggle and debate around often painful questions, not just in South Africa but elsewhere in the region and transnationally; sensitivity to nuance, contradictions, and silences that speak volumes; alertness to the need to avoid simplistic dichotomies as well as essentialization of whole groups of people or discrete communities; consciousness that "gender" is not synonymous with "women" but must also include how men and masculinities shape the exercise of power over women; commitment to finding the voices of, and naming, individuals from marginalized groups, especially women; care to avoid knee-jerk attributions of blame rooted in received wisdom from bigger cities in South Africa; and consciously seeking out the perspectives of African intellectuals rather than relying on the established canon for theoretical framing. The key terms of reference (i.e., "eco" and "health") were understood expansively to include the social determinants of health (physical, mental, emotional, etc.), and the living/natural, social, and built environments.

All of us working on the project were attuned to the crisis of service delivery and governance unfolding in the municipality around us. How not to notice the distinctive pop of tear-gas canisters on the walk to the archives, as happened to me on a couple of occasions in the city centre? Or that one of Edendale's rare public sporting venues was being torn down to make space for a new mall? Or that the municipality's highly respected medical officer of health, acclaimed for a path-breaking initiative on HIV/AIDS, had been sidelined to managing a wholesale food market, among other grotesqueries of poor governance? The wider contemporary context also gave much piquancy to our concerns and questions. The controversies around HIV/AIDS during Thabo Mbeki's presidency notably demonstrated how quickly and bitterly insinuations of anti-African racism could enter into debates about science and health (see Nattrass 2012 and Robins 2008, for example). My own admittedly unscientific observation over the years is that public discourse has become even more racialized and accusatory in tone since Mbeki, and that inflammatory, essentializing rhetoric has become commonplace in mainstream media.[18] To be fair, the most unsettling demagoguery tends to unfold from the national level. Local politicians, however, are clearly also adept at riling up apartheid-era grievances for short-term political objectives. I observed, for example, the mayor of Msunduzi open the historic Edendale Land Summit in 2006 by first sharply castigating the principal organizer for speaking English to the audience, thus demonstrating his obvious allegiance to the colonialist past. Both triumphalism and factionalism within the ANC have meanwhile spilled into the heritage field in ways both crude and sophisticated. It may take a sharp eye to discern, but, to give but one example, seemingly mild and mature criticism of Nelson Mandela in the Mandela Capture Site museum near Howick ("Mistakes and Contradictions") is in fact a scornful dig against his successor, Mbeki, and a none-too-subtle promotion of the following president, Jacob Zuma.

It is wise not to overreact to such provocative insinuations. Nonetheless, having spent some years in Harare on the cusp of Zimbabwe's "Third Chimurenga," and while acknowledging some quite sophisticated historiographic reflections even in government heritage propaganda (Sithole 2014), I find this rhetorical turn deeply disconcerting. Careless or overhasty attributions of blame by scholars in this context may abet the polemics and hence contribute to what Terence Ranger has called "patriotic history" (Ranger 2004 and 2007). Ranger was reflecting in part upon his own role in Zimbabwean historiography when he raised this concern. He describes a dangerous oversimplification of the colonial past that was once important for mobilizing opposition to an oppressive regime but today lingers on to obscure the failures and crimes of the current ruling party. A task Ranger sets for historians today is to challenge patriotic history with more rigorous analysis of the complexities of the relationship between colonizer and colonized. Such a challenge is not to gloss over the cruelty and harm of colonialism or cultural racism. Rather, it is to encourage historians to be alert against unwittingly perpetuating blanket generalizations about the past that were politically astute at the time (i.e., the 1960s to the 1980s) but may now muddy the often-fraught initiatives to address present-day sanitation, service, and other crises, imbued as they frequently are with existential questions around individual, ethnic, and national identity.[19]

Many South African scholars are indeed revisiting the anti-apartheid scholarship, not to dispute the trauma that apartheid caused but in light of the mixed record of subsequent governments, the disappointment or disillusionment that many people currently express about democracy, and the growth of comparative scholarship from other parts of the continent or the Global South more broadly. Alexius Amtaika's survey of the challenges facing local authorities in South Africa today underscores the complex factors behind maladministration and that some of them – including "laziness, incompetence, inefficiency, and irresponsiveness of public officials," "fruitless and wasteful expenditure," political interference, ignorance, apathy, and corruption – have in many cases grown markedly worse since the early years of transition (2013, 159, 170, and passim).[20] A provocative piece by Jacob Dlamini (2009), to give another example of rethinking the apartheid legacy, goes so far as to speak positively about township life under apartheid for some people and even has positive things to say about Afrikaans, the language. William Beinart, one of the pioneers of environmental history in South Africa and a cogent critic of racial capitalism, has recently wondered if the earlier critique of the Bantustan and betterment schemes wasn't overstated (Beinart 2012). Also pertinent to my project here, Sean Field worries about "the domination of narcissistic nationalist discourses" in the post-apartheid era, with the consequent need for much more sophisticated approaches to oral history than tended to be the case in earlier "romantic," anti-apartheid approaches (2012, 16, 5). I find myself agreeing as well with Mbembe and Nuttall (2008, 12) in their critique of scholarship that is "sutured to a political agenda (the critique of the apartheid state)." Insightful and passionate as it can be, that agenda feeds into a tendency towards South African exceptionalism in the scholarship (which ignores comparisons to places outside the country and especially reference to the rest of Africa), a disproportionate emphasis on political economy (which undervalues culture), a tendency to assess every decision through the lens of race, and the neglect or submersion of diverse local perspectives within a grand national narrative that tends to lionize individual great men.

It was a challenge for me to begin writing up many years of research at a moment of especially intense reflection among scholars, business leaders, and politicians, even within the ruling party. The years 2013–14 saw, among other things, profound anxiety about the meaning of twenty years of democracy, the longest and one of the most violent industrial strikes in South African history (Marikana), and the re-election of the ruling party after a campaign of often breathtaking demagoguery.[21] Because the political, economic, health, and other context has changed so radically in the two decades since the democratic transition, new conceptual tools are needed to make sense of it. "Ecohealth," when framed as described above, offers one such set of tools and ways of asking questions that address the contemporary multi-layered crises. Indeed, the ecohealth project in Msunduzi produced a wealth of data that push us to reassess many taken-for-granted assumptions. Perhaps most significantly, as Goebel (2015) demonstrates in her elegant monograph, gender relations emerge as a core factor explaining differential health outcomes of urbanization. Gendered welfare practices, among other interventions, have actually introduced new forms of marginalization for women and children in the impoverished informal settlements that still encircle the formal city like a "black belt," eighty plus years after the term was coined.

My penultimate point of reference (for now) helps to explain the format I have chosen to present my case. Julie Parle's presidential address to the South African Historical Society noted how scholars in South Africa are not presently rewarded for authoring books. Rather, they gain academic merit and funding by publishing in peer-reviewed journals ranked according to their impact factors (2012, 168). Of course there are many outstanding monographs by South African scholars, but the current system gives primary incentive to the production of disjointed articles or small, "popular" books with sometimes indifferent editing. Parle thus specifically cites the need for the "sustained and cumulative exploration" of historical questions that is possible only in monographs. Fortuitously for me (since this is my preference in any case), the latter are rewarded in my home institutional system in Canada. My hope is that the structure I have chosen of relatively discrete component chapters within a monograph allows them to be read and debated on their own merits as well as for the cumulative arguments as they unfold over the longer text.

When appraised through the "archives and justice" lens elaborated by South Africa's chief post-apartheid archivist (Harris 2007), and read beside studies of contemporary urban stresses elsewhere or the gathering critique of the perpetual economic growth paradigm, my sources posed questions to me that were unexpected given the secondary literature on Edendale and environs. In the process of thinking about them, I became aware of the gaps or stretches in the historiography of Msunduzi noted above. I also developed an affection for Msunduzi as a fascinating, beautiful, exasperating, and potentially great place in which to live. Indeed, thinking long term and bearing in mind the climatic, demographic, technological, and economic changes that the world as a whole is facing, the city is surely full of potential for a new kind of urbanism that contrasts with the present highly problematic form. My growing conviction over the years was thus that Msunduzi deserved attention both in its own right and for how its history might engage scholars elsewhere in the region to reflect on the transdisciplinary and applied methodologies that an ecohealth approach enjoins. I am not going to claim that such an approach calls for radical revisions in the scholarship. On the contrary, on key issues an ecohealth history of Edendale supports and enriches the arguments of critical scholars across several disciplines. In other cases, however, that history does call for greater alertness to inherited but stale or unhelpful concepts, overgeneralizations, essentialisms, teleologies, eurocentrisms, androcentrisms, anthropocentrisms, and other blindspots. To be sure, this is an academic volume, with all the limitations that that implies for "knowledge translation." However, it is my hope that this book can contribute to such alertness and so to provide support to those who are working towards environmental justice and ecohealth in Msunduzi and elsewhere.

≈

The title of this book is deliberately cheeky. It inverts a common practice still very much in use that places Pietermaritzburg at the centre of the analysis and that obscures Edendale within anonymous, generically poor environs. My primary focus, by contrast, will be on Greater Edendale as a highly complex and creative place. The themes referenced in the subtitle are key to the historical narratives I have reconstructed, with the proviso that understanding environment, health, and gender requires the consideration of many other intersecting issues (race, class, governance, economic growth, and human rights, notably – but a title can only be so long). The reference to "an African city" is also deliberate, in that it echoes three previous notable publications. Ingram (1898) and Hattersley (1938) described Maritzburg exactly that way, although in an unintentionally ironic manner, in that Africans were entirely invisible to them (as, for the most part, were women, Indians, and coloureds). In this way, I am calling attention to major blindspots in the historiography of the place up to the 1980s. The reference also pays respect to Laband and Haswell (1988), whose "new portrait of an African city," for all its quirkiness, nonetheless contains the first published scholarly studies of the black communities upon which earlier colonial fantasies depended. Over a quarter century of momentous change has happened since Laband and Haswell's publication, during which time Maritzburg has become recognizably "more African" both in terms of its demographic makeup and the problems it faces as an urban area.

Chapter 2 offers an overview of the ways in which the environment in the district surrounding Edendale has been conceived, researched, and represented. It becomes clear, and is much in line with observations made elsewhere on the continent (see, for example, Leach and Mearns 1996; Wolmer 2007), that several highly problematic tropes have been embedded in the process of knowledge production about the local environment. The imposition of gendered imagery to represent the land, for example, sometimes melded closely with justifications for colonial conquest (as McClintock 1995 found in her powerful analysis of the work of H. Rider Haggard). Another key finding by environmental historians in the region has been that scientists commonly made fundamentally incorrect assumptions about African soils, climate, flora, and fauna, and overstated the wildness of what were, in fact, populated areas. They also critically understated or ignored not only indigenous African perceptions of, and methods for managing, the environment, but also, and especially, African women's knowledge. Poor science and condescending or outright racist attitudes towards African knowledge then underpinned development initiatives that in many cases exacerbated the very environmental problems they sought to fix. These included, among others, the introduction of exotic species or epizootics that wiped out local subsistence economies; erosion controls that increased erosion; and "improvement" or "betterment" schemes that marginalized and further impoverished the poor in general and women in particular. The language of conservation against threats such as erosion or desertification often barely disguised a widespread desire by whites to appropriate land from Africans who supposedly did not know how to husband it, at least compared to wise European scientists and settlers.

This history of paternalism or worse underpins the highly dichotomous debates between green (conservationists) and brown (social justice activists) that often poison political discourse today. The racialized nature of the discourse in South Africa bubbles quickly to the surface to derail reasoned discussion upon how to avert or mitigate evident, gathering environmental harms. Among innumerable recent examples, white Democratic Alliance (DA) councillors in Richards Bay raised "green" concerns about a gathering of tens of thousands of Shembe church members in a protected forest without provision for cooking, toilets, or parking. The black minister of agriculture and environmental affairs shot back: "The DA must learn that blacks can think. They should stop this thing of saying black people would just destroy nature" (Mbjanwa 2014).[22] In an even more tense exchange, the DA vigorously distanced itself from the very real racism of one of its members masquerading behind her concern for clean beaches (DA 2016). Complicating matters further is that stubborn gender ideology that associates concern with "Mother Nature" with femininity, and technology, development, and progress as masculine (Merchant 1989).

Can the critique of colonial practices and knowledge production specifically about nature in an urban setting be applied to break that green/brown impasse? Many intriguing ideas are being raised by black intellectuals in South Africa (and beyond) in this regard. They wrestle not only with the material but also the cultural and psychological legacies of racial capitalism (Ramphele 1991; Bullard and Wright 2009; Maathai 2009; Ramose 2015, for example). While the South African government and multilateral agencies continue to talk of growth as an imperative, albeit qualified by vague adjectives such as "green," "inclusive," "resilient," and "sustainable," UN-Habitat (2015, 32) is frank in its call for a "profound shift in the imagination of African elites." My aim in chapter 2 is to set the stage for such a shift and for "decolonizing" or "reterritorializing" writing about Msunduzi, liberating it inherited paradigmatic traps like those noted above. Understanding how the local environment has been used for political purposes, and often seriously misrepresented in the process, is an important first step to reconstructing the ecohealth history of the city.

Chapter 3 makes a second small gesture towards that goal of "decolonizing" knowledge about Msunduzi. Radical historiography at present tends to be heavily preoccupied with race relations and is framed around the idea of the nation as an entity in the painful process of birth. Key dates in the Natal case – 1838, 1893, 1910, 1913, 1948, 1983, 1994 – mark the ebb and flow of the political struggle for democracy, which subtly entrenches the white/black and green/brown dualisms. But what if we were to adopt a periodization that shifted the focus from racial politics towards moments of dramatic changes in human impacts upon the local environment and changing practices around how to manage them? What if, rather than focusing on the great men who led the great political struggles, we looked instead primarily at how environmental and health stresses unfolded for the majority population in a specific place? The political economy of race will necessarily remain a central concern but, I argue, important, generally understated factors like gender and sexuality would come more to the fore. Chapter 3 lays the groundwork for the more focused discussions to follow, with an overview of key environmental and health factors that are typically obscured by the storms of the *mfecane* (the wars and migrations of the 1820s), colonialism, and capitalism as they swept over the valley. It covers the period up to the first systematic attempt to deal with the human health consequences of environmental degradation, in the early 1940s.

Chapter 4 tackles a recurrent concept not only in the historiography of Pietermaritzburg but throughout urban South Africa, and even farther afield. Maynard Swanson first coined the term "sanitation syndrome" in his study of the origins of racial segregation in Durban (1976). He posits that scientific knowledge about contagious disease was co-opted, sensationalized, and exploited by racist whites – sometimes unconsciously – as a rational-sounding, humanitarian cover for the politically difficult goal of physically removing Africans from the city into tightly controlled, racially segregated native locations. The concept struck a chord among radical historians not only because it problematized science as metaphorical, and hence political, but also because it challenged a long-standing, self-congratulatory staple of liberal South African historiography. In effect, the concept shifted blame for the antecedents of apartheid away from rural, conservative, and unabashedly racist Afrikaners onto urban, self-styled "progressive" British officials and voters. British officials were said to have promoted segregationist policy for a variety of reasons disguised under the rubric of an inflated health scare (infection of whites by diseased blacks), which was happily accepted by selfish white voters. In chapter 4, I ask whether the empirical evidence from Maritzburg supports such an analysis and examine the history of debates around the creation of a "native village," debates that lasted more than eight decades before producing any results and that involved a wide range of voices, including those of African elites. I highlight Edendale women's first foray into the public political sphere in the early 1920s, long before the usual starting point for African women's political activism.

The next two chapters focus on a period roughly analogous to what D.A. Low and John Lonsdale (1975) termed the "second colonial occupation" in British East Africa. This involved stepped-up state intervention in development through a wide range of projects and programs designed to improve nutrition, health, soil conservation, water management, housing, and the sense of citizenship and pride of place among the poor majority. It was an "experiment" on several levels, which included ecological rehabilitation, training in local governance, and social medicine or primary health care.

Chapter 5 is particularly interested in the implications of that experiment for gender relations. African women's health had become an implicit concern of the state in Natal as early as the 1880s, with the emergence of shebeens (illegal drinking establishments) on the fringes of the colony's growing cities. By the 1920s, several notorious informal settlements stretched southwest of Pietermaritzburg and had begun to undermine the integrity of old Edendale itself. So-called loose women sold beer and sex with detrimental effects on their own health and that of their children, as well as on the health, discipline, and productivity of the male labour force. Attempts to clean up the area foundered for a long time over jurisdictional disputes. Less well understood, however, was African opposition to health care interventions by the state. Women in Edendale played a significant role in this opposition, from evasion and physically attacking city officials to political challenges at higher levels of government. A central part of chapter 5 focuses on one of the challenges arising from a scandal provoked by the first-ever attempt to systematize research on maternal health, in 1944. It reflects on how that scandal problematizes some of the staples of contemporary scholarship on sexual health and sexual rights in southern Africa, such as oversimplified invocations of the legacy of racism to explain high rates of HIV/AIDS or gender-based violence.

Chapter 6 aims to pin down the answer to the question, "Who killed the experiment at Edendale?" I want to be careful not to romanticize the social medicine regime of the Local Health Commission or to make its early years into a "liberal Camelot idea of a magic moment," as Bill Freund (2014, 595) phrases what he sees as a dangerous tendency in the historiography led by Shula Marks. It is an often overlooked fact, however, that from 1942 to the mid-1950s, the LHC committed itself in word and deed to a relatively holistic redevelopment of the wider Edendale area into a place where people of all races could live in dignity and good health according to the principles of social medicine, a very progressive if not radical concept in the world at that time. It included moves towards a multiracial form of local authority, an unprecedented political commitment in South Africa and rare anywhere on the continent.

According to Shula Marks (2014, 185–7), while social medicine was quickly suffocated in South Africa itself, the model was inspirational to the development of community health centres in the segregated South of the United Sates, to the rediscovery of primary health care by the World Health Organization in the 1970s, and, belatedly, to the present government of a democratic South Africa. This experiment in applied, pragmatic liberalism deserves to be remembered and critically assessed better than has tended to be the case. Indeed, if not ignored altogether, the LHC has typically been dismissed as hypocritical pie-in-the-sky or doomed-to-fail paternalism that can safely be lumped together with all other white initiatives in the same irretrievably racist boat (and Africans who tried to work with it as mere stooges or dupes). By examining closely some of the key decisions that ultimately led to the abandonment of the liberal promise in favour of so-called separate development and racial zoning, which began to happen definitively in 1960 (not 1948 – another important distinction that is often overlooked in assumptions of a coherent, undifferentiated "apartheid era"), I hope to contribute to a better understanding of both the violence that followed in the 1980s and the HIV/AIDS epidemic that followed soon after. Of particular interest here are the cultural construction of racial and ethnic identities, the "rise and fall" of LHC idealism, and the role played by H. Selby Msimang, a scion of the old Edendale elite who served on the LHC advisory board for over two decades.

Chapter 7 examines the period from 1974 to the present. These tumultuous decades are normally periodized into two very distinct eras: late apartheid and the liberation struggle up to 1994 followed by democracy/globalization. The break is quite rightly celebrated as a political triumph, even a "miracle," that extends to almost every aspect of life, including environmental activism. *Inkosi* Zibuse Mlaba used precisely the word miracle to describe a negotiated peace between the organizers of the Dusi canoe race and the chiefs along the Msunduze River valley in 1995. This led him to become one of the first Africans to participate in the hitherto white-dominated and insensitive, if not provocative, sporting and conservationist event. Mlaba subsequently wrote, "a new realization is that we must respect and care for nature," and "we" (meaning South Africans of all races but implying black South Africans especially) needed to develop "new ways of seeing animals" (Mlaba 2007, xv). New ways of seeing took place within an emerging constitutional and policy framework that appeared to place South Africa at the forefront of the sustainable development movement in the world.

The transition from apartheid to democracy is particularly stark in Edendale, where conditions after the end of the LHC regime deteriorated to all-out war in 1990 and the near complete breakdown of health and environmental services. This was followed after 1994 by quite radical changes for the better: democratic elections, the end of civil war, the merger of old apartheid jurisdictions into a single municipality, and explicit commitments by various spheres of government to sustainable development, gender equality and environmental justice. I do not at all question the importance of those achievements and, indeed, I draw attention to some notable successes. But I also want to consider some of continuities across the 1994 watershed and to examine forces at work that have not only significantly undermined the achievements of the democratic years but may be laying the groundwork for long-term environmental and other injustice. Some see this contradiction as being inherent to the elite nature of political transition from 1990, the seduction of black South African elites by neoliberal development discourse and the fetishization of economic growth as the panacea for South African problems (Patrick Bond is a leading voice among many who make this point). Growth discourse, in this view, including its purportedly "sustainable" or "green" versions, is in practice a highly sophisticated form of entrapment to atomized and patently unsustainable consumerist values. For rhetorical purposes I call it the "third colonial occupation." Resistance seems futile against such a global force. Worse, as we have witnessed repeatedly over the past decade throughout South Africa, people's frustration at the impasses that the fetishization of growth creates sometimes finds its expression in nihilistic and xenophobic violence.

Chapter 7 follows the threads of industrialization, environmental activism, and health as they have played out over four decades across the whole of the Maritzburg and Edendale polities. A particular focus is upon the self-described revolutionary Greater Edendale Development Initiative (GEDI), or GEVDI as it is now known in an apparently token gesture to Vulindlela.[23] GEVDI is the latest official plan to redevelop the area as an "urban Pietermaritzburg place of high quality" – that is, a recognizable suburb or edge city and a healthy place to live (and invest!). I am not going to comment on the technical elements or modernist aesthetics of the GEVDI plan, the feasibility of its financing, or its relationship to other powerful planning bureaucracies and business groups that have manifestly competing interests in the area (although these all call out for critical attention). Nor am I going to assess whether GEDI/GEVDI is truly as moribund as it appears to have been since at least the 2010 budgetary crisis. My focus, rather, is on the tensions between environmental health, social justice, and the assumed imperative of economic growth. Seen through a socialist ecofeminist, feminist political ecology, or degrowth analytic lens, this history allows us to discuss some commonly overlooked but deeply unsettling continuities between the late apartheid era and the GEDI/GEVDI plan.

It is important not to be naive about the power of history to inspire, to change minds, or to commend good policy to decision makers in the present. The notion that history can ever heal psychological traumas, as South Africa's Truth and Reconciliation Commission, GEDI, and many oral historians in Africa seem to believe, has been sharply questioned (see, for example, Posel and Simpson 2002, Field 2012, Denis, Ntsimane, and Cannell 2010). We need to be all the more alert to this critique, given how powerful forces in South Africa today purposefully refute historical accuracy and fairness, actively disparage a vaguely defined liberal intelligentsia or other "clever" people, and seek to co-opt memory to serve a national myth or an illusory "green capitalism." Nonetheless, a long tradition in South Africa sees learning from the past as a way to give guidance or inspiration to present-day struggles for social justice. Whether it was called "respecting the ancestors," whiggishness, or radical history, that tradition remains alive, however attenuated. I offer the following both in that varied and honourable tradition, and for the plain fun of uncovering some neglected pieces of Africa's urban heritage.

# TWO | Colonial Tropes and Traps

Early histories of Edendale and environs centred almost exclusively on Pietermaritzburg and tended to exude what John Laband called an "elegiac colonial fragrance" (1988, xxxi). They not only paid scant attention to the non-white population of the city but also wrote as if human achievements and conflicts played out against a steadily "improving" environment that reflected the particular genius of British colonialism: barren grasslands turning into beautiful and productive forests; thriving farms and neat plantations replacing wasteful native scratching; flowering gardens and tree-lined streets steadily taming a choking dust bowl; a stinking open sewer giving way to a recreational river; the hum of industry bringing prosperity without destroying the city's charming architectural heritage and natural playgrounds. The two big worrisome exceptions to this happy state of progress were the smoggy conditions of the city in winter and the evident deterioration of the environment just outside city limits – the so-called black belt. However, the former could be explained as a fluke of geography, while the latter could mostly be ignored and in any case was somebody else's fault and responsibility.

This writing tradition roughly aligns with the conservationist or green side of contemporary environmental politics. It is somewhat unfair to cite Alan Paton as poster boy for this way of seeing and thinking about old Maritzburg, but he does invite raised eyebrows by opening his memoir with a paean to "the lovely city … My hometown was paradise … Access to unspoiled nature was immediate" (1980, 1–2). Romanticism of this sort was in fact mostly long dead for South African scholars by the time Paton wrote. Many of the authors discussed below had for years attested to how Paton's illusion of a local paradise significantly depended upon mass population removals, manufactured deserts, horrendous pollution, and desolate urban spaces located just beyond his view. Fellow Maritzburger Ian Player's encounter in 1951 with the Msunduze River scarcely a kilometre downstream from Paton's childhood home is evocative in that way: "from the stench we soon realized we were in the sewerage farm … A hundred yards further on we came to a packed and solid mass of logs piled thirty feet high. The water simply disappeared. The logs had dammed the river into a pool – a trap for every conceivable type of garbage. Rotten oranges, cabbages, carrots, dead cats and other foul smelling refuse lay in a thick scum on the still water" ([1964] 2007, 12). Player was so moved by the experience, and specifically by the sight of only two small duikers during his six days of canoeing on the river, that he went on to become a key figure in the rise of the modern conservation movement in southern Africa.

The end of apartheid witnessed a raft of new legislation to protect the environment while addressing the injustices that underlay Paton's and Player's disparate impressions. It also saw a new form of progress narrative emerge in official discourse. In this view, democracy, social justice, and balanced environmental management ideals triumph over apartheid cruelties and environmental crime. *Balanced* here means both protection of natural features and biodiversity (the green side) as well as providing jobs and houses for the poor, black majority (the brown or justice side of the debate). In practice, the emphasis by politicians has tended to be on the brown, to the serious detriment of environmental health in many cases. Indeed, although couched in vaguely defined terms like sustainability, the brown agenda has tended to be pursued through pro-growth strategies that largely conform to the expectations of the economic bosses of the previous regime plus foreign advisors, investors, and tourists. Such growth sometimes explicitly involves intolerance or impatience with the green ideals now expressed in the constitution and law. The national minister of housing put it this way when warning against delays caused by the legally mandated environmental impact assessments for a proposed massive housing development on the thornveld approaching the city from the Durban side: "We can no longer be held hostage by butterflies' eggs" (Lindiwe Sisulu, quoted in Macleod 2006).[1] Not-so-subtly coded here is blaming privileged white conservationists for black poverty.

Few dispute that growth so conceived will impose costs upon the environment. But loss of habitat, increased carbon emissions, higher levels of pollution, and so on, are simply seen as the price of progress. Such growth may also require some building or marketing that addresses the expectations of established but deeply problematic international investment and tourism. Tourists mostly want to see "the big five" in an imagined pristine wilderness or to experience a "real Africa" heavily redolent of colonial nostalgia. Msunduzi cannot provide the former but it has rich potential for the latter. This potential creates tensions, which are abundantly evident in municipal documents. One hardly expects sober analysis in tourism bumf, but does this (ANC-approved) city propaganda today not sound much like Paton's paradise: "boasting a benign climate and natural splendour, the city is growing into a popular leisure destination," with its "treasure trove of buildings" and "an array of master-built Victorian mansions that increasingly are coming into vogue as choice residences" (Msunduzi Municipality 2006, 65, 67). The showcasing of select gentrification sits poorly with the ethic of social justice, but we are presumably invited to accept it as a small price to pay for jobs under currently prevailing economic conditions.

To be sure, a niggling sense of doubt does sometimes creep in among proponents of the "balanced" view. Various crises anchored beyond the city's or nation's borders (global climate change, the "great recession," demographic explosions elsewhere in Africa), and stubbornly low levels of concern for – or even active hostility towards – the environment among the majority population, for example, give occasional cause for worry. The study by Roberts, wa Kivulu, and Davids (2010, 193) has especially disturbing implications for a place like Msunduzi. Their survey of public opinion found "environmental concern" to be lowest among urban informal dwellers, black Africans, low-income earners, people with low education, and people who believe the Bible is the literal truth. All of these characteristics are overrepresented in Msunduzi's population compared to other cities in the country, and they portend significant political challenges to making sound environmental choices. The prevailing assumption among brown environmentalists, however, is that such obstacles will eventually be solved through poverty alleviation, education, technological innovation, and respectful consultation with the people.[2]

A third progress narrative paradoxically focuses on the negative impacts of economic growth and the seeming irreconcilability of green and brown, as conventionally conceived. This narrative sees us on the cusp of a defining moment of environmental crisis. That crisis is not only causing great harms but is also engendering new social movements to fight them and to imagine healthy and just alternatives to hegemonic ideas about progress, growth, and modernization. Player ([1964] 2007, xxiii), for example, expresses great optimism with respect to a burgeoning environmental consciousness linked to the transformation of South Africa into non-racial democracy. Others have praised the achievements, courage, and vision of grass-roots, African feminist-inspired groups like Abahlali baseMjondolo (the South African shack-dwellers movement). Eco-socialist networks in alliance with a global anti-capitalism, post-development, or degrowth movement hold promise to reconcile green, brown, and red (politics explicitly and primarily focused on the erasure of wealth disparities). This would entail the radical re-measuring, reconceptualization, and redistribution of wealth that promotes environmental justice and human well-being.[3]

My own heart pulls me somewhat in this direction, and I will argue that case in chapter 7 below. My rational brain, however, quails at the fragility of the efforts when stacked beside the enormity of the tasks. Indeed, notwithstanding electrification and other welfarist or brown provisions since 1994, many today argue that polluted living conditions for the poor in South Africa have in many ways actually gotten worse over the past twenty years. The fact that more homes are now hooked up to electricity means little to the people who live there but cannot afford to pay for the service and so still burn coal, kerosene, tires, or scraps of particle board and plastic. For ample supplies of the latter, the municipal dump makes for good pickings, with at one time as many as a thousand trespassers on the property each day rummaging for scrap and discarded food (Hallowes and Munnik 2008, 157). For others, however, the spontaneous combustion of garbage that frequently pours acrid smoke from the dump over the neighbouring suburbs is a highly visible sign of a wider environmental crisis that can be directly linked to the failure of post-apartheid governance within a neoliberal economic framework.[4] The RDP houses that have colonized the hills of the area known as France are meanwhile not only smaller and more shoddily built than the old matchboxes of Sobantu, Ashdown, and Imbali townships, but they are prone to high levels of indoor pollution, broken toilets, and poor water drainage, and their inhabitants to social anomie (Greater Edendale Environmental Network 2002; Savage 2007; Goebel 2015).

Harmful legacies of environmental injustice can be more subtle, and intractable, than visibly fouled rivers and smoking mountains of garbage. Glen Elder (2003) coined the term "malevolent geographies" to suggest how

2.1 France: RDP homes spread across the former white farmlands

the design of urban spaces in South Africa created pathological and often violent social relations that will impede the development of environmental and social justice consciousness for many years to come. Similarly, Jennifer Beningfield (2006) analyzes the enduring imprint of white male "imaginative entitlement" upon the physical landscape of South Africa through urban segregation, Bantustanized rural areas, idealized (and discreetly militarized) wilderness areas, and a host of totalitarian and other architectures of inequality that fuelled alienation and epidemic diseases among the majority population. The "trauma that inheres in our landscape," as Albie Sachs puts it in his preface to Beningfield's volume (2006, x), is frighteningly manifest in farm murders/militias, land invasions, and the politics that have roiled neighbouring Zimbabwe since 2000. It is evident as well in the way in which environmental conditions favourable to the transmission of communicable diseases have been highly concentrated in certain areas of the city, and remain so. A conservative estimate in 2003 placed the number of people living with HIV/AIDS in Msunduzi at 88,000, of whom at least 90 per cent were black and up to two-thirds female (BESG 2003a, 7).[5] This aspect of Msunduzi's malevolent geography clearly threatens to cement, through ill health, some of the most grotesque inequalities of the apartheid era.

Creative ideas on how to transcend inherited traumas without creating new ones, and for challenging the green/brown and other polemical dichotomies, are coming from a wide range of planners, scholars, and civil society organizations around the region, sometimes reaching back to precolonial or indigenous knowledge about environmental management. From my observations in Msunduzi, unfortunately, awareness of the history of the local environment is scanty. Basic facts – such as that maize, donkeys, gum trees, and wattle are not indigenous or that South African scientists were once among the best in the world in environmental management (as were health professionals) – tend to get lost in the prevailing political struggle or economic underdevelopment narratives. Could looking back at history with fresh eyes help us see past blindspots about the environment and hence free our imaginations to create a more sustainable and just re-ordering of urban space? I would suggest that the answer to that question is yes – even with an understanding of the limited ability of *any* form of scholarship to affect change. To get there, let us first look at how knowledge about the environment in the region has changed over time and how different theories of human-nature interactions can guide us to a richer appreciation of what these might mean for the politics of health and healing in Msunduzi today. The first section to follow examines popular culture, while the second focuses on critical scholarship as it emerged to explain gaps or contradictions in the evidence.

## SEEING AND MAKING THE MIDLANDS

The environment and climate featured strongly in writing about, and justifying, empire in southern Africa. They provided a seemingly scientific fig leaf for a certain type of white supremacist ideology, viz., temperate climate and heavy soils in northern Europe gave rise to clever, disciplined, agro-industrialist people, fit to rule over slothful, lascivious tropics dwellers or incompetent, dreamy, war-like pastoralists. This notion in fact long predates colonialism in Natal. The very first specific description in English of the people living there (by "Hamilton," 1683, recounted by John Bird) focused on the slothful side, opining that "the natural fertility of those countries … made the natives lazy, indolent, indocile and simple" ("East Indies" in Bird 1888, 25).[6] Such views became a staple of promoters of and apologists for British imperialism by the early twentieth century. Alfred Bryant, notably, was the pre-eminent historian of the Zulu people in that era. He conjured pre-*mfecane* Natal, and particularly the Midlands area surrounding Pietermaritzburg, as "a Black man's arcady smiling in the joy of peace and plenty and perpetual sunshine" ([1929] 1965, 236). "Nature's own children" lived there in timeless harmony with their domestic animals so that, while numerically abundant due to the bounty of the land and their own natural fecundity, they were obviously never going to amount to much. Indeed, their paradise was peremptorily destroyed by Zulu invasions from 1820, "polluting the virgin sward with gore and putrid corpses" (381, 557) and opening the way for supposedly fitter people to occupy it.

For Bryant this imagined prehistory of colonialism served two purposes. First, it disproved claims by Afrikaner trekkers and certain British settlers that they were occupying a country that had been empty of prior settlement (and that contemporary African inhabitants thus had no historic claims to be there). Second, it justified British intervention both to protect the hapless African population of Natal from their enemies (Afrikaner and Zulu) and to promote the development of Natal's natural assets for the benefit of all humanity. The Zulu, as cattle-obsessed savages, had proven easy enough to manage by the time Bryant wrote, the Boers less so. Indeed, while Afrikaners may have come from promising stock in Holland and France, too long living as pastoralists in the African veld had sapped their cultural vitality. The British takeover from them thus enabled Natal's reconstitution into a productive garden from what would have become, as the first premier of the self-governing colony put it, a "paradise of sluggishness and stagnation according to Boer ideas and aspirations" (Robinson 1900, 70).

Like the Boers, Africans were seen as being in dire need of British industriousness and benevolence. As Thomas Bulpin (1966) explained in his chapter on the Africans in the Pietermaritzburg area in the early colonial era, "the debilitations of tropical diseases, bilharzia and fever, had left them a legacy of cultural lethargy … They were floundering about, making a hopeless mess of agriculture" (221). Clearly, more and better supervision, with predominant ownership of land by British settlers, was required for Africans' own welfare.

An imagined primordial Africa peopled by incompetent naifs with debilitating tropical diseases played a significant role in the early conservationist movement internationally (Grove 1989; MacKenzie 1988). Down to this day those images remain a staple of developmental and tourism discourses, framing debates with wild animals, deep jungles, and pristine landscapes, on the one hand, and a long-standing Malthusian "crisis narrative" about desertification, deforestation, relentless population growth, and sick, ignorant, or exotic African peasants/Bushmen on the other (Leach and Mearns 1996, notably). Such imagery was commonly invoked to justify authoritarian or militaristic conservation programs in Africa, often to the frank disregard of local African perspectives and well-being, let alone empirical evidence (Maritzburg is not in the tropics and has no endemic tropical diseases). While in South Africa this is most acutely felt in the big game parks (Meskell 2012), one finds the imagery intruding in the literature on Msunduzi and area as well, often within stories ostensibly about something else. Note, for example, the photograph in Laband and Haswell (1988, 254), in a chapter that discusses the origins of the Dusi canoe marathon as a means to raise awareness of local environmental issues. The photo depicts a naked black boy lying on the sand while a white sport/environment enthusiast whizzes by in a kayak. "*Yeka lamandla agangayo*," the caption has the boy wondering – "What a waste of strength!" The incomprehension of "tribal Zulus" downstream of the city is also taken as a point of humour in Graham Linscott's fawning biography of the race's founder and pioneering conservationist, Ian Player (2013, 28). Ironically, Player's own vaunted spirituality around nature anchors itself in an imagined pre-colonial conservationist ethic with its roots in the aristocratic, militaristic concept of King Shaka's "royal hunting grounds," for which there is scant basis in Zulu history (Brooks 2009).

An Africa that was feminized in romantic popular imagery could also be invoked to justify or normalize masculine, imperial subjugation and ordering of space (McClintock 1995). Tellingly, for example, both Bryant ([1929] 1965) and Robert Russell (1911) apply the term "Arcadia" to pre-*mfecane* Natal, invoking an idyll from Greek mythology that was populated by naked female nymphs and over-sexed forest sprites. The Midlands landscape in pre-colonial days lent itself well to that gaze, its softly curvaceous hills rising up to lap against the stern cliffs of the Drakensburg. The article by former American consul to South Africa Russell Millward, "Natal: The Garden Colony" (1909), also provides a revealing illustration of this trope from an emerging and admiring imperial power. Following immediately after a long article that lionized President Theodore Roosevelt's hypermasculine hunting exploits in east Africa, Millward's contribution contains numerous photos of bare-breasted African women and girls in rural and village settings. It contains not a single clearly identifiable townscape. In his text, meanwhile, even the big city does not escape the feminizing treatment, Durban being "the prettiest and most coquettish" city in the South African colonies (278). Tellingly, even the science of biology was feminized in popular discourse, and one of the commemorated contributions of the pioneering biologist of Natal, John William Bews, was how he instilled a sense of manly adventure in the pursuit of environmental knowledge (Gale 1954, 61). Nostalgia for the days when men were men and Africa was metaphorically a woman to be wooed or conquered recurred well into the 1980s. One hagiography put this bluntly with reference to pioneering white settler Joseph Baynes: "Just as a man takes unto himself a lover, so Joseph possessed Maritzburg" (Pearse 1981, 32). Critics of colonialism also sometimes employed gendered language, equating the process to rape, with the metaphorical heterosexual violation of the African landscape a defining aspect of white South African culture.[7]

This trope of feminization dovetailed closely with another conceit of the colonial era, viz., in addition to protecting Mother Nature from Africans, British men would (eventually) emancipate African women from the unspeakable barbarism and drudgery they experienced at the hands of African men. Some admiration for the African patriarch was allowed, provided it was contained within romanticized natural settings. Referring to an area just beyond they city's borders in 1925, for example, the *Natal Witness* described an area of Zwartkop that was already starting to feel the impact of migrant labour and peri-urbanization as "The Beautiful Valley of a Hundred Kraals: Africa's Barbaric Children." The "pastoral peace" was made flesh in the person of Chief UmVova, that "magnificent specimen of the finest Zulu physique" (*Natal Witness,* August 25, 1925). Most authors, however, quickly reverted to the trope of African men's lasciviousness or brutishness when broaching the topic. Periodic moral panics around the "Black Peril" to white women in the late nineteenth century belied a taken-for-granted notion that black men's oppression and abuse of black women would need to be tempered by the hand of European civilization. Much of the rumination upon the Native Mind that prefigured the rise of ethnopsychiatry as a science was really about African men's gonads and how best to discipline them.[8]

A number of other fascinating tensions emerge in the early writing about Natal. On the one hand, a strong motivation among authors was to entice immigrants from England with glowing descriptions of the land and climate. Propaganda from the late 1840s deliberately concealed the limitations of the environment (from a commercial productivity point of view) and conjured in its place the image of an extended "nobleman's park" open for easy development (Emery 1985, 5). Its scenery, as described in James Methley's opening paragraph, "immeasurably surpasses England." "What an immense extent of hill and dale is spread around. What myriads of gorgeous flowers, that glow like gems amidst the verdure of this grassy ocean: the vast primeval forests, where the huge straight trunks of the trees are wound around with gigantic vines and blossoming creepers" (1850, 1). Maritzburg itself was "surrounded by hills, the sides of which *are mostly clothed with wood* and present, the year round one continued appearance of freshness and verdure" (12, my emphasis). To counter stereotypes of inhospitable Africa, pioneering trader Henry Francis Fynn portrayed the region as an inviting enclave, "its soil, climate, productions, and aspect, which differ essentially from those of the surrounding countries: so widely, indeed, that it would appear as if Nature itself had set boundaries to that district" (Fynn 1839, reprinted in Bird 1888, 74).

Less tendentious descriptions, by contrast, disparaged many aspects of the indigenous environment. These describe the hills surrounding Maritzburg as desolate and barren, covered by "rank and useless vegetation" (in the words of the Lt. Governor Benjamin Pine in 1850, quoted in Hattersley 1950, 252–3). The prevailing treelessness was a "deformity" (quoted in Emery 1985, 4) that called out for afforestation with useful exotic species. Such observations underpinned the deep reluctance of the Colonial Office to assume responsibility for a place "too worthless to justify throwing the burden on our national resources" (quoted in Guy 2013, 138).

Even after the commitment had been made to make Natal a viable colony after all, the worthlessness trope still served a useful purpose. It strengthened the improvement narrative, which centred upon the introduction of supposedly more beautiful and useful exotic species, the progress of plantations and fences across the Natal countryside, the transformation of the dusty townscape into garden city, and the emergence of Christianized, market-oriented Africans at Edendale who shared the aesthetic. Not unlike the admiration for magnificent Zulu men, nostalgia for the lost herds of big mammals and expanses of open lands did creep into some accounts (Robinson 1900, 73–4, for example). Yet on the whole this genre of writing presented the transformations as a positive thing, demonstrating the affinity of British imperialism for making nature better. In some cases this myth making went to the extent of retroactively erasing the memory of the real historical geography. John Ingram's paean to Pietermaritzburg settlers, for example, illustrated "the wilds of Africa" that the settlers had supposedly tamed as a dark and dangerous-looking jungle, rather than the far more common grasslands of the area (1898, 4). "A Natalian" went even further

2.2 Maritzburg, ca 1900. Not even Town Hill (background) is yet clothed by the trees that today are commonly regarded as part of its "natural" beauty

in his 1897 memoir, retrospectively covering Mbube mountain "black from base to summit with forest" in the late 1850s (1897, 18). In fact, paintings and photographs from as late as 1880 depict it mostly grassy or with only pockets of forest on the lower reaches. With the mountain almost completely bare of trees, one painting from 1880 already hints at soil erosion gulleys, presumably from rough-hewn wagon tracks (Wedderburn 1991, 229–30).[9] "A Natalian" was evidently overzealous in imagining the magic of British colonial afforestation.

In numerous other cases over a long period of time, the Midlands' beauty was associated with the degree to which the landscape had been shaped by the colonial hand to approach romantic ideals of a more familiar place. "And yonder, as we open out the valley, lies Edendale. But surely this is not South Africa at all. This is a piece of Yorkshire" (Statham [1881] 1969, 199). As late as 1933 Plessislaer resident Frank Lee defended the area against claims that it was a slum: "the peaceful surroundings, lovely walks and views with beautiful river scenery remind me forcefully of rural England (*Natal*

*Witness*, 7 March 1933)." Indeed, even today, tourism promotion evokes meandering through the Midlands' green and pleasant, shire-like ambience, highly distinctive from the usual Africa brand.

Dissident opinions did occasionally enter the discussion. Louisa Grace Ross, notably, was in the 1860s a rare voice extolling the beauty of indigenous forests ("primeval patriarchs"), naturally grassy hillscapes, and the sparkling streams of Edendale. Her view of the spread of exotic trees is worth repeating in that it was so far ahead of her time: "Hateful civilization! How I loathe you ... What Vandalism thus to mar the earth and rob her of the beauties God has given her" (Ross [1864–65] 1972, 101). Robert James Mann, superintendent of education in Natal in the 1850s and otherwise a fervent booster of British immigration, also conceded to a dim view of early settlers' wasteful destruction of scarce indigenous trees: "sordid colonists ... thus stripped the soil of its garniture" (Mann 1859, 101). Such views, however, were not taken especially seriously until well into the twentieth century, when H. Raymond Burrows's authoritative survey cited the colonialists' "ruthless destruction" of indigenous forests as a legacy Natalians would need to deal with (1951, 73). Far more typical of the late nineteenth century was Lady Barker's assessment: "Gum-trees seem the nurses of all vegetation in a colony: they drain the marshy soil and make it fit for human habitation" (1877, 149). Among other things, the health benefits of such plantings seemed obvious. As one letter writer to the *Natal Witness* put it, "a plantation of gum trees on the flats – formed of vegetable deposits – around the town would very largely extract the poisonous vapours which, diffused as at present, induce fever" (*Natal Witness*, 17 December 1886)."

Scientists eventually began to cast doubt on such health and environmental claims (Bews 1913, notably 521, although even he judged certain indigenous flora as "worthless"). Policies were then gradually put in place to temper the promotion of exotic flora over indigenous. Yet, for many, the strong presumption of an inexorable, triumphant value in exotic (that is, modern) flora lingered well into the late twentieth century (if not still). The first indigenous trees to be planted in the city for aesthetic purposes (some rows of Natal mahogany) were not planted until 1920, but these remained anomalous for decades thereafter (McCracken 1988, 60). The city's engineering department stated the aesthetic quite baldly as late as 1983 when praising its colonial predecessors' wisdom: "Pietermaritzburg's reputation as a city of beauty stems largely from the surrounding tree-clad [that is, with introduced Scotch pine and wattle] hills" (Meineke and Summers 1983, 90).

Highly technical, positivist, or instrumental approaches to geography that were in vogue through much of the twentieth century contributed to the perpetuation of problematic tropes, blindspots, or essentialisms in popular culture and political discourse. Academic practice thus obscured the ways in which the racialized political economy affected both the land itself and the gaze of those that the system privileged enough to afford to do research and to have a voice. The discipline's "prostitution to science" (Maharaj and Narsiah 2002, 89) in effect blinded researchers to the existence of manufactured swathes of poverty and land degradation within and beyond South Africa's borders (such observation would have been suspect as "political"). There was also an unmistakable bias towards the study of rural or wilderness-seeming environments to the neglect of the urban. As Maharaj and Narsiah point out, throughout the entire period between 1917 and 1980, the *South African Geographical Journal* published exactly two articles focused on black townships (89). Edendale was not one of them. Indeed, somehow Edendale and the middle Msunduze valley even slipped under the gaze of the father of grassland science in South Africa, John William Bews. Other than a single photograph of an uninhabited section, and a list of plants that inhabit the rocky hillside formation, Edendale is absent from his otherwise meticulous and historicized description of the Midlands' flora (1913, 527 and plate XLIV).

Writing about health and healing in Natal evinces similar patterns and tensions. As Karen Flint (2008) describes, early colonial impressions of African health in the Midlands were generally positive. The strenuous outdoor life, balanced diet, the low alcohol content of traditional beer, the relative absence of infectious diseases and parasites, the abundance of fresh water, and the culture of wifely submission and parental discipline were all thought to contribute to the robustness of the "Zulu race." This general good health existed in spite of a culture of healing that Europeans mostly found risible or positively dangerous. While giving some credit to herbalists, who could identify plants or other naturally occurring ingredients with effective pharmacological qualities, and allowing for a placebo or psychosomatic factor in some ritual practices, they tended to dismiss African beliefs in sorcery, witches, imps, ancestors, and other metaphysical explanations of ill-health as ignorant superstition. Lady Barker's condescension towards female diviners (*isangoma*) is typical, inviting them for tea primarily for the amusement of her guests and reading public (Barker 1877, inside cover and 169). Bryant ([1909] 1966), while casting himself as a friend and defender of the native, was positively racist in his contempt for African ideas about medicine. The colonial state took steady steps to repress traditional beliefs and practices in that regard, first banning witchcraft and witchcraft accusations (1847) and then rainmakers and healers (1862), and then instituting a licensing system for herbalists (1891). In tandem with missionaries and private health practitioners (see, for example, McCord 1946), Europeans sought to propagate a Western biomedical model to replace Africans' approach, much as they had promoted useful exotic fauna to replace the worthless indigenous species. When in doubt, as in Bulpin (1966), authors simply fabricated a pestilential past that proved the benefits of progress.

European authors were by no means to only ones to express such contradictorily judgemental views. Although explicit depictions of landscapes by Africans are exceedingly rare from this early period, one can easily infer from oral testimony (Webb and Wright 1986), novels (Dhlomo 1928), and published sketches for an assumed European audience (Savory 1965) an implicit nostalgia for the rural life. R.R.R. Dhlomo's *African Tragedy*, notably, contrasts the happy village of "Siam" (Siyamu, neighbouring upon Georgetown) with corrupt and bleak city life. Yet it also asserts that African healing practices were "digusting and humiliating" (Dhlomo 1928, 28). Tensions between nostalgia for the rural past and the desire to be modern also played out in popular music (Ballantine 1993). How those tensions were refracted in representations of the environment in isiZulu sources over the next few decades has yet to be researched. It is telling, however, that one of the first rehabilitations of Shaka as an African nationalist icon (by R.R.R. Dhlomo's younger brother Herbert) added environmental consciousness to his edifying character. The great tyrant of colonial infamy became, in Dhlomo's hands, an "experiment-minded progressive" who "inspired the study of nature," and hence stood as a pre-colonial rebuke to colonial conservationist lies about Africans (H.I.E. Dhlomo 1954).[10]

Dhlomo's fictional "green" Shaka was an outlier. African invisibility or culpability remained by far the dominant tendency in representations or discussions of the environment, amply established in paintings, literature, and official analysis by whites throughout this period in a variety of ways. As Laband (1988) and Wedderburn (1991) detail, Africans in Pietermaritzburg either appear as little more than a token of the exotic locale in an otherwise domesticated (and amplified) land and cityscape, or they are entirely absent. They are even almost invisible to the camera:

2.3 Africans in the frame: the village of Georgetown nestled in among the hills, seen from the east, ca 1890

of ten catalogued photos of Edendale in the provincial archives, only one captures Africans in the frame, and even then they are too far away to discern their faces in a shot that emphasizes the panorama. In other cases in the literature, Africans are depicted as the principal cause of their own implicitly deserved downfall. Blaming Africans for the consequences of a century of environmental racism is a theme I will return to, but we can note here that it is even evident in studies that are otherwise self-consciously "liberal" in the South African context. Brookes and Hurwitz's survey of the native reserves of Natal is a case in point. While strongly condemning the negative environmental impacts of the migrant labour system and recommending the promotion of qualified black professionals over whites in reserve administration, Brookes and Hurwitz ultimately blame food shortages and soil erosion on Africans themselves for their "conservatism of attitude" (Brookes and Hurwitz 1957, 129).

## ENVIRONMENTAL TO ECOHEALTH HISTORY

In the previous section I described how artists and intellectuals played an important role over the first century and a half of Natal's history when they portrayed the environmental and health impacts of colonialism as either desirable (such as the flowering of the city and the afforestation of the hills around it), invisible (outside the purview of scientific research), or regrettable but primarily Africans' own fault. Environmental history and historical geography as serious modes of academic enquiry had to overcome such cultural baggage. They grew out of a fruitful engagement with archaeologists and the observations that, however obscured by racist ideologues or modernist presumption, environmental factors *were* often demonstrably central to historical conflicts and that landscapes were closely entwined with processes of social, political, and economic change. In pre-colonial southern African history, these factors included periodic drought, overgrazing around population concentrations, and competition for scarce valued commodities like gold and ivory, all of which could be linked to the rise and fall of states such as Mapungubwe, Great Zimbabwe, or the Zulu "empire" of the early nineteenth century. Southern Africa in this respect was profoundly affected by global exchanges long before it became directly ensnared in the emerging world capitalist system. Indeed, a compelling case has been made that the current ethnic make-up of the whole region of southern Africa – including the fact that isiZulu is the predominant African language spoken in Msunduzi – owes much to the importation of maize from the Americas by the Portuguese in the sixteenth century. As the relatively nutritious and productive plant spread throughout the region from ports in Mozambique and Angola, it sustained steadily growing populations. Maize, however, is susceptible to drought, and when a prolonged one affected southeastern Africa in the late eighteenth century, hungry people began to raid their neighbours for survival. This in turn gave rise to militarized states, sometimes notoriously aggressive and aggrandizing (as in the case of Shaka's kingdom), sometimes primarily defensive (as in Moshoeshoe's). Ethnic identities (Zulu and Basotho in the aforementioned cases) adhered around these successful political units. They then spread through migration (the Ndebele, Kololo/Lozi, Nguni, and so on) as far afield as western Tanzania through the so-called *mfecane*.[11]

Care was taken in such analyses to avoid environmental or climatic determinism and moral judgements or sentimentality about different types of environments. The new historiography analyzed the environment as a contributing factor to socio-political change and cultural forms rather than a defining or determining one. As popularized by such authors as Robert Bullard in the United States (for example, Bullard and Wright 2009, which develops the concepts of environmental racism and injustice) and Jared Diamond (whose 1997 *Guns, Germs, and Steel* includes a chapter memorably entitled "How Africa Became Black"), this analysis often begins with a critique of racism as a factor in various areas, including technological development, the location of toxic dumps, cultural change, and political struggles over natural resources. For example, from the beginning, environmental historians of Africa were concerned with understanding the role that the environment played in the construction of racist colonial ideologies and disciplinary structures. They demonstrated how the commonplace imagery of Africa as an essentially wild place – and of Africans as essentially ignorant and incompetent to manage it – constituted an important strand within an overarching framework that justified a whole range of violent interventions against Africans. In Charles Mather's phrase, the environment was an important rhetorical "weapon" in the hands of white South African farmers, who used scientific-sounding language about the threat of supposedly destructive black farmers to strengthen their case to have Africans evicted from the land (1995, 231).

Another theme emerging from environmental histories from the 1980s onwards was that well-intentioned efforts to protect or "improve" the environment were frequently based on faulty science and in many cases actually contributed to environmental degradation.[12] Africa's unique ecologies were poorly researched and/or inappropriately assumed by European scientists to be analogous to more familiar ecologies elsewhere in the world. Hence came soil erosion control campaigns based on Tennessee experiences, afforestation based on Scottish precedents, and the introduction of exotic species of flora and fauna ostensibly to solve development problems in Africa as had presumably been done in Australia or South America. In other cases there may have been a Machiavellian element in the initiative. Karen Middleton (1999), for example, suggests that French officials in Madagascar may have conspired to abet the propagation of an insect pest that conveniently hastened the extinction of the prickly pear, a backbone to the local subsistence economy that had long enabled Malagasy men to resist French plantation owners' recruitment of low-wage migrant labour. The coerced population movements and concentrations to make room for parks, plantations, and white farms that often ensued from poor science had observable negative impacts on the land and on African peoples. So-called betterment schemes and conservation programs consequently often became a focal point of African political resistance to colonialism and apartheid. That resistance in turn fed back into alarmist portrayals of conservative, ignorant, lazy, and ungrateful Africans who caused environmental crises.

Generally opposed to that crisis narrative are studies that illustrate the diverse ways that Africans participated in and positively shaped debates about the natural world and "progress." For example, Luig and von Oppen argued from the ethnography that, while "African languages have no proper terms of 'landscape' or 'nature' in the abstract sense of the European equivalents," African societies did have a rich metaphorical or symbolic language that captured meanings of difference, danger, harmony, and healing in relation to the natural world (1997, 21). Very often this entailed a concept of guardianship over the land under the watchful gaze of ancestral or other spirits. Physical processes such as drought, migration of animals, lightning, and fire interacted with metaphysical forces and were in turn closely imbricated with political power and process. The Rozvi ruler or *Mambo* in ancient Zimbabwe and the Rain Queen of the Lovedu in Limpopo, to use two famous examples, emphasized their mystical power over life and death, fertility, rain, preservation of natural resources, and the protection of their people from enemies. In other cases, a ruler's ability to prevent environmental harms such as loss of soil fertility or the desecration of sacred forests was a central aspect of his or her political cachet. Close local knowledge about natural resources and environmental management enabled sometimes relatively dense populations of people and domestic and wild animals to co-exist in stable ecosystems. Bringing religion and culture into the analysis, as Terence Ranger demonstrates in his study of racialized struggles over the Matopos area of southern Zimbabwe, can help to explain some of the frustrations and disappointments encountered by formal scientific environmental management methods in Africa (Ranger 1999; see also the wide range of case studies examined in Sheridan and Nyamweru 2008).

As with H.E.I. Dhlomo's and Ian Player's writing, and more recently Mogobe Ramose's thoughts on *uBuntu* as a potential path to a sustainable society (Ramose 2015), there may be a tendency in this type of analysis to romanticize African philosophies towards the natural world or to overstate their "green" credentials in order to critique Western attitudes and capitalist practices. A cautionary note is thus advised. For example, pre-colonial society was strongly patriarchal and gerontocratic, and oral traditions reflect this by remembering the exploits of great men and rendering the poor, the young, and women mostly invisible. In that sense, colonial ways of thinking, representation, and learning could be (indeed, demonstrably were) oppressive but also revelatory and liberatory. Moreover, what may once have been sustainable or environmentally astute – like sacred groves and respect for mermaids – rarely functioned well under colonial and capitalist conditions, and are unlikely to do so in the foreseeable future. Traditions became garbled, or transformed under the pressures of restrictive political borders, market production, rapid population growth, new technologies and exotic species, and the material benefits that expert knowledge and the profession of Christianity promised to certain classes of people. With respect to the latter, Diana Jeater (1995, 2007) has shown how missionary translations of chiShona erased subtleties and contradictory meanings from the language while conveying positivist messages that suited the interests both of the colonialists and a small class of African "progressive farmers." Likewise, some Africans stood to gain by aligning themselves with and promoting the culture of blame against supposedly primitive African agricultural practices and cruelty to animals. As Jacobs (2003) shows in her study of the "great Bophuthatswana donkey massacre" in the 1980s, this could result in the same or even more extreme levels of violence enacted by African leaders against African farmers or herders in the name of conserving the environment.

Environmental histories elide easily into the history of health in Africa.[13] Path-breaking studies of the epidemiology of specific diseases in fact often began with accounts of the sudden and dramatic environmental changes introduced by colonialism and capitalism. Exotic animals introduced exotic parasites (rinderpest, notably, which spread from Somalia to devastate African peasant economies at the end of the nineteenth century, with powerful ripple effects on gender and intergenerational relations). Dams created stagnant pools of water, which contributed to malaria and schistosomiasis; colonization of tsetse fly–infested zones exposed people to sleeping sickness; underdevelopment of the labour reserves gave rise to diseases of malnutrition such as pellagra; and the migrant labour system fuelled epidemics of sexually transmitted infections. The concentration of large, often undernourished and overworked populations in unsanitary camps or "native locations" also enabled a host of epidemics, most fearsomely, in the early twentieth century, the bubonic plague. This in turn led to what Maynard Swanson famously called "the sanitation syndrome," whereby colonialists cited Africans' ill health and supposedly low standards of hygiene as an imperative to physically segregate them from whites. The erroneous belief that African children were the main vector of malaria was similarly cited as justifying laws to keep them (and by implication their mothers) out of the urban areas, consolidating with poor science the self-interested economics of a cheap, migratory male labour force.[14]

An analysis that links health, social inequalities, and sexuality comes full circle to environmental history through the concept of "new variant famine" (de Waal and Whiteside 2003). In that sense, HIV/AIDS is an environmental as well as health calamity. When people are desperately ill with a shameful disease, they tend to abandon fields and the maintenance of conservation works and not to take part in the co-operative labour that is normally needed to avoid harmful environmental impacts of subsistence farming or simply to keep the neighbourhood tidy. When people die unexpectedly in what should be the prime of their life, they also tend to leave behind much unfinished business that further complicates housing, health, and other development initiatives: orphans, above all, but also tangled property titles, incomplete household repairs, and abandoned pet dogs (that now roam some southern African neighbourhoods and villages in feral packs). And does anyone know what are the long-term impacts upon the ecosystem of millions of people consuming (and excreting traces of) powerful anti-retroviral medicines?

Despite the growing sophistication of research linking the study of human health and the environment, several decidedly under-investigated areas remain. A notable blindspot lingers in environmental history with regard to women and gender. Finding and representing African women's perspectives is unquestionably methodologically daunting, despite the long-recognized central role of African women in environmental management at the micro level. The easiest research task in that respect is to disaggregate people by the gendered division of labour and to examine their differential impacts on the environment. Studies using such an approach have focused on new crops, animals, labour impositions, and the expansion of the cash economy under colonialism. Often these were unambiguously negative for African women and children: men's migrant labour in German East Africa removed them from their traditional task of hunting and led to a population explosion of wild pigs which decimated women's subsistence crops (Sunseri 2009); women's compulsory cultivation of cotton in Portuguese East Africa led to exhaustion and malnutrition (Isaacman 1996); the loss of women's usufructory rights to land under codified patriarchal "custom" or private land tenure regimes, and both market and state incentives that promoted men's control of production, led to, among many specific examples, the neglect of women's sophisticated *matengo* system of terracing for subsistence crops in Tanganyika in favour of men's soil mining for commercial crops (Mackenzie 1995, 102). In other cases, however, the gendered changes in production had positive, or at least mixed, results. The introduction of pigs ("women's cattle") in colonial Lesotho and donkeys (the equivalent of women's horses) gave women access to meat, transport, and draught power they otherwise were not traditionally allowed, a huge asset, albeit one that came with some serious disease and erosion effects (Epprecht 2000). Female-led environmental protests have also gained attention in the scholarship as distinct from, and in some cases more effective than, male-led movements (Steady 2009, for example).

Feminist analysis is sometimes faulted for understating conflict and differences between women, or for reasserting the degradation narrative in new but equally problematic forms by essentializing women as closer to "Mother Nature" than men, with men and masculinity then taking the blame for environmental crises. Feminist political ecology and socialist ecofeminism seek to guard against such dead ends by insisting upon a materialist analysis of gender roles and stereotypes in relation to other shifting cultural constructions like race.[15] Glen Elder's (2003) application of queer theory to a micro-urban environment (former men-only hostels in Johannesburg) also suggests some very interesting potential to make hidden normative assumptions visible to researchers. *Queer* in this instance is understood not as a politicized homosexual identity but as an approach to research that destabilizes all taken-for-granted identity categories or analytic concepts including *man*, *woman*, *city*, *country*, *black*, *white*, *sex*, *natural*, and so forth. A queer approach would emphasize the contingency, fluidity, and subtlety by which sexuality pervades human interactions and is mutually constitutive of such matters as architecture, state formation, labour, land tenure regimes, and human health. Mark Hunter (2010) eschews the term *queer* but captures the shift in approach in his study of the "geography of intimacy" that fuels the HIV/AIDS pandemic in KwaZulu-Natal.

This scholarship, and other postcolonial critiques under rubrics like Africanity, postdevelopment, transnationalism, and ecohealth, constitutes a remarkable improvement in how we as scholars ask questions about and see the relationships between the environment and humans in Africa. It also raises the bar for us as researchers to a level that can, in a practical sense, be extremely difficult to reach. A key concern of ecohealth theory (Forget and Lebel 2001; Goebel, Dodson, and Hill 2010), also recently intimated by eminent environmental historian Jane Carruthers (2014), is that the research challenges are so vast that a transnational team is imperative – to access resources, to bring diverse intellectual and political traditions to bear, and to share lessons from potentially analogous histories that borders, of all kinds, often render invisible. The team must also be explicitly committed to transdisciplinarity with regard to sources, methodology, and heuristic lenses. Attentive to the wide-ranging debates and perspectives across scholarly literatures and academic conventions, such researchers hold that the ultimate objective of the knowledge they produce is to motivate individual consciousness and responsibility, community engagement, and policy interventions that will improve both people's and environmental health in a manner that is fair and truly sustainable.

Aspire as we may to these ambitious goals, it remains deeply challenging for researchers to find and digest African experiences and non-positivist epistemologies around environment, gender, and health. Sometimes this oversight is quite bluntly admitted, and without acknowledgement of postcolonial critiques of research in Africa. Harald Witt (2005), for example, opens his history of wattle cultivation in Natal by stating that Africans as well as Europeans adopted the exotic crop but that he will not be discussing the Africans. Although he does not explain this choice, we can probably assume that it is a reflection of the eurocentric nature of the archival record, and eurocentric assumptions about what constitutes reliable evidence (oral history presumed to be less trustworthy than written documents or "hard" scientific data, notably). In other cases, even when oral evidence is sympathetically collected from African informants, the difficulties of incorporating it into Western analytic models are evident. That tension can be seen in Kate Showers's critique of erosion control in colonial Lesotho. Showers is scathing in her assessment of colonial initiatives and insists that "the Basotho are *not* responsible for the massive erosion gullies that scar their landscape … The Basotho were victims of an untested experimental technology" (2003, 256–7). This could well be true. Yet when she allows her Basotho informants to speak directly to us, we hear that they quite strongly *do* hold themselves responsible. They explicitly blame themselves for not following the scientific advice they were given by colonial officials whom they often explicitly admired: "I wish we really went on doing them [anti-erosion works], but we were lazy" (229); "Yes, it [my field] used to have contours, but through carelessness, I destroyed them" (226); "People are like animals, they need a foreman [like the British]" (215); "They [the Basotho of those times] were blind, and did not know anything about soil erosion" (229); and so on. The glaring absence in Showers's analysis of metaphysical explanations to explain soil and fertility loss – such as witchcraft or ancestral disapproval – also raises suspicions about what may have been edited out, perhaps by the informants themselves so as to appear suitably modern to their interlocutor.

Showers's work poses some difficult questions. Which colonialism is the biggest worry, and, hence, most in need of our critical attention? Is it the actual colonialism of the British, long gone but living on in a bureaucratized development state that aspires against the odds to be a nation? Is it the hegemonic culture that robbed Basotho of their former self-assurance and lingers on as "common sense" through positivist images about the land and people? Or is it the academic practice that does not hear or dismisses Basotho perspectives when they do not fit the preferred model or ascribed role? Meintjes (1988b, 42) alludes to the latter problem as well, noting how self-doubt and self-recrimination among African informants justified her decision not to pursue oral history too vigorously. It was a fair call, and there is no doubt in my mind that marginalized informants' self-blaming very much remains a serious concern. To self-blame, one might also add concerns with nostalgia. One could thus argue that the researcher has an obligation, first, to accurately perceive and then somehow offset the culturally colonized or victimized, self-blaming, and/or nostalgic mentality of African informants (particularly women, poor people, and so on). Yet it is not difficult to see how this can lead to paternalistic judgements, censorship, and the disempowerment of the very people said to be in need of empowerment.

How then exactly to decolonize scholarship, as South African geographers M.F. Ramutsindela (2002) and Jennifer Robinson (2006) insist, and to develop new ways of seeing the environment that can transcend mutual suspicions across racialized and gendered divides like green/brown? Many intriguing ideas are emerging from various strands of post-development thinking, including ecological accounting, citizen science, transition towns, slow food, *uBuntu* capitalism, re-indigenization, and much more. For me, these were anticipated by Carol Merchant's call for a "global ecological revolution" (1989), but now come together in the concept of *degrowth*. D'Alisa, Demaria, and Kallis (2015, back cover) succinctly defined degrowth as a burgeoning global movement to "repoliticize public debate colonized by the idiom of economism. It is a project advocating the democratically led shrinking of production and consumption with the aim of achieving social justice and ecological sustainability." Degrowth for now is generally thought to apply to developed or "overdeveloped" countries to bring down greenhouse gas emissions and other environmental stresses in a managed, co-ordinated manner while improving the quality of life in non-consumerist terms. The assumption is that degrowth in the Global North will allow the Global South to catch up in the medium term in building the basic infrastructure and social and health safety nets that are needed to reduce inequalities. This would then allow mutual, fair, long-term degrowth towards a steady state economy (with no depletion of non-renewable biophysical resources); high levels of local, autonomous, democratic governance within a global framework of enforceable environmental stewardship; and a "convivial" society (where human relations are mediated primarily by sharing, community well-being, and emotional fulfilment, rather than expert-driven, consumption-oriented commodification and industrial productivity). I wonder if the concept might already be helpful in a struggling South African city and perhaps in other so-called emerging markets in the Global South. Could degrowth provide a roadmap for imagining strategies out of the current dangerous trajectory and the racialized impasse between brown and green perspectives?[16]

Reaching back in time to reconstruct ways of knowing that predominated before political colonization, the scientific paradigm, the market economy, the English language, Christian ideology, and professional knowledge production is clearly not tenable as a degrowth (or any other) strategy. IsiZulu was itself a colonizing language and culture in the region prior to the arrival of Afrikaans and English. And there is no escaping the fact that science, even if it was the "handmaiden" of colonialism and apartheid (Wylie 2001, 234) and today abundantly serves neoliberal agendas, will be needed if we are to address the many environmental, health, and other challenges we face. What, then, would a decolonized science for degrowth look like? Or a decolonized history of Msunduzi? Or any other form of knowledge and activism informed by Enlightenment traditions and expressed in English (including Black Consciousness, Marxism, and feminism)? How does Africa degrow without inflicting further suffering on people already marginalized by ill-conceived growth, particularly if people in the Global North do not contribute fairly to the effort, as is abundantly the case so far and, from past practice, is eminently predictable for the future?

As a historian, I am not equipped to answer those questions, although I will be pointing out aspects of both traditional culture and earlier attempts at modernization that restrained consumption and may, as such, be adaptable for degrowth under future conditions (e.g., communal sharing; organic, local, and urban agriculture; and commuter trams and other public transport). It is also obvious even to a historian that any local initiatives will need to be supported by policy and attitudinal changes at the national, regional, and global level in order to survive – a tall order. I leave such initiatives for others to strategize and create. For the purposes of this chapter, and without meaning to be ironic, I will simply conclude that there has been remarkable progress in understanding relationships between humans and nature since people first began writing and depicting them. Positivist and other blindspots, colonialist tropes, and epistemological traps remain, but our ability to see them with insights gained from decades of critical scholarship is unquestionably improving. This opens the door to the needed "profound shift in the imagination of African elites." In the next chapter, I would like to build on that history of knowledge production by examining the causes and trajectories of specific environmental and health changes in the Edendale area over the *longue durée* up to an important turning point, 1939.

# THREE | Sketching the Environmental History of Msunduzi to 1939

What happens if we periodize modern South African history without emphasizing political transitions (pre-colonial, conquest, age of segregation, apartheid, democracy) or economic development (mineral revolution, Great Depression, racial Keynesianism, crash of the rand, neoliberalism), as is the current norm? What if we focused instead primarily on changes to, and struggles over, the environments in which people lived? Some of the environmental and political economy "turning points" would certainly closely coincide. The arrival of white settlers, notably, immediately introduced a wide range of radical disruptions to the ways people related to their living environments and quickly transformed much of the built environment as well. Other turning points, however, would diverge. For example, 1893 is often taken as a key moment in Natalian history, the year that white settlers attained responsible government and began imposing more explicitly racist legislation than hitherto under direct British or Cape Colony control. Yet 1896 – the beginning of the rinderpest trauma – was far more immediately significant to the lives of the majority population. Similarly, despite the political significance of the formal establishment of apartheid in that year, 1948 had very little meaning for people's daily lives in a place like Edendale. The years 1939 (the introduction of mandatory culling of cattle in the reserves), 1943 (the advent of the Local Health Commission), 1954 (the roll-out of anti-tuberculosis drugs and the opening of Edendale hospital), or 1963 (the actual start of population removals under the Group Areas Act) were all more visible, impactful, and memorable turning points at the local level. And, without denying that 1994 was a political watershed with enormous potential for long-term ecohealth improvements, a focus on the environment reveals more continuities than ruptures with the late apartheid era. In terms of environmental thinking, a more meaningful marker of change would be 1990, the year that the city scored its first direct victory against environmental crime and jurisdictional barriers to environmental health began to be torn down over the wider urban area.

Shifting the emphasis given to dates this way may seem trivial. I see it, however, as symbolically contributing to the "decolonization" critique I introduced in the previous chapter. It slightly shifts our attention from a narrative that is highly susceptible to patriotic or great man tendencies to one that reminds us of the view from below. Focusing on the year 1943, for example, shifts us from obsessing about white male politicians in Pretoria to remembering the complex and often conflicted people of Edendale who demanded development on their terms. By the same token, 1954 prompts us to remember how people began to see and positively welcome the dramatic improvements in health (and job prospects) that came to the valley that year. The shift in focus gently reminds us of young girls who collected polluted water or who bicycled around a still largely peaceful community, of poor women who struggled to put food on the table and to nurse the ill, and of fathers who aspired to slaughter a beast for their sons' weddings. The shift from speeches and dramatic gestures to the everyday lives of inhabitants acknowledges that the concerns of highly literate and politically engaged people in big cities, astute as they may have been, were not necessarily shared by the majority of the population, and that we might gain insight from the latter, however tenuously we can access their views.

Different parts of the city obviously developed, or underdeveloped, at different paces in relation to the broad changes to the political economy and the impacts of changing technology, flora and fauna, and demography. And this fact presents a challenge to revised periodizations. It is nonetheless possible to sketch rough periods of time in which a clearly dominant tendency can be identified over the whole of Msunduzi. In the first part of this chapter, I make that sketch for the first three such periods: 1) before white settlement began in 1838; 2) the era of rapid ecological, economic, and social transformation under the colonial system to roughly 1880; and 3) the decades of underdevelopment up to 1939. Underdevelopment did not, of course, stop in that year, and in later chapters I examine how economic and governance structures have continued to limit or undermine human potential and environmental health to the present day. But 1939 was the beginning of a marked shift in policy by the state towards a more interventionist, Keynesian/Virchovian role in addressing manifest environmental and health crises in the African population, marked in the Maritzburg area by the start of aggressive anti-erosion works in Zwartkop and by forceful recommendations of the Thornton Committee on the urgency of the health crisis in Edendale. The final section then draws out key themes from this history. I thus introduce the parameters of the debates around development that cohered in the 1940s and that still, to a remarkable degree, shape contemporary initiatives to redevelop Edendale into a healthy place to live.

## BEFORE *ABELUNGU*, TO 1838

We need to begin by challenging the Africa-as-Eden trope, a "black Arcadia" (Russell 1911, 127) that, in romantic interpretations of South African history, was lost sometime between the wars and migrations of the Shakan era (the *mfecane*), the arrival of whites (*abeLungu*), and the "improvements" (or ravages, depending on perspective) inflicted by subsequent colonialism, population growth, and/or industrial development.

The area of present-day Msunduzi does span a remarkable range of geological formations and biological niches richly populated with diverse flora and fauna. It also has a climate that tends to be pleasant, provided one has shelter from winter frosts, blazing summer heat, frequent violent thunderstorms, and flash floods. It was not, however, a timeless natural paradise. Archeological evidence reviewed by Tim Maggs (1988) points to a natural process of steady, relatively rapid erosion – he estimates the land surface to be on average fifteen meters below what it was 100,000 years ago (14). Archaeology also indicates the ebb and flow of forests and grasslands in relation to vicissitudes in the regional climate such as periodic long-term droughts and the varying intensity of winter fire season. The migration or extinction of fauna took place in relation to those shifts as well as, in part, the impact of the area's human population. Scanty as their numbers were, and as modest as their consumption, Stone Age hunter/gatherers in the district are likely to have contributed to driving some larger indigenous mammals into extinction between 7,000 and 15,000 years ago (16).

Sedentary agricultural communities came relatively late to the area. Gavin Whitelaw dates the earliest Iron Age settlements to 400-450 CE, scattered along the coastal areas stretching southwards of Msunduzi. Higher elevations were not significantly settled until several hundred years later, with the earliest Iron Age settlement on the Msunduse River dating from around 1100 and permanent (stone-built) villages appearing on the highveld above the present city only in the fourteenth century. The people who built these villages, the ancestors of today's Sotho and Nguni speakers, introduced cattle, sheep, and chickens, mining and metallurgy, long-distance trade, and the rudiments of political organization to the area. The vagaries of rainfall and temperature contributed to an often-marginal existence, given the level of technology, and there is some evidence of violence between communities to defend (or enhance by theft) their occasionally meagre food surpluses and cattle, the primary currency of a healthy community. From the accounts of Portuguese travellers in the late sixteenth century, populations remained small and there was extensive vacant land between villages. A big "kingdom" might be fifteen to twenty kilometres across in area, while many chiefs held sway over no more than three to five villages (Whitelaw 2009, 55).

The "Black Arcadia" trope was not completely fabricated. Poor as they may have been, the ancestors of today's majority population had a diverse and balanced diet, lots of exercise, and a built environment that inhibited epidemic disease. Settlement tended to be scattered, rather than concentrated. Households comprised several discrete huts rather than an overcrowded single-family dwelling. Moral claims over land inhered in the lineage, not individuals, and as such the *inkhosi* (chief) had the discretion to reallocate fields or move the whole village according to its health. He (in most cases the chief was male) would have been advised on that question by the ancestors through observable changes in the weather, flora, and fauna, the fertility of the soil, and/or certain manifestations of stress among the people as interpreted by diviners and healers.

The household itself clustered around a central cattle byre or kraal, which served to protect the animals at night, as the site for ritual slaughter, and as the men's court or place for public socializing. Living in such close proximity to domestic animals introduced its own health hazards – lice and tick-borne diseases, notably. But it also gave rise to a culture that drew strong connections between humans, cattle, and the surrounding biophysical world. Indeed, a strong "public health" or sanitation consciousness was present in the concept of *ukuzilungisa*, translated as "balance" or "moral order," in the relationships between people (living, deceased, and yet-born) and the environment and the metaphysical forces – often from away – that could pollute or disturb symmetry (Ngubane 1977, 27). As Shirley Brooks (2009) warns, we need to be wary of some of the bigger claims about a traditional, mystical conservationist ethic sometimes asserted in the "royal hunting grounds" discourse. It does appear, however, that a culture linking human health across the living/dead/unborn divides to the surrounding environment that was needed to maintain the health of the cattle herd cautioned people against long-term destructive or exploitative land-use practices. The philosophy is captured in the term *uBuntu*, which Dent and Nyambezi (1969, 441) translate simply as "human nature; good nature," and which Ramose (2015) interprets as a sustainable relationship between people, biosphere, and the land.

The centrality of cattle in the traditional *Ubuntu* culture throughout the region can hardly be overstated. A key aspect of the emerging political ecology from at least a thousand years ago was *ilobolo* (bridewealth), a gift of cattle from the family of the groom to the family of the bride.[1] Ownership of many cattle enabled men to acquire multiple wives for themselves and their sons through the exchange of *ilobolo*; many wives brought many children, who contributed labour to the household (including sons to look after expanding herds of cattle) and eventual further networks through marriage. Many daughters brought more cattle as incoming *ilobolo*. This combination allowed a successful patriarch to offer hospitality, beasts for loan, or brides to other men in exchange for their labour and political loyalty. Going back to the concept of *ukuzilungisa,* however, it is important to emphasize the dangers of too many wives for maintaining a harmonious balance. Some of the metaphysical forces threatening to disrupt the moral order and to introduce illness, infertility, or other calamity accompanied wives who came from away to join a family. Conjugal relations, including sexuality, thus had to be very carefully regulated by ritual balancing ceremonies and medicines, not simply for the immediate family but for the good of the whole community (Ngubane 1977, 42–4).

As noted in the previous chapter, the Portuguese unwittingly sparked a major transformation to the regional environment by introducing maize early in the sixteenth century. Whitelaw posits that wider and wider areas of land fell under hoe cultivation as the new crop was generally adopted and a better-fed population grew steadily from around 1700. At the same time, the Portuguese colony at Delagoa Bay (modern Maputo) created a demand for meat, ivory, and, to a small extent, slaves. Thus, elephants, people, and expanding herds of cattle were drawn into the global commodity nexus, giving rise to partly militarized, albeit still small, polities throughout the region. When the climate turned drier around the end of the eighteenth century, competition to supply the export trade and for pastures to feed local populations led to the consolidation of these small polities into larger and larger ones, and eventually (beginning around 1815–18) to the wars and refugee migrations of the *mfecane* period. The Msunduzi district was on the periphery of these changes, with even the biggest "kingdoms" in this period still amounting to only "a few thousand people inhabiting an area of a few hundred square kilometres" (Wright 1988, 18), in a constant state of flux and migration. The only chieftaincy from that time that still survives as a place name today was kwaMacibise, named after a female chief who ruled a marginal section of the Nqondo kingdom.

Farming, and consequent changes to the landscape, remained limited by the constraints of the hoe and human (mostly female) muscles, although expanding human settlement likely did have a deleterious effect on the forests. For example, the construction of a traditional beehive-style hut required between 100 and 1,000 saplings, and large amounts of charcoal were used in smelting iron. Meanwhile, the common practice of winter grazing of cattle in the forests killed additional trees (McCracken 1986, 24). Jeff Guy ([1979] 1994, 9) has also speculated that the cultural imperative of ever-expanding herds of cattle was already having harmful impacts on fragile soils in the more densely populated parts of Zululand prior to the 1820s. Shaka's creation of female age regiments, which delayed girls' marriage for several years, may have been a purposeful strategy to restrict population growth and so protect the environment within the core of the kingdom (Guy 1980, 116). Whether birth control was either the primary intention or successful result of such regiments, it did have an ironic long-term demographic impact in neighbouring Natal. Zulu kings' meddling in the sex lives of their subjects, including not only the female regiments but also a ban on the practice of *hlobonga* (pre-marital sex play) and the insistence on the king's power to approve all marriages, was a major grievance that fuelled a constant stream of people escaping from Zululand as late as the 1870s (Atkins 1993, 31–2).

During the *mfecane* itself (most of the 1820s for the Msunduzi region), various uprooted groups fled through the valley to escape the centre of the power struggles to the northeast. They stole food, cattle, and people on the way. Those who remained in the valley after these violent intrusions took to sheltering in deep forests and caves, and reverted to hunting, gathering, and very small-scale farming rather than keeping or rebuilding their herds of cattle (which would only have attracted the unwanted attention of more powerful or desperate neighbours). By the time the first British traders arrived from Port Natal in the late 1820s, followed shortly thereafter by Afrikaner trekkers coming down off the highveld, settled life was only just beginning to be re-established to the north of the present-day city and on the slopes surrounding Mbube (Zwartkop) and Table Mountain. Elsewhere, abandoned fields and resurgent forests and wildlife gave newcomers the impression of entering virgin or ownerless lands. We know it was politically convenient for white settlers and historians to overstate the generalized destruction supposedly caused by Shaka's "wasting hordes." In this instance, however, there is an element of truth in the old trekker mythology that they were staking claims to lands in the Msunduze valley that were apparently empty of permanent African settlement. An area that today is home to perhaps half a million people and still has extensive open spaces (Vulindlela and Edendale) was then home to perhaps a few hundred.

One final legacy from this period needs mention. Shaka is generally renowned for the military prowess that laid the foundations of expansive Zulu political, cultural, and linguistic influence in the region. Another of his reputed innovations, however, was to have a long-term health impact on the nation as a whole. Historians assume that his ban on the traditional practice of male circumcision had a political/military function that can be traced back to the environmental stresses of the times. In its stead he instituted *amabutho*, or age sets of celibate conscripts who could achieve adulthood (that is, marriage) only after they had proven themselves in battle. In this way Shaka brought young men's symbolic transition to manhood directly under the control of the state in order to forge a disciplined, professional standing army that assured a flow of tribute to the centre. The ban was widely emulated among other African peoples throughout the region over the following decades and thereafter strongly encouraged by Christian missionaries. Yet whatever the military, political, or cultural advantages that intact foreskins were thought to have had at that time, they made men more vulnerable to sexually transmitted infection. By the late twentieth century, the widespread lack of male circumcision was a strong contributing factor in making KwaZulu-Natal among the worst AIDS-affected jurisdictions in the world (Halperin and Epstein 2007; Timberg and Halperin 2012).

## COLONIAL ECOLOGICAL REVOLUTION, 1838 TO THE 1880s

The early years of European settlement and growth radically transformed the human/natural environment in a very short period of time. Carol Merchant's typology of colonial and capitalist "ecological revolutions" in North America,

3.1 A "traditional kraal" in Elandskop, at the top end of Zwartkop, ca 1900. A.W.G. Champion may have had such an image in his mind when he worried about the natives from Bulwer learning to disrespect their social betters

which Jacobs (2003) also employs in her reconstruction of Karoo history, is highly applicable to the history of colonial Midlands Natal. The basic contours of Msunduzi's four distinctive, racialized sectors were established in this era.[2]

The colonial transformation of the Midlands and today's Msunduzi involved, first and foremost, the rapid slaughter for food and for sport of native animals, plus the systematic fencing off of potential commercial agricultural or pasture lands from ever-shrinking pockets of indigenous forests and wildlife. As early as 1841 the Afrikaner Raad (town council) rued the "reckless destruction" of wildlife by its burgesses and imposed a fine on those who killed animals without a "proper" use in mind (Ellis 1998, 42). This does not seem to have abated the killing – the last elephant was seen in the district in 1848, a mere three years after the British established their first civilian administration. Indeed, among the selling features of Natal in the early period of British rule was the fact that there were "no game laws" (Hattersley 1951, 42, citing an advertisement for settlers from the period). A rogue buffalo was reportedly chased from downtown and killed in the Slangspruit in 1851 (Hattersley 1950, 332) and, from Bulpin's colourful description (1966, 215–16), the city remained an outpost for "hunters and adventurers" hawking trophies and captured wildlife from the highveld up to the end of the 1860s. By the time game conservation laws were put in place in 1866, most preferred game in the immediate district was extinct, including edible fowl (Ellis 1988, 32). One important, if ill-studied aspect of this slaughter is that much of it was carried out not by rifle or traps but through saturation of the land by poisons, strychnine above all, with unknown long-term effects upon human health.[3]

Perhaps surprisingly for a colony that was intended to be primarily agricultural, the early British colonialists also brought with them an impatient if not rapacious attitude towards the soil. Robert James Mann hinted at this when he wrote to promote immigration from England to the Pietermaritzburg and Midlands districts: "Manure is not actually needed for the greater part of the lands in this district. There are many choice spots where successive crops have now been taken from the soil year after year, without any sign of exhaustion or diminished fertility" (1859, 130). This attitude, together with poor match between climate and the initially preferred cash crop of cotton, partially explains why so many immigrants failed at their first farming schemes and that Natal as a colony urbanized relatively quickly. Half the plots on Joseph Byrne's first settlement scheme on Slangspruit farm just outside Maritzburg, as an example, were abandoned within two years of being staked out (Schnurr 2013, 120). The disappointment of this early wave of British settlers then opened the door on both the west (Slangspruit) and east (New England) sides of the city for small numbers of Africans and Indians to purchase parcels of the abandoned freehold land.

Undercapitalized settlers had other unintended impacts on the environment. They tended to take the easy way – fire – to suppress the thick growth of coarse, indigenous grasses, and to replace them with species more amenable to imported cattle and sheep. Fire was of course nothing new to the grasslands, but the overloaded wagons that fanned out across the district to tap into the emerging commercial market up-country were. In combination these "created havoc" on the exposed, delicate soils of the fire-swept veld (Hattersley 1950, 265, 278). Scars in the red soil from the wagon traffic up and down the escarpment are still clearly visible today.

As for trees, it may be that some deforestation had already been caused by pre-colonial African communities. The scale of destruction, however, enormously increased with the advent of white settlers. By the 1870s

Pietermaritzburg was estimated to be consuming over 6,000 tons of timber per year, mostly out of Karkloof forest about forty kilometres to the north and from ravines in the nearby Zwartkop and Table Mountain reserves (McCracken 1986, 34; Ellis 1988, 29). Trees were used as fuel, and lumber gave rise to the first manufacturing industries – wagons and furniture. The most valuable indigenous hardwood trees like yellowwood were harvested and exported through the 1880s to even more timber-hungry Kimberly and Transvaal.

While all this destruction was taking place, the settlers introduced their own favoured species from away. In town itself, the settlers enthusiastically planted various exotic shade and ornamental trees like oak, eucalyptus (blue gum), and syringa (lilac). This was followed by decorative exotics like jacarandas from South America, Australian chestnuts, and plane trees from Europe. As the town grew in population, so did the size of urban forest. By the end of the nineteenth century, brightly flowering trees were starting to climb up the slopes of Town Hill and, in the first decades of the twentieth century, they began to colonize the thornveld to the southeast through new suburbs like Scottsville. In the surrounding countryside, meanwhile, black wattle, pine, and gum spread rapidly to cloak the hillsides. Wattle, first imported from Australia in the 1860s, became a popular commercial crop by the 1880s. By 1904 there were over four thousand wattle plantations within Pietermaritzburg county, often on hitherto agriculturally marginal lands whose African occupants were evicted to make room. Wattle thus became the first of several monocultures or "green deserts" to demarcate the limits of the city (Witt 2002, 2005; Ellis 1998).

The white settlers in this era had another indirect impact on the environment by stimulating African production for the market. Following the end of the *mfecane* disruptions, Africans who had fled the Natal area returned in numbers. They, and others who fled continuing strife in Zululand through to the 1880s, were desperate to rebuild their herds, families, and means of subsistence as quickly as possible, what Keletso Atkins terms a "crisis of reconstruction" (1993, 27). White settlers provided new technologies, new crops and animals, and new opportunities. Villagers from Edendale and Zwartkop profited in particular from the sale of wood, maize, forage, and animal products to the city and white-owned farms around it. African productivity was actually a selling point for colonial officials hoping to attract new settlers from England. As Mann put it, "many parts of the midland districts are thickly peopled with Kafirs, who grow very large quantities of

Indian corn and Kafir corn (millet), and dispose of their superabundant produce at comparatively low prices to the white settlers … Many settlers are now beginning to reap good incomes, by purchasing Indian corn of the natives in somewhat remote districts, and conveying it to the markets of the capital and Port, ground or unground, in their own wagons and carts" (1859, 135–6). Increased production by Africans was facilitated by their widespread adoption of the iron plough over hoes and the turning of virgin soil to supply the market. Africans also embraced wattle as a convenient, fast-growing source of fuel for use and sale, and it proliferated around "traditional" homesteads. Similarly responding to the growing market generated by white settlement in the Midlands, Indians freed from indenture on the coastal sugar plantations began migrating to Maritzburg in the late 1860s to stake out densely cultivated gardens and orchards at the lower end of town and its eastern periphery.

Edendale and Zwartkop began to diverge quite strongly in physical appearance and culture in these decades. The *amakholwa* built their village of Georgetown on European principles, including single-family dwellings laid out in straight lines concentrated around the central church. The large commonage was managed in theory by the Edendale Trust but in practice by the inhabitants themselves through an informal system of moral suasion and of shaming and sanctions for environmental abuses. Otherwise, private ownership of property meant that decisions about land use were reserved to the individual patriarchal households. The nature of Christian patriarchy also implied a change in land use. Indicative of the shift from polygynous families married through the exchange of cattle (as remained the norm in Zwartkop), the population of cattle in Edendale actually declined over the first few decades of settlement (Meintjes 1988b, 129). Large numbers of scrawny beasts had little social value in Edendale whereas smaller numbers of healthy oxen and dairy cows were an economic asset in the emerging capitalist economy.

The city of Maritzburg had a somewhat rocky start to its development, with the abandonment of the colony by most of the original Afrikaner settlers after 1845, the aforementioned failed cotton farms, and rampant land speculation (Guest 1988). Another almost intractable brake on the white settler economy was implicit in the successes of African peasant production. Profits from market production enabled the growing African population to demur with respect to providing the cheap labour that white settlers needed to realize their ambitions. All these factors contributed to a very deep depression that hit the city in the late 1860s. Many burgesses left, and virtually all

3.2 Downtown Georgetown looking towards Maritzburg, ca 1900. The Wesleyan church (1866) is on the left, beside the monument to "Queen and Country," Troop Sergeant Major Simeon Kambula, and the Edendale contingent of the Natal Native Horse brigade that served with honour in the Anglo-Zulu War

capital works ceased. Even the chief constable and inspector of nuisances was retrenched to save costs (Meineke and Summers 1983, 24). Through to the end of the century, visitors often disparaged the town's unkempt appearance, insufferable dust or quagmires, and stinking river courses. It remained, in general terms, a mess: "shabby assemblages of buildings," "not a single handsome building in the whole place," and "grass-grown streets," as Lady Barker put it (1877, 59). The *Natal Witness* was even more scathing in frequent upbraids over the decades, as in this lead editorial from the mid-1860s: "Our streets are literally the receptacles for all the filth and rubbish of the town – the most disgusting heaps are to be observed festering at any corner. Pig-styles, cattle kraals, shambles, filthy putrefying heaps of stable manure are the ornament of each vacant piece of land throughout the city. Our water courses are the homes of ducks, geese, dogs and whatever else lies to wallow in them" (1 January 1864).

The need to solve the labour shortage crisis sparked some of the most vociferous debates in the colony's early decades, including around the wisdom, the size, and the location of reserved lands for Africans. The tendency over time was for government to impose ever more severe restrictions on Africans' ability to own land and hence to be able to opt out of wage labour, high taxation requirements, and corvée (*isibhalo*). By the end of the 1870s, such policies were showing mixed but promising results for the colony as a whole. *Isibhalo* produced a network of roads, albeit of such notoriously shoddy quality that the merchant trade they stimulated took a heavy toll of erosion on the land. Small-scale industries arose to service both the surrounding farming districts and the "overberg" (including the diamond boomtown of Kimberly after 1879). When the railroad from the coast arrived in 1880 it provided an additional boost to a wider commercial export economy (dairy products and tannin from wattle plantations, notably). Growth in scale of production tended to consolidate small holdings into larger, scientifically managed farms. A series of wars and the suppression of African rebellions around the fringes of the colony (in 1873, 1879, 1880–81, 1885, 1899–1902, and 1906) added another stimulating effect upon the local economy, typical of a military base situated well behind the frontlines.

The city's growing prosperity began to show through the flowering of grandiose public architecture, not least of which the largest brick structure in the southern hemisphere and most expensive town hall in the country, completed in 1903. Social pretentions were also apparent in the emergence of the Midlands' "squirearchy," a notoriously cliquish colonial sub-culture centred in Maritzburg and dominating the political life of the colony (Morrell 2001). Pressure then began to grow to clean up the appearance (and odour) of the city to match the pretensions of its elites. After the worst of the depression of the 1860s, considerable expense was devoted to fixing messes inherited from earlier poor or non-existent planning, notably by putting in the rudiments of a modern sanitation system and covering open watercourses and hundreds of cesspits that had become a notorious health hazard (Meineke and Summers 1983, 81). An initiative to develop a whole new suburb as a healthy living option for white citizens (Scottsville) was supported by a public electric tram system, opened in 1904 (Wills 1988b, 138).

Lastly, the colonial ecological revolution involved the introduction of many new pathogens to the region, and a built environment that favoured their propagation through human and animal populations. Some new diseases had spread through trade contacts with the Portuguese even before whites had settled in the area. Smallpox scars observed in the 1830s, for example, suggested an epidemic in Zululand as early as 1816 (Flint 2008, 45). Afrikaner and subsequent settlers who employed African domestic servants in close quarters increased this transfer of disease from Europe and Asia to southern Africa. In addition to a more virulent form of smallpox than was indigenous to the region, the settlers brought whooping cough, plague, polio, tuberculosis, pneumonia, leprosy, cholera, and syphilis. As long as the majority of the African population remained in the rural areas or relatively dispersed within towns, and bearing in mind the generally if fitfully rising prosperity of the age, the burden of new diseases remained low. But in the decades after the 1880s, more and more people were exposed to environmental and health risks, with rippling impacts on the whole political economy.

## THE DEVELOPMENT OF UNDERDEVELOPMENT, 1880S–1939

Initial periods of relative prosperity in the main African communities around Maritzburg did not last long, although in each case economic downturns (and the environmental and health degradation that ensued) happened for different reasons and were drawn out unevenly over a prolonged period. The Edendale economy, for example, suffered a hard blow from the depression that affected Maritzburg in the mid-1860s, with bankruptcies among leading landowners and artisans who then struggled to recover through the 1870s. But the seeds of ecological and health collapse in Edendale were already being sown in its early period of relatively successful market production. While private property and the profit incentive in theory motivated good husbandry, in practice landowners admitted their own inability to protect the environment as early as 1870. In that year they requested the government to draft formal land-use regulations for them, to provide them the right to collect local taxes and to allow them to elect advisors to the Edendale Trust for managing the commonage (Cape Colony 1883, ii, 58). Lady Barker also hinted at disaster in 1877 when she noted how "for more than twenty years the soil has yielded abundantly without an hour's rest or manuring" (199). For all his high estimation of Edendale as its ultimate bureaucratic overseer, Theophilus Shepstone conceded shortly afterwards that the land itself no longer had any value, presumably because of soil exhaustion (Cape Colony 1883, ii, 58). One of the rare African testimonies on this issue subsequently confirmed the worst. As *amakholwa* farmer Cleopas Kunene told the Native Affairs Commission of 1906–07, "owing to lack of manure and defective methods of cultivation it [Edendale] had become much less productive" (Natal 1907, 916).

Expanded herds of environmentally harmful grazers like goats and donkeys, plus a steady influx of migrants from the reserves bringing their own animals, also began to strain the commonage. In 1888 the Trust found twelve families living on the common with illegal permission from some of the leading families of Georgetown. Then, and in subsequent attempts, humanitarian concerns prevailed to stay the squatters' eviction, in effect condoning private subsistence use of the most ecologically sensitive public lands.[4] Moreover, by the turn of the century the fence that theoretically kept Zwartkop's scrub cattle out of protected pastures and guarded improved milk cows in Edendale from disease and impregnation was long gone. The first proper census was conducted only in the 1950s, and it found the area to be 1,000 per cent overstocked. Combined with the spread of eucalyptus and wattle trees throughout the valley (which had a profound desiccating effect on the soil), livestock overgrazing and trampling had resulted in "fantastic" levels of erosion damage.[5]

Another transformation for Edendale in this period came through the proliferation of undercapitalized industries in and around the community. Notably, Edendale's hand launderers, the *amaWasha,* took in the bulk of the laundry from Maritzburg, rinsing copious suds and other cleansing agents straight into the Msunduze River (Atkins 1993). The Edendale tannery opened in 1890, commencing more than a century of dumping lime, blobs of rancid tallow, and other poisonous effluent into ponds that spilled into the spruits and eventually made their way into the city, leached into the surrounding soil, and a created foul stench and fly infestations over a wide area (Zuma 2002). Better-capitalized factories such as Sutherland's Tannery and South African Wickerworks may have been less directly noxious but they did little to mitigate growing poverty in Edendale, with all that that implied for environmental stress. On the contrary, industrial employers in Edendale generally preferred to bring their African workers in from Zwartkop. They regarded the "tribal" migrants as more disciplined and hard working (for less remuneration) than Edendale residents (University of Natal 1951, 79).

The arrival of the railroad in Maritzburg in 1880 exposed the vulnerability of the Edendale economy by facilitating the importation of cheap grain from overseas and driving down the price of locally produced maize to

3.3 Gum trees and the commonage being "eaten" as Greater Edenale today sprawls up and outwards towards the fence that once marked the border with Zwartkop

economically unviable levels. In that same year the city also took aim at the hand-laundry industry by opening an industrial facility in town (Atkins 1993, 136). Further blows fell in the 1890s as plagues of locust and cattle diseases like lung sickness and rinderpest struck both peasant production and the transport business. Many *amakholwa* moved out to establish new farms elsewhere in the region, but after 1903 this option was effectively closed when government froze the sale of Crown lands to Africans. Many Edendale landowners then either turned to renting for income or sold out to Indian and European farmers and non-resident landlords. The latter, also known as shack or "kaffir farmers," maximized their profits by crowding desperate emigrants from the increasingly impoverished reserves into decrepit wattle and daub structures. The 1913 Land Act generated a further flood of evictees from white-owned farms. In this context, the moral economy that had once tempered neighbours' over-harvesting of wood or stream-side cultivation, lost its ability to shame or ostracize. In 1929, the Edendale Trust conceded to losing its ability to control matters such as the "irregular use of dongas" (erosion gulleys).[6] The city declined to help on that and numerous subsequent occasions through the 1930s. Indeed its deportation of hundreds more Africans into Edendale from its border areas in the 1930s only worsened overcrowding, pollution of watercourses, and soil erosion. When the Thornton Committee into uncontrolled urbanized areas visited in 1938, it received evidence that the once lovely farmland of Edendale had become an "insanitary death trap."[7]

The state, it must be emphasized, remained largely aloof from, if not hostile to, calls for action to address the crisis. As early as 1880, leading Edendale citizens approached the colonial government to establish a new form of local government that would enable the *amakholwa* to assert control over health and environmental threats to the community: "the time has come for the adoption of some suitable measure which shall secure better sanitary arrangements, the maintenance of roads, the preservation of forest or bush land, and the due reservation of commonage, and for this object the aid of Government may be required."[8] This request was rejected. In another important case, leading *amakholwa* men in 1902 launched a long and detailed denunciation of Chief Stephen Mini, mostly focused on his prolonged absences, exploitative lending practices, and other incompetences that were allowing the village to deteriorate. The Native Affairs Department (NAD), upon the advice of the Wesleyan missionary in Edendale at the time, ruled in Mini's favour.[9] Yet the NAD subsequently ruled against him in another case. In 1914 it rejected a decisive vote by landowners led by Mini to have Edendale declared an exclusively native area with NAD powers to protect against the problems of an urbanizing, highly transient population.[10] The common thread right through to the creation of the Local Health Commission was for the NAD and other arms of the state to avoid expensive entanglements or difficult political questions about governance in an increasingly multi-ethnic and multiracial district.

The nearest local level of government (Maritzburg town council) was also notoriously remiss in accepting responsibility for its role in creating those problems, including directly polluting Edendale. In 1905 the problem was the city contracting waste disposal to private companies who simply dumped their loads outside city boundaries. A petition by residents of Edendale took the matter to the colonial government: "stench arising from the carcasses buried on the land, in close proximity to our properties constitute[s] a grave danger to the Public. The dead beasts and human excrement for 'Fort Napier' and elsewhere are buried in the midst of a large population of Europeans,

Indians, and Natives. Enteric fever caused through the unsanitary conditions of the District in the past was very rife."[11] Chief Mini also appealed to the province to intervene to stop a different type of menace to his community coming from Pietermaritzburg: "Soldiers of the Fort have been molesting my people at Edendale. Several cases have occurred lately in which they have forced entrance into houses for immoral purposes."[12]

Neighbouring Zwartkop meanwhile charted a pattern of underdevelopment that became characteristic of communal tenure areas throughout almost the whole of southern Africa.[13] The persistence of traditional mores in the reserve meant that rising income from market production tended not to be invested in technologies that would improve productivity or protect environmental assets. Rather, they went primarily to increasing the quantity of the main customary indicator of wealth: cattle. The practice of exchanging *ilobolo* had been around for perhaps a thousand years and had been a contributing factor in the consolidation of African states in the pre-*mfecane* era. But in the context of rising population and restricted borders, wealth measured this way quickly became unsustainable. When rinderpest swept through the region in the late 1890s it provided some respite to the pressure on pasturelands. By virtually wiping out wealth invested in cattle, however, rinderpest left families with little choice but to send their sons out to look for employment, principally in Edendale, Pietermaritzburg, and Durban but for a growing number as far away as the goldfields of Witwatersrand. The "masculinity ratio" (the proportion of males to females) eventually stabilized between 1934 and 1944 at around 70 per cent (Brookes and Hurwitz 1957, 68), but this was an unhealthy stability that implied a far higher gender imbalance among adults of working age. Moreover, the wages remitted could not compensate for the loss of men's labour from the subsistence economy, and indeed, as care of the herds and fields fell increasingly to women and children, labour short cuts took an inexorable further toll on veld conservation. Even wattle turned from asset to burden in the absence of so much male labour. Without constant attention, copses of wattle around homesteads ate into arable land and became so densely overgrown that the trees choked themselves to commercially inefficient levels (McCracken 1986; Witt 2005, 105). Like gum trees, they were calamitous for natural wetlands (Bews 1913, 495).

Taking all these factors together, the proportion of the maize crop produced by the reserves over the whole colony dropped from 80 per cent of the total to 38 per cent between the 1880s and 1904 (Lambert 2009, 215). Migrant labour did channel cash into the reserves to enable the purchase of food and other consumer products that somewhat offset the decline in subsistence production. Much of the remittances, however, was dedicated to rebuilding the herd size, and the population of so-called scrub cattle in fact made a rapid recovery rummaging off abandoned fields through the 1920s (MacKinnon 2009). As a result, by the time the Native Economic Commission toured Natal to examine conditions, it found overstocking and weed growth in some districts to be "as bad as the worst parts of the Ciskei," that is, an area the commission ranked as the country's most object lesson in despoliation and denudation (South Africa 1932, 43, 32).

Zwartkop's unusual proximity to town and relatively favourable climatic and soil conditions protected it from the worst impacts experienced in many of Natal's other reserves. Nonetheless, diseases of malnutrition like pellagra followed closely upon the poverty and deteriorating living conditions. Subtle but possibly important social and cultural changes resulted as well. Anthropologist Judith Gussler (1973), for example, posited a connection between pellagra, male outmigration, and the increasing numbers of malnourished or starving women coming out as *isangoma* (spirit mediums) in the reserves. "Hysteria" (to European observers) or spirit possession (to Africans) was one of the only means for women to express their anguish (and get food), with unknown consequences for the traditional healing system. We know that hunger also drove young women in particular to take risks with their health to get food – the risk of pregnancy, sexually transmitted infection, and gender-based violence among the most widely observed, and rued. And we can surmise that the loss of indigenous flora to exotic monocultures in the surrounding farms removed an important source of traditional medicines that might have mitigated the impacts of deteriorating diet and health in the reserves.

Maritzburg was not immune to these trends. Its prosperity in the 1880s rested on a relatively fragile base and soon proved unsustainable. By the early twentieth century the city had begun to experience a type of underdevelopment that belied its pretentions to grandeur. The steady impoverishment of Africans in Edendale, Zwartkop, and other surrounds, and the consequent erosion of that consumer market, provided one of the biggest hits to the economy. The completion of the railroad from Maritzburg to Johannesburg in 1895 was another. It allowed trade from overseas arriving at Durban on the way to the burgeoning industrial centre of the Rand to bypass the city as an entrepot. The Act of Union in 1910 preserved Maritzburg as the administrative capital of Natal province, and that year also saw the incorporation there of Natal's first university. But the closure of the military base in 1914 ushered in a long period of economic stagnation relative to Durban. Indeed, between 1905 and 1944, only six new industrial establishments were set up in Pietermaritzburg (Seethal 1993, 33). One impact of unemployment is alluded to in a Natal-wide nutritional survey in 1928, which actually found a higher level of malnutrition among white schoolboys than among Africans (21 per cent versus 14.8 per cent respectively), with "several deaths" of white children from marasmus or starvation (Dyer 2012, 140; see also Morrell 1992). A survey of Maritzburg in 1935 found that whites made up just over 1 per cent of the total population then living in slum conditions on the west side of the city (36 of 3,064, mostly in Camp's Drift) (Pietermaritzburg [1904–90] 1935, 93); a similar percentage had to be rehoused after slum demolitions began in 1937 (1939, 122).

None of this is to excuse the lamentable track record of the city's white population in treating its non-white population with fairness and dignity, its obsession with maintaining expensive showcases of colonial values, or its steadfast refusal to take any responsibility for the health and environmental crises emerging just outside its borders. However, the fact was that the white population declined by nearly a fifth in the first decade of the twentieth century, that white-owned homes in the neighbourhood surrounding the university remained on the bucket system of sewage removal until the end of the 1920s, and that white families could still be found living in "appalling crowded conditions," including wattle and daub shacks and squatter camps, as late as 1952 (respectively, Pietermaritzburg [1904–90] 1917 and 1929, 21; *Natal Witness*, 11 August 1925).[14] Indeed, while Pietermaritzburg by the late nineteenth century had unquestionably been fashioned into a pleasant spot for property owners, educators, and civil servants, for a section of the white working poor it offered dismal opportunities for many decades to come. An important piece of public infrastructure that helped mitigate the effects of poverty was closed down at the end of 1936 when the city, citing the need to cut costs, terminated its tram system (Wills 1988b, 138). A further sign of the times may be taken from the response to a national decision in 1937 to locate the regional South African Railways workshop in Durban. A near-riotous crowd of two thousand white men came onto the streets of Maritzburg in what the *Natal Witness* sympathetically described as a "spontaneous uprising" (*Natal Witness*, 29 April 1937; see also Pietermaritzburg 1904–90 [1937], 36). Council's refusal to impose new expenditures upon its declining number of ratepayers, which became a recurrent theme in discussions between city officials and higher levels of government over the issues of poverty, health, and pollution in the Edendale valley throughout the 1930s, has to be understood in this context.

## THEMES IN THE EARLY ECOHEALTH HISTORY OF EDENDALE AND ENVIRONS

The year 1939 is a convenient point to end this overview. Years of economic depression and the looming prospect of world war focused people's minds on the need for decisive action to address the poverty and injustices that had become so entrenched in South Africa and that were shamefully visible in the "black belt" that had emerged around Pietermaritzburg. Relatively radical ideas about social medicine, welfare, and state-led development projects to address the issues were beginning to find expression in national policy. The visit of the union's chief health officer, Sir E.N. Thornton, to Pietermaritzburg in June 1939 in particular foreshadowed a profound shift in the relationship between the local state and the people of Edendale.

Several themes can be discerned running throughout the history sketched above, some of which are counter-intuitive or diverge from conventional historical narratives. First, Msunduzi's environmental and health problems cannot be understood without reference to global influences. Globalization in a broad sense has been having a profound impact upon the region's physical environment for at least three hundred years, in the effects of maize from the Americas, immigrants from Europe, India, and St. Helena, epizootics from the Middle East, the expansion of "green deserts" to serve consumer tastes far and wide, as well as financial and political crises in the imperial centre. For all the talk of a "sleepy hollow" somehow bypassed by hustle of the wider world, developments that shaped the environment of Msunduzi today have long been closely tied to ebbs and flows of the world capitalist system.

Second, the cash economy often drove people to make short-term decisions without regard to long-term environmental impacts. Soil mining and deforestation for commodity production were perhaps the most obvious harms, but so too was afforestation with inappropriate exotics. For the *amaWasha* and many of the city's small-scale industries, the river was the only way to dispose of sometimes toxic waste. The super-exploitation of African and Indian labour, meanwhile, including through *isibhalo* on badly constructed roads or sub-subsistence wages in formal employment, passed housing and other survival costs on to people who could least afford them in sustainable and healthy ways. Powerfully symbolic of this was the proliferation of wattle and daub shacks and emaciated cattle on the steep slopes and along streams that had once been protected commonage. Greed, of course, played a role in this, but so too did poverty. Poor whites, while small in number compared to Indians and Africans (and to the poor white population in other provinces), were visible and vocal enough to influence the political process against costly investments that might have addressed the growing environmental crisis on the city's fringes.

Third, the state has often been the city's most visible and egregious polluter. In the early years this was as crude as illegal dumping or incompetent disposal of sewage and other effluents by the city or the military. Sometimes it was state-owned industries like South African Breweries and the railways, whose shunting yards and poorly fired coal-burning engines spewed black smoke over the city until well into the 1970s. Institutions, notably the hospitals and prison, further contributed to air and water pollution. The city administration did put in place various initiatives over the years designed to reduce such pollution in the city centre, but, as we shall see in the chapters to follow, it also commonly blocked attempts to enforce controls by other levels of government if those initiatives were thought to endanger investments or to expose the city's own neglect of and incompetence with respect to environmental concerns.

"The state" had a further indirect role in abetting environmental harms through its very incoherence and structural weaknesses. Bad (indeed, virtually non-existent) governance allowed poor farming and forestry practices, unsafe waste disposal, and unregulated construction to go uncorrected in Edendale for so long that large sections of a once-thriving community turned into a slum. Bad governance was evident in the many injustices inflicted upon Africans and Indians and driven, in large part, by the inability of the Colonial Office in London to restrain the racist cupidity of the local settlers. Bad governance was further manifest in the hodgepodge of local authorities that resulted from race- and rate-obsessed decision making. Numerous distinct local authorities administered the present area of Msunduzi, some governed by duly elected bodies, some by local bureaucrats, some by "traditional" chiefs, and some, as in Edendale proper, by an unwieldly meld of elected chieftainship, church, and trust. Rival jurisdictions had little incentive or ability to coordinate policy between them, let alone to intervene in environmental management concerns across boundaries.

Changing technology and infrastructure are another important theme in environmental history. Several infrastructural developments and technological advances in Msunduzi's history were unambiguously positive in terms of health and quality of life. One of the area's first major development initiatives was to cover the open furrows or "sloots" that Maritzburg city founders had constructed to deliver fresh water for household consumption. When constructed, and protected from the encroachment of settlement and pasturage, the Henley Dam supplied some of the naturally cleanest water in the nation both to the city and (eventually) to a growing distribution network throughout the Edendale valley. Similarly, the tarring of roads from the late nineteenth century immeasurably improved the quality of life for citizens, who had previously lived with the effects of choking clouds of dust or quagmires for much of the year. Another major public health victory was achieved over the 1920s when, after the discovery of a connection between cattle and enteric fever, dairies and privately owned cattle were restricted within the city limits. Where enteric fever had been the major killer of whites from infectious diseases up to that time, by the 1930s it had been reduced to insignificance in that population, with the eradication of other such diseases following over the next two decades (Brain 1988, 194). Yet other innovations were of mixed benefit. Going back to the first period of economic boom in the 1850s, the almost universal adoption of the plough by African farmers enabled increased production and a period of considerable prosperity. Ploughs cut much more deeply into fragile soils than traditional methods of hoeing, with the effect of vastly speeding up natural soil erosion. Similarly, new exotic crops, trees, and animals often brought direct, short-term benefit to poor communities and, specifically, to marginalized members within communities. The unforeseen long-term impacts, however, could be "devastating," to use Witt's term regarding the uncontrolled spread of wattle in the reserves (2005, 122).

And some technologies may have had unknown effects that may be revealed only in the course of time by yet-to-emerge technologies. For example, the Academy of Science of South Africa (2015, 61) alludes to a study linking high levels of intersex births in Limpopo province to high usage of a notorious endocrine disrupter, DDT, which may have affected unborn children's physical sex, sexual orientation, and gender identity *in utero*. The new science of epigenetics, for another example, has shown that environmental and psychosocial stresses experienced by one generation can cause damages at the chromosomal level that are inheritable by the next generation. Dowsing large swathes of Natal with strychnine and arsenic undoubtedly made the province primed for commercial agriculture but could it have inflicted genetic damage that may still affect generations to be born? Such questions are worth pondering, as unsuspected stresses arising from apparently benign technologies are either ongoing or bound to happen again.

Finally, gender and sexuality are an important, generally understated theme in environmental history. The gender division of labour, gender identities, rituals, and roles, and sexual mores played a key role in mediating African relations with the biophysical world pre-*abeLungu*, and in subsequently shaping African responses to the colonial state and the emerging markets in land, labour, and produce. Many of the state's most ecologically damaging and politically costly interventions were rooted in its perceptions of traditional African patriarchy, notably, that Africans, and Zulus in particular, needed strong paternal figures to guide them. This view underpinned the Shepstone system of indirect rule through managed "traditional chiefs" and the migrant labour system that significantly depopulated African reserves of their adult male population. It was a view so entrenched in white Natalian culture that, in the 1930s, when Maritzburg city council elected an experienced white woman to be a member of its Native Affairs Committee, one male councillor resigned in protest. It was, he asserted, simply not "in the best interests of the native" to have a woman in a position of authority over them (*Natal Witness*, 19 September 1933). Also, as we shall see in the chapters to follow, African and European men alike often blamed African women for the breakdown in traditional discipline that had formerly sustained a generally healthy community. Yet ironically, Zulu conservatism around gender relations was widely blamed for the emerging crisis that the Native Affairs Committee was facing, above all the ruination of the reserves and consequent influx of migrants to the city and environs supposedly caused by the Zulu obsession with cattle for *ukulobola*.

I will be returning to this theme with reference to some specific areas of contention in Edendale and Maritzburg. Let us first step back in time from the decisive hearings of the Thornton Committee on Edendale's future to consider an earlier debate that has shaped the historical geography of so many African cities: where, under what principles or regulatory guidelines, and at whose expense, to house the growing population of African families in and around the city?

# FOUR | The Native Village Debate in Pietermaritzburg, 1848–1925

A number of contradictory clichés have been used to describe Pietermaritzburg: "Sleepy Hollow" (evoking the slow pace of municipal decision making and lack of cultural vivacity), "the last outpost" (due to its loyalty to the British Empire and patrician values), "the city of flowers" and "the healthiest city in the Union" (which by the 1930s was likely true for most white people). Its burgesses, while tending to be socially conservative, nonetheless voted quite consistently to make it a bastion of United Party, Progressive, Liberal, and other (in the context) moderately left-leaning politics almost to the end of the apartheid years. Its liberal reputation rested in part on "reasonably relaxed" racial attitudes, as Meineke and Summers put it (1983, 133), and progressive city policies towards its black population. As an example of the latter, well into the 1980s the city prided itself on Sobantu (or the Model Native Village, as it was known when first established in 1928). The city touted the village as one of the best-laid-out and best-managed African urban locations (townships) in the wider region and hosted study visits by officials from as far away as Kenya, Cape Town, and Rhodesia. Middle-class Africans in other townships also cited it as the standard to aim for – "the best houses offered in Natal," as the Ashdown Advisory Board put it in protesting their own conditions on the Edendale side of town.[1] Indeed, scholars otherwise critical of city governance concede the good quality of housing and high level of electrification that the city provided to Sobantu compared to other South African townships of the era (Peel 1987). Not least among its assets – recommending it well even

4.1 Sobantu, twenty years after democracy: improved homes in the township below, with a formerly coloured neighbourhood above and new neighbours informally occupying the old buffer zone

today – is that it is an easy stroll from Sobantu to both the downtown business core and one of the main industrial zones of the city, thus offering nearby employment opportunities.

In contrast to this liberal image, several historians have focused on a cruel, insidious streak of racism that underlay the municipal decision-making process. Sobantu, in this analysis, is evidence not of the city's enlightened native policy, but of "the sanitation syndrome." Prima facie evidence of the latter was the whites-only referendum in 1925 that overruled strongly expressed African opinion on the question of where to locate the proposed village. White voters decided in favour of putting the city's showpiece African housing project next to the "conservancy depot," that is, an open pit where most of the city's night soil removals and animal carcasses ended up. For this reason, and many other indicators of racial segregation built enduringly into the urban geography, Pietermaritzburg has been bluntly described as a typical, even "quintessential apartheid city" (C. Merrett 2009, 240).[2]

As noted in the introductory chapter, Maynard Swanson (1976) first coined the term *sanitation syndrome* in his study of the origins of racial segregation. A subsequent article on Cape Town and Port Elizabeth found that the bubonic plague offered an especially compelling pretext to overcome opposition from those white employers who stubbornly preferred the status quo for their African employees (either on-premises barracks or small backyard shelters known as *kias* or *khayas*). The invocation in scientific terms by white officials of a dire health threat from Africans was ultimately successful in justifying the shift of hundreds of African families to Ndabeni and New Brighton, many kilometres out of their respective cities, notwithstanding the fact that Africans were neither the main vectors nor major victims of the plague (Swanson 1977). Swanson later read the concept a bit farther back in time to explain the movement to segregate Indians in Durban, a goal ultimately achieved through colony-wide restrictions on Indians' ability to purchase land. In effect, white citizens' fears of economic rivalry from Indian traders were disguised under the cloak of protecting the health of whites from "nests of Oriental dirtiness," as one contemporary phrased it (quoted in Swanson 1983, 407).

Swanson's most important achievement was to illustrate how sometimes subtle racist or eurocentric assumptions about Africans and Africa or Indians and "the Orient" imbued the production and application of scientific knowledge in pursuit of narrow political or economic goals. Numerous studies from around the continent have since corroborated this.[3] In short, the language of science had a powerful conjuring effect that enabled good people to make and to justify bad policy decisions. Those decisions have lingered as facts on the ground long after the science was debunked. Citing Swanson's work, for example, Howard Phillips describes mass evictions around the country in the wake of the plague (or "epidemic expediency" as he terms it) as having laid the foundations for South Africa's stubbornly racialized urban geography: "From small germs do mighty townships grow" (Phillips 2012, 66). Moreover, the importance of science as a metaphor actually increased over time as overt expressions of racial prejudice grew less socially acceptable. On the latter point Diane Wylie makes a persuasive case that "cultural racism," understood as an overweening hubris among whites arising from European scientific achievements, significantly broadened political support among whites for National Party rule in the early 1950s (Wylie 2001). The "malnutrition syndrome" she describes was a tendency to blame Africans themselves (rather than low wages and land dispossession) for their poor diet choices, inappropriately slow infant weaning and harmful agricultural practices. The language of nutritional science thus helped to lay the groundwork for the implementation of grand apartheid by winning sceptical liberal voters over to the cause of separate development. This would partially explain how the pro-apartheid National Party finally managed to win its first parliamentary seat in Pietermaritzburg in 1977.

A first glance at the documentary record does seem to suggest that Maritzburg conforms to the sanitation syndrome. James Just Niven, inspector of nuisances, was the most vocal proponent of removing Africans from the city in terms that precisely reflect this syndrome in the second decade of the twentieth century: "I consider that the housing of natives, with their native conditions and customs, in the heart of the City, and surrounded by Europeans, a distinct menace to the health of the whole community" (Pietermaritzburg [1904–90], 1915, 119). The mayor in the same year stoked fears by referring to a case of non-sexual transmission of syphilis from a female African domestic worker to a European child (16). Subsequent debates over what to do about peri-urban sprawl were also frequently characterized by the exactly kind of hyperbolic language that Swanson found in his studies. In the words of the city's medical officer of health during a crucial municipal debate in 1941 on whether to incorporate Edendale into the city so as to clean it up, Dr M. Maister described that community as the "thin crust of a volcano ready to spout an epidemic of disease," with Pietermaritzburg "right in the path of the lava that will overflow."[4] One Maritzburg writer even anticipated Swanson's argument when he defended Africans in the surrounding "black belt" against detractors whose political agenda was barely disguised by the health argument: "To suggest incorporation [of peripheral African areas] in the borough on the grounds of 'insanitary conditions' constituting a 'public menace' would be a ridiculous bluff" (F. Berger, "Incorporation," *Natal Witness*, 10 March 1933).

Yet Pietermaritzburg also presents some obvious anomalies to Swanson's theory. How powerful could the sanitation syndrome really be when Sobantu was established so close to the then white suburb of Mountain Rise and farms of Bishopstowe (immediate neighbours) and to the city proper (about two kilometres to the eastern edge of the central grid)? When the main African opposition had proposed an alternative site that was farther away from the city on the flood plain at the most polluted section of the river as it flowed out of Edendale, and had their proposal vetoed? And when as late as 1970 (twenty years after the proclamation of the Group Areas Act) almost half the African population still lived in the town proper, either in large hostels near the heart of the central business district or, more commonly, directly on employers' property (Thorrington-Smith et al. 1973, 32)? How can a syndrome be responsible for "dictating" (Lester 1998, 87) a major urban planning decision that took more than seven decades of hapless debate before it was partially settled? Were the syndrome so politically important, would it not have been a major part of election campaigns and visible in popular discourse? Yet, it was not; indeed, not counting the annual report of the medical officer of health, one has to very patiently trawl through the public record prior to the 1930s to find *any* references to African health, menacing or otherwise. It scarcely featured even in the 1925 referendum, which itself was virtually a non-issue in the municipal elections of that year. Although his family played an important role in that and subsequent debates around health and housing, it pointedly does not warrant a mention in H. Selby Msimang's (n.d.) history of Edendale.[5]

Moreover, astute as Swanson's analysis was back in the 1970s, subsequent iterations have clearly tended to lose his early sensitivity to the complexity of debates and political relationships. Swanson's original unpublished argument placed sanitation as but one concern among many that led Durban to adopt its native location policy: that version also showed how Africans' fear of infection by whites was a compelling part of the argument (1965, 387). As well, while presenting ample evidence of repugnant racist sentiment among government officials, Swanson was initially careful to show that most whites did *not* support formal racial segregation in the colonial period, and did *not* perceive Africans as a particular danger to their health, worth the risk of higher taxes and wages that proper residential segregation implied. The evidence from Durban suggests that this view prevailed. Far from being sudden or reactive to an epidemic health crisis in 1902, the establishment of a "native village" took decades (Lamontville in Durban was laid out in 1936) while the eradication of lingering freehold communities near the inner city such as Mkumbhane/Cato Manor took decades further (1959–63). How then could the limited quarantines of 1902 be described as a "definitive step" towards residential segregation, as Swanson subsequently maintained (1977, 396)?

Discontinuities and ambivalence around sanitation over the decades have in fact often disappeared in the historiography, subsumed under harsh and sweeping judgments asserted with reference to Swanson about the relationship between science and segregation or the seamlessness of whites' racism over time. Historians, including in a later article by Swanson himself, have claimed that interventions to address Africans' ill health were merely "diversionary" or "meliorist" (Wylie 2001, 199, 124; Swanson 1996, 276), and that perhaps there was no real health crisis among Africans even as late as the 1930s (rather, only a "putative" one, according to Swanson [1996, 296]). To Nuttall (1984, passim) blacks who worked with the state to improve sanitary conditions in their communities were "collaborationists." Scientific expertise, in this view, was "a muffler and a blinder," the "handmaiden" of apartheid (Wylie 2001, 242) that obscured the real problem – "white panic, racist paranoia and squeamishness," as Maylam put it when questioning the merit of this analysis (1995, 25). Without at all denying the insidiousness of white racism and paternalism, I do wonder whether these judgments are fair in all cases and over many decades of dizzying social and political change. Is it sufficient to make such claims by citing Swanson only, in lieu of any direct local empirical evidence? We might also usefully ask whether Africans were simply and always at the receiving end of such a syndrome, or might they have actively participated in its construction as a social metaphor for their own possibly diverse purposes?

With these questions in mind, this chapter traces the history of debates around African health and housing in Pietermaritzburg and surrounds. The focus is on three specific moments or issues where one would expect the sanitation syndrome to have played a noteworthy role: the multiple decisions *not* to create a segregated native village prior to 1925; Pietermaritzburg city council's responses to specific contagious disease and water pollution threats in the period in which the syndrome was held to be driving segregation in other cities; and, finally, the infamous decision to place the village beside the conservancy depot. Does the evidence support the model? And how much does the model explain the evidence?

≈

When the British took over Maritzburg in 1843, they made it the main base for their colonial army and a staging point for ambitious British settlement and agricultural schemes. The town's role as an economic and population magnet for Africans increased in tandem. Ironic as it may strike us today, white settlers for decades commonly referred to Africans in the Midlands area as "immigrants" owing to the fact that they mostly moved there from Zululand or the southern fringes of Natal.[6]

Conflict over land was inherent in the rapid growth of the African population and the colonial government's desire to market Natal as an attractive destination for immigrants from England. The colony's first "Diplomatic Agent to the Native Tribes" and eventual secretary for native affairs, Theophilus Shepstone, devised a system to manage that conflict while at the same time securing a steady supply of cheap African labour for the benefit of white settlers with minimized costs and responsibilities accruing to the government (McClendon 2010; Guy 2013). Towards those multiple objectives, in 1846 Shepstone ordered the demarcation of native reserves, including Zwartkop, on a stretch of rough and tumble land in the "mist belt" along the escarpment overlooking Pietermaritzburg. The sale of land within reserves was prohibited in favour of communal land tenure, with disputes adjudicated by loyal and dependent chiefs exercising formalized customary patriarchal law under Shepstone's watchful guidance. A similarly governed but much smaller reserve was also demarcated on the east side of the city around Table Mountain.

The so-called Shepstone system of indirect rule drew on earlier segregationist ideas from the eastern Cape and elsewhere in the empire. Africans within the colony were supposed to have a protected land base where they could live according to their presumed ancient, more or less healthy ways and with the costs of maintaining social harmony and justice primarily borne by themselves. Zwartkop and other reserves in the system were neither big enough nor well-placed enough to disturb the expansion and the cost-efficient operation of European farms. On the contrary, because male reserve dwellers were required to pay tax, they were motivated to seek employment or sell goods for whatever the farmers had to offer. Hence the reserves helped to stimulate the growth of a white-owned commercial agricultural and forestry sector in the areas surrounding them. After rendering their services, and perhaps sustaining injuries or communicable diseases in the course of their labour, the men would retire back to the reserve to be taken care of by their wives, children, and chief in a salubrious rural atmosphere.

Soon after Zwartkop was created, another group of African "immigrants" established their own discrete community nearby on almost exactly the opposite principles: private property, commercial production, and Protestant values of monogamy, reward for hard work and educational merit, and self-sufficiency. The *amakholwa* at Edendale saw themselves as the vanguard of an African elite who could compete on an equal footing in the settler economy. As such, Edendale was never formally reserved for Africans prior to the application of the Group Areas Act beginning in 1960. On the contrary, over time many *amakholwa* landowners began to rent accommodation to migrant workers including Indians, or to put their property up for sale. An increasing number of Europeans, Indians, and coloureds became landowners in that way from the late nineteenth century. Non-African settlement picked up in the second decade of the twentieth century so that, by the time of the first official census in 1936, the wider Edendale area comprised one of the most racially diverse semi-urban communities in the country, with 277 Europeans, 119 coloureds, 1,283 Indians, and 7,011 Africans.[7] To be sure, people tended to live and socialize in discreet enclaves, with fences and hedges for those who could afford them. With the exception of some extremely dense pockets of settlement, the population remained spread out over a large area with a semi-rural ambience. Yet homes were close enough that children played together, and, as I will discuss in the next chapters, residents eventually came together to form associations that led to a rudimentary multiracial local authority.

Pietermaritzburg during this period was also never formally reserved for whites only. On the contrary, from the beginning it became home to traders and labourers from many racial groups. City Councillor J. Boshoff in 1855 referred to an unregulated "village" of Africans having sprung up on (and eventually been removed from) empty land on the south side of the river where it flowed past the town centre (*Natal Witness*, 19 September 1855). A painting from 1870, a time when the African population was estimated to be around 2,000, depicts a Zulu beehive hut in what is now near the heart of the central business district (Wedderburn 1991, 249). Africans with the means remained free to purchase their own property in town. This they did with the establishment in 1865 of an *amakholwa* "colony" in the New Scotland/Topham Road district, just across the Msunduze River from the central grid and a short walk from the city's premier park. This neighbourhood was eventually sufficient in population to warrant the creation of a titular (elected) *amakholwa* chief for Maritzburg to represent black residents in their relations with the Native Affairs Department (NAD).

The first attempt to quantify African property ownership in 1898 found that Africans occupied ninety-six and owned forty-two "buildings" in the city (Natal [1892–1908] 1898, B71). The district surgeon clarified this a few years later as meaning that some "possess houses of their own, and live in a more or less civilised manner" (Natal [1892–1908] 1903, A79). By the 1920s about fifty Africans owned sometimes substantial brick homes in the mixed-race neighbourhood, often renting to diverse tenants, while an estimated

4,000–5,000 Africans lived as families scattered around the city on town lands or pockets of freehold farms.[8] At least one prominent African family (the Gumedes) owned a residence on fashionable upper Berg Street right in the town itself (Brown and Aitchison 1988, 209). A small but relatively diverse coloured population (then still considered more or less white) and, after the 1870s, a rapidly growing Indian population also found residence in town, primarily in the lower east end of the grid. Around the turn of the century, a new Indian "colony" was established on the west side of the city on the main road to Edendale (Pentrich/Camp's Drift). A 1952 census found that about 300 Africans lived in Pentrich, mostly renting rooms from Indian landlords but also owning twenty-five acres or 5 per cent of the area.[9]

That said, and much like almost any city in the world in that era, people from different cultural, linguistic, and class backgrounds tended to self-segregate or cluster into more or less discrete neighbourhoods. Housing and land prices added strong incentives for this pattern, particularly as the great majority of African "immigrants" did not conform, or aspire to conform, to colonial notions of proper urban life. The first generations of non-*kholwa* Africans in particular tended to regard the city as a stepping stone towards a prosperous life back in the rural areas, a temporary residence rather than a cultural home. This attitude quickly led to conflict with property owners (and voters) over a range of issues, notably noise, fires, animal smells, bare breasts, and other aesthetic questions. The first proposal, in 1848, to establish a discrete and properly managed native town already anticipated these issues. The so-called Gibb scheme put forward by the colonial government envisioned the purchase of 1,000 acres of land two to three miles from the city centres of both Maritzburg and Durban. The first fifty of those acres would be allotted to "advanced natives" who would form the nucleus of a stable, monogamous, and appropriately attired African workforce. As their anticipated prosperity became evident, other Africans would be attracted to the lifestyle and "reclaimed." Far from proposing to expel the desired intelligent and industrious Africans from proximity to the city, Gibb (and Shepstone, who also backed the proposal) wanted to bring them nearer. "Thus the European and Native population would be brought into more intimate and closer connection with each other, and under more favourable circumstances than can apparently in any other manner be brought about … A nucleus of natives will thus be formed to foster and disseminate enlightened views and principles and the first great difficulty in their improvement be overcome."[10]

This proposal fell through for a variety of reasons, no doubt including "want of funds" (Atkins 1993, 118) and perhaps unstated anxieties about the precise meaning of "intimate." The founding of Edendale in 1851 and the growth of another *amakholwa* station to the east (Ekukhanyeni) also undermined the initiative by promising to fulfil one of the main objectives without any cost to the colony's purse – that is, providing a place for "advanced natives" to settle and inspire their compatriots. However, Edendale and Ekukhanyeni were still too far away to be of much use to city employers, who wanted cheap, readily available labour. The city thus stepped forward with a proposal of its own in 1855. Its native village was conceived as what we would now call a "site and service" project – that is, the city would provide the land for free and give the tenants one year to build on it or lose their claim. In the first draft regulations for the village, we can see it imagined in spacious, semi-rural terms, with one-acre plots that would allow subsistence cropping and domestic animals as a supplement to the low wages that African workers received, the latter being an assumed necessity both for their employers and for Africans' own moral welfare.[11]

During the first town council debate on the issue, supporters explained that their main concerns with the unregulated status quo were fourfold. First, there was the noise ("psalm-singing till late hours in the night was a perfect nuisance," as Councillor Hoffmann put it; *Natal Witness*, 14 September 1855). Closely related to this was the presence of squatters and unemployed visitors, who not only contributed to the racket but were felt to be a threat to public safety, as they had no means of support except, presumably, petty crime. Third was a desire to protect legitimate African employees from the exorbitant rents that they were required to pay to live in town. For those whites who were not landlords, there was obvious self-interest in this concern. Rents of up to £3 a month exceeded the average worker's wage and, since wage hikes were ruled out, virtually required Africans to top up their legal income by other, implicitly criminal, means. Cheap, well-managed accommodation would avert this problem while reducing pressure on employers to pay their workers better.

Finally, there was the disruption to service that resulted when male African workers whose families still lived in the rural areas made constant demands to return to distant homes or simply absconded when family called. As the *Natal Witness* put it in expressing its support for the village, the city could solve that aspect of the labour problem by using its most abundant resource (land) to "foster a new class of labourers, and to endeavour to add to their number by holding out inducements for them to settle comfortably with their families near to our labour market" (21 September 1855). Other councillors pointed out the opportunity for the citizens of Maritzburg to fulfill their Christian obligations by providing decent homes for their employees and families, while at the same time conveniently providing the city with a steady new source of rental income (14 September 1855).

The main opposition to the motion focused on the potential that such a village had to harbor "lazy, dangerous and mutinous vagrants," as letter writer "Alpha" put it. His other main fear was that a village would undermine individual white employers' ability to discipline cheap labour under their direct supervision on private properties. There was an implicit political anxiety in this as well: "no encouragement should be given [to Africans] to isolate themselves, or band together in large masses or stations" (*Natal Witness*, 6 April 1855). Councillor Boshoff also spoke against the motion but in more liberal terms. If the various sites out of town such as Edendale were indeed not appropriate, he argued, then the preferred solution to the admitted problems was to allow Africans freedom to "build houses and live properly," dispersed throughout the city on its abundant supply of privately owned but empty plots as determined by the real estate market (14 September 1855).

The terms *sanitation*, *hygiene*, or *disease*, and the idea that these occur among or are caused by Africans, are strikingly absent from the debate at this stage. Certainly, there were isolated comments at the time upon the "wretched" living conditions and "filthy huts" for Africans living in town (Holden [1855] 1963; *Natal Witness*, 30 November 1855). Blame for this, however, was generally cast upon the employers rather than any innate slovenliness among Africans. Shaming whites for their blatant disregard for Africans' well-being was a common theme among letter writers, although the *Natal Witness* itself cast the problem in less moralistic terms. Where poor accommodation existed, it stemmed from a desperate shortage of artisanal labour and capital available to whites to enable them to do the right thing for their employees. That shortage prevented the construction of proper rows of cottages for Africans to rent in the preferred "ordinary" manner, as in England or as did the white working class (*Natal Witness*, 14 and 21 September 1855). In the case of some shoddy-looking houses at the Edendale station, Bishop Colenso also exonerated Africans during his visit there in 1854: they had no spare time for home improvements since every effort was being channelled into cash crop farming so as to pay off their debts for the purchase of the land as quickly as possible (Colenso, 1855, 50).

Draft regulations for the proposed village make it clear that town councillors scarcely regarded sanitation as an issue. Of fourteen proposed regulations, only one (the very last enunciated) touches on the subject: "No broken bottles, bones or other nuisances, to be allowed in the streets of the village." Detailed specifications meanwhile provided for minimum house size (24 x 12 feet), number of rooms (2–3), a chimney, and a garden enclosed "with a neat and substantial fence." They thus ruled out traditional style homesteads and informal structures. They also ruled out certain behaviours – "no heathenish rite," polygyny, uncontrolled dogs, and so forth (*Natal Witness*, 5 October 1855). But the regulations were completely silent on the question of privies.

The motion to create a village where Africans presumably took adequate care of their own sanitary requirements was approved by a vote of 5–2. That decision, however, was quickly overshadowed by a much more heated debate around whether to "secure the sanatory [*sic*] safety" of city dwellers by imposing residential building standards in Pietermaritzburg itself (*Natal Witness*, 12 October 1855). Some councillors probably saw the proposed regulations as a discreet method to push Africans out of town and hence produce a population for the newly approved native village. Councillor Boshoff, however, pointedly insisted that "the question should be looked at without reference to the natives" (30 November 1855). Rather, the central question was how to get white property owners to respect their neighbours by maintaining adequate structures and tidiness so as to inhibit the spread of diseases such as cholera and smallpox.

Compelling the clean-up of private property quickly proved to be a far more sensitive topic than the native village, particularly since lax English-speaking citizens (by far the main property owners at this stage) were the obvious prime targets of the proposed regulations. To Councillor Leathern it was "selfish, arbitrary and calculated to check industry," "not fair to natives" (not explained but presumably because it would push them out of town), and "discreditable" to the assumed inherently high standards of English dignity and democratic traditions (*Natal Witness*, 30 November 1855). The motion nonetheless eventually passed in a compromise form. Notably, it declined to exclude livestock and associated outbuildings from within the city boundaries. With regard to African tenants, the approved standard specified only that African-occupied dwellings be limited to one for employees per property (theoretically leaving the door open to large barracks and family-sized *kias*). While limiting the *quantity* of African residences per property, however, the city tellingly did not interfere with owners' freedom with respect to *quality*, leaving the standards of employees' on-site dwellings completely unregulated. Incredible as it may seem, the city did not even require that employers provide a privy for their employee accommodation until 1923.[12]

Irrespective of sanitation matters, the approved native village never happened. The fact was that open land remained plentiful enough within the city's boundaries to push isolated extreme nuisances out to the periphery (and thus out of sight) in an informal, ad hoc way using available police powers. The bush along the river, the spruits, and the fringes of town was meanwhile thick enough to allow procrastination on the privy issue. Indeed, with procrastination such an easy option, there was insufficient consensus among voters to support a budget for a more formal approach not only to native housing but to the most basic needs for workers in town.[13] A deep economic depression lasting from 1865 to 1870 confirmed that the city was in no position to take on such a bold venture. It could not even afford to pay its white employees, let alone start building homes for a hoped-for advanced class of Africans. A mooted private native location at this time, again backed by Shepstone himself, also appears to have been abandoned in the face of the financial crisis (Parle 1988, 136). An even more radical proposal by the city's inspector of police and of nuisances in 1873 was to expel *all* (his language) Hottentots, Kafirs, and Coolies from the city and, citing crime, sickness, sexual menace, and property values alike, to place them in separate locations two miles beyond the city limits ("Inspector of Nuisances," *Natal Witness*, 7 March 1873). However, this idea sank without a further trace.[14]

A further anxiety persisted well into the twentieth century that undermined enthusiasm for the village idea among most white officials and voters. This was the fear that an African village, however much it was intended to encourage the development of a stable working class, would inevitably attract a new population of "undesirable" and "won't work" residents. This fear may partly explain the failure of yet another village proposal in 1875. This time it was for a site on the town lands rising to the north (agriculturally marginal but today among the city's most valuable real estate). It was well known that Africans squatted in the nooks and crannies of the Town Bush Road area. Why not formalize them and starting earning rental income? This village proposal was again approved but then almost immediately abandoned in favour of a counter-proposal for barracks for single men on the lower side of town (Wills 1988a, 40). Council believed that the Fitzsimmons Road barracks was preferable to a village in that it better suited a tightly controlled but transient male labour force and was far cheaper to build and manage.

Over time, the colonial government's and city's ad hoc approaches to the non-European presence in the city resulted in a raft of policies and by-laws that somewhat policed the movement, settlement, behaviour, and attire of Africans and Asians. The colonial government, with Shepstone's influence, notably sought to impose a coordinated policy by introducing the *togt* system in 1874. This required Africans to register for labour in all urban jurisdictions on a daily basis and to wear a badge to display their legal status. In addition to the *togt* requirement, Pietermaritzburg city imposed a curfew on Africans and instituted laws prohibiting the sale of home-brewed liquor in order to keep the noise down at night. However, there was no end of creative ways to adapt or get around the rules by all parties, and in practice none of these restrictive laws was particularly well enforced. Swanson notes, for example, that many employers in Durban were uncooperative with the *togt* system and that never more than one-third of African employees at any given time actually possessed *togt* passes as required (1976, 165). As late at 1886, the *Natal Witness* denounced the *togt* system as "laissez faire" and tentatively condoned panicky calls to get tough on Africans' de facto freedom of movement and accommodation (14 December 1886).[15]

Cool heads prevailed. The need to secure African labour (and *amakholwa* loyalty to the Crown) far outstripped the risk of alienating it by harassment and unnecessary restrictions. Aside from the above unusual instance of moral panic, a certain laxness towards Africans in town is suggested even in official discourse around African health. As Umgeni District Surgeon D. Campbell Watt observed in 1898, "the health of the Natives in this Division would appear to be very good indeed … Most cases are Chicken-pox, brought into town by Kraal Natives, who under the present Borough Regulations, are allowed to come into town, and invade your premises at their own sweet will, calmly taking up their quarters for the night in your Kafir house" (Natal [1892–1908] 1898, B27). As for liquor laws, an indirect way of restricting the size and activities of the female population in town, these were branded "utterly futile" which was likely just as well. As the *Natal Witness* presciently argued in its case *against* the registration of African workers and in favour of tolerance of discrete, often female-run shebeens in town, strict laws would only push the problem out of the borough boundaries to form "a sort of belt of outer darkness," far more difficult to manage (9 December 1886).

In early iterations of the village debate, the feared undesirables that a proper village would supposedly attract were male "loafers" and "spongers." There was also the problem of "impudent" laundrymen known as *amaWasha*.

Their guild-like stranglehold over a critical service industry gave them the economic wherewithal to ignore city attempts to impose restrictions on either their movements or their ability to facilitate entry into the city for their *togt*-less kin (Atkins 1993, 131–6). By the last decade of the nineteenth century, a further problem, in the form of "loose" African women, had become increasingly apparent. Radically divergent views on what to do about them further stalled movement on the question of a native village. Put simply, if the police and property owners could not control women and illicit brewing even in their own literal backyards in the heart of the city and in middle-class neighbourhoods, then how could a native village be anything other than an open door to uninhibited drunkenness, thieving, and immorality? But taking a pragmatic view, the district surgeon in 1903 proffered a powerful health argument that the status quo, however unsatisfactory, was preferable either to aggressive enforcement of existing laws or expensive innovations: "Many Native girls and women wander from their kraals in the country and take up the profession of prostitutes in the City ... Anything like a wholesale banishment of these women to the country is sure to be followed, in my opinion, by an epidemic of what has been termed the 'social pest.' I am disposed therefore not to be too severe on these women ... and not to trouble so much about those whose relatives make no application for them" (Natal [1892–1908] 1903, A80). In other words, better to contain the health risk in a known area that was large but relatively manageable (Pietermaritzburg) than to disperse it in rural areas or overconcentrate it in a formally segregated village.

One other opinion on the matter of women in town was only obliquely documented at the time: that of African chiefs in the neighbouring reserves. Heads of households and lineages in the reserves were deeply concerned about the breakdown in traditional social hierarchies that city life implied. According to Benedict Carton's sympathetic account, heads frequently complained to the government that it was supplanting their patriarchal authority over young men and women, albeit so ineffectively that it was exacerbating gender and generational conflict. The Bhambata rebellion, or "war of the heads," in 1906 was thus as much a struggle between young and old Africans as between Africans and white settlers, with elders looking to the state for help in shoring up their eroding authority while youth turned to violence to assert their own. Chiefs in the reserves around Pietermaritzburg stayed out of the Bhambata fray. However, a joint statement by local chiefs in the run-up to the fighting laid out their grievances with youth insubordination, which they believed conditions in town inflamed. They rued sons who went to the city attracted not only by the possibility of independent income but also of irresponsible sexual exploits with town women. The chiefs wanted government to "hound" the women (the "treacle") out of the town. But where then would they go? One chief, Ziboni, shared the government's anxieties about deporting them back to the reserves and proposed a unique solution. As he put it, "Can't these bad women be put into a kraal … all by themselves?" (cited in Carton 2000, 82).[16]

There was, in short, a bewildering muddle of impractical and self-contradictory opinions that seemed to shift with the breeze of every new labour demand or moral panic. Did, as Swanson argues, the bubonic plague clarify matters enough to swing the argument in favour of formal segregation? Recall that it was not that the disease killed many people but rather that fear of infection *from whites* sent African workers packing for presumed safety at their homes in the rural areas – business came "almost to a standstill" as a result (Swanson 1965, 387). At the urging of principally Durban officials and major employers, the colonial parliament passed the Native Locations Act, whose primary intent was thus not to protect whites' health or morals but rather to stabilize the city's African workforce. Tellingly, the idea of offering African workers decent, subsidized, serviced, and plague-free accommodation was greeted with open derision by the *Natal Witness*. It was not so much because the editors disagreed with the location idea. On the contrary, they praised Cape Town's native locations, which they noted had lately been "improved" in response to the plague threat in that city (5 June 1903, 5). Their difficulty with Durban council even considering a similar policy in Natal was in the practicalities. Natives in Natal, they pointed out, even urbanized Christian ones, generally continued to practise polygyny and *ukulobola* – marriage in exchange for cattle. This implied a cultural incompatibility with proper urban life, which was not the case for the Cape's mostly coloured or "Hottentot" populations, and this factor would make a truly urban native location unmanageable.

It was ultimately not the plague, but another issue altogether, that laid the groundwork for deciding to build a native location. The Natal Native High Court ruled in January 1908 that home-brewed beer (*utshwala*) was *not* an intoxicating substance. This ruling removed the power of the police to arrest Africans who brewed beer for sale and whose backyard shebeens were the focus of many of the noise and prostitution complaints of the past decades. However spottily effective the law had been before, the court's ruling opened the gates to a surge of shebeening within the city and with the traditional

4.2 The Edendale Beer Hall was strategically situated – not in Edendale itself, but in Plessislaer (Sutherlands) – to intercept African men on the way home from work in the city. The city's monopoly on beer sales here and elsewhere underwrote its contributions to "native administration"

brew doctored to make it truly intoxicating (*isishimeyana*). An unsatisfactory but basically tolerable situation suddenly became much worse. Chief Constable of Police Fred W. Moore put it this way in his annual report: "The outstanding feature of the year is the Increase in Drunkenness among Natives." Without the easy target of beer brewers, he continued, he now faced "the indiscriminate and promiscuous quartering of such numbers of natives in private yards and locations within the City and suburbs, under such circumstances that it is quite impossible to exercise any proper supervision or control over them" (Pietermaritzburg [1904–90] 1908, 45). A tightly regulated urban location would provide such control, and Moore consequently threw his political weight behind that proposal.

The native beer ruling created a similar crisis in Durban, where politicians mobilized to push a bill through the colonial legislature that would enable the reassertion of control over African brewers and drinkers (La

Hausse 1992). Passed as the Native Beer Act of 1908, it provided the means to realize the location idea in practical terms, at least as far as alcohol was concerned. First, it allowed municipalities to declare for themselves a monopoly on the production and sale of beer. Then it restored police powers to arrest not just brewers *up to five miles beyond* the town boundary, but also anyone transporting home-brewed beer into the city. Any adult male consuming beer off the premises of the newly established native beer hall committed an offence, as did any African woman or person under the age of fifteen. Applied to Maritzburg, the majority of people arrested for transgressing the law in its first year of operation were in fact women, and on that basis the city declared success in "stamping out" the shebeen problem within its borders (Pietermaritzburg [1904–90] 1909, 34). Another success was tallied the following year as the city's monopoly on beer sales generated no less than £4,300 in revenue (1910, 21).

Much of the new revenue went straight to the police. In the short run and from the perspective of the voting population, repression removed the perceived main need for a properly managed native location. With shebeens driven out of sight, mostly up the valley towards Edendale, the political pressure to create a native village temporarily abated. In the long run, however, the police focus on liquor violations obviously did nothing to address poor housing and associated ill health among African workers, let alone the peri-urbanization of Sutherlands, Macibise, Edendale, and other emergent slum areas. On the contrary, enforcement of the beer law within the city and its five-mile radius simply pushed offending properties and practices out to the periphery. They may have been out of sight, and out of jurisdiction, but they were also directly upstream from the city on the Msunduze River, which duly delivered their effluent back to the good people of Maritzburg. Some of the beer revenue was therefore banked towards the anticipated costs of getting a properly managed native location established if and when conditions on the periphery deteriorated to truly intolerable levels, or other events necessitated. The decisive initiative that finally demonstrated the need was to come not from Maritzburg or even Durban, but from Johannesburg.

≈

Before looking at the debates leading to the creation of the Model Native Village in the 1920s, it helps to assess attitudes towards and decisions related to specific public health issues as they played out in relation to questions of racial difference over the preceding decades. To begin with the concept of sanitation, official documents and popular press alike make it clear that the term was understood primarily to mean securing the city's water supply from pollution and improving the disposal of sewage. Eight of the 114 regulations laid out in the first document proposed to govern urban life in British Natal were devoted to water quality, including the stern admonition that "any person who shall bathe in, wash in, or cast any filth or rubbish into any public dam, watercourse or canal, or in the water above the dam, or in any other way willingly soil the water therein, shall forfeit a sum not exceeding two pounds, nor less than one pound sterling" (South Africa 1960, 186, draft of regulations, 1 and 8 October 1847). In this way, councillors were attentive to emerging scientific knowledge about the connections between dirty water and disease. As noted above, for example, cholera was specifically cited in the 1855 motion to introduce residential building standards in Pietermaritzburg – the discovery of its connection to contaminated water had been made in London only the year before (*Natal Witness*, 30 November 1855). Dr Mann warned that inaction on protecting Pietermaritzburg's water supply would result in "a deadly mist" spreading fevers, cholera, and contagious dysentery (cited in Hattersley 1955, 48). Other more tangible contaminants were also commented upon with concern. Mrs S.E. Lamond, for example, recalled collecting drinking water (and fishing for crabs) at her home near the town centre as a child in the 1860s. "It was not uncommon to see a dead dog or cat in the sluit," she wrote, an observation that made visitor William Nicolson's caution about the water – "excellent when passed through a filter" – seem a tad understated (cited in Hattersley 1936, 122 and 147, respectively).

Water, rather than conscious or even "metaphorical" town planning, thus accounts for the first de facto racially segregated neighbourhood in the city. The original layout of the city had channelled naturally pure water from a stream (the Dorpspruit) into open furrows, called sluits or sloots, leading down each street in the grid. For those at the top of the grid, this was a wonderful free asset, and those who could afford to buy property did so on that side of town. The same sluits received domestic and industrial waste, not to mention animal excrement and other run-off from the streets, as they flowed through town. Hence, the lower you were on the grid, the dirtier your source. During heavy rain, the sluits amplified floods, while during drought plots (erven) in the upper part of town passed precious little water on to the lower. Erven in the lower, southeast corner of the grid thus had been largely unpopulated before the early 1860s, when the first wave of Indian immigrants arrived in search of affordable property to purchase. The availability of cheap land was also likely the main reason to site the city's first barracks for African workers in the same area (Fitzsimmons Road in 1874).

The above debates were framed by almost fundamentalist presumptions of private property rights. Liberal opinion later came to be associated with state-led development, but in the late nineteenth century it could be vehement in its defence against state infringements upon private property. This produced some remarkable flourishes of rhetoric and (in retrospect) twists of illogic with regard to public health. The *Natal Witness*, for a noteworthy example from 1886 (a time of "moral panic" over interracial sexual relations), came out strongly against a proposed bill that aimed to relax punitive measures against prostitution in favour of a preventative public health approach. Most people conceded by that time that prostitution was a problem, particularly since it involved a significant number of white women, who presumably catered mainly to the military base.[17] City council proposed a shift away from evidently ineffective policing to giving suspected prostitutes medical examinations and treating them for sexually transmitted infections, if needed. The *Natal Witness* opposed this initiative in favour of a harsh crackdown on the women, which it partially justified by reference to property rights: "There is … [a] principle with regard to which the English people are sensitive. They object to imposing, for any reason, any limitations upon the liberty of any class of Her Majesty's subjects; and still more do they object to place in the hands of any authorities, police or other, powers which may be used for purposes of the most detestable persecution" – that is persecution of property owners rather than "immoral" women (leader, *Natal Witness*, 24 November 1886).

Africans did not feature in these early discussions about water, rubbish, or the relationship between health and private property. Although inferior accommodation and bedraggled Western clothing led to concerns about the health of blacks in town, whites generally regarded the Zulu as a robust and clean people. Empirical observation seemed to bear out such views. As late as 1903, for example, mortality from typhoid was lower among rural Africans than it was among whites in the city (Dyer 2012, 100). This point also provided an important ideological underpinning for the Shepstone system of "traditional" reserves. The strongest racism expressed by city officials on issues related to sanitation and health over the course of the decades was not directed against Africans but, rather, overwhelmingly against Asians. As the district surgeon bluntly put it, "the Indian, in spite of sanitary inspection, still wallows in his native stench and filth" (Natal [1892–1908] 1893, B57). To Magistrate Charles Barter, the "petulant Indian … has made some of our villages well-nigh uninhabitable, and has seriously increased the difficulties of maintaining sanitary regulations in our towns" (1894, B2). It was thus the predominantly Asian side of town that was the focus of the first serious complaints about insanitary accommodation within the city. These led to a proposal for a "Coolie location" (unsuccessful) and legal restrictions on the sale of property to Asians (successfully enacted by the colonial government in 1893).[18]

Africans living in town first entered the sanitation debate rather indirectly. By the 1870s, drought and growing population had exposed the inadequacy of the Dorpspruit to supply the city's needs. Exacerbating the problem from the perspective of white citizens was the stranglehold that African *amaWasha* had on the city's hand-laundry business. A virtual monopoly on this gruelling job gave them the power to "extort" high prices and free accommodation from desperate employers in town (Atkins 1993, 136), with all the unruliness that that implied. City council thus had labour, housing, and discipline questions, as well as more narrowly conceived sanitation concerns, firmly in mind when it approved the construction of the first modern waterworks in 1880. These centred around a dam on uninhabited town lands above the city near the border with Zwartkop. Clean water so collected was piped directly into ratepayers' homes. The city also built an industrial laundry facility that year, which allowed some source point control of pollution. Perhaps more importantly to white customers, the steam laundry together with home delivery of piped water promised to undercut the *amaWasha* monopoly over the laundry business.

These big investments soon proved to be stopgap measures. By the start of the new century, the city's growing needs for clean water turned council's attention to inhabited lands farther upstream – Zwartkop. Securing land there for a new, bigger dam required negotiation with the NAD, a bureaucracy whose raison d'être was precisely to protect its subjects from the encroachments of urban life. The NAD, however, being headquartered in Pietermaritzburg, was apparently open to persuasion that, in this case, the natives would need sacrifice for the greater good, and it eventually agreed to sell a thousand acres of its trust over Zwartkop to the city. This land was both for the proposed dam itself and to create a fenced and afforested buffer zone that would keep Africans and their animals away from the precious resource. But city officials fretted that one thousand acres was insufficient to the task. This led them to make their first secret proposal for a mass population removal. With construction already underway on Henley Dam, the mayor discreetly approached the NAD in April 1902 with a delicate request: transfer the entire population of nearly 50,000 Africans from Zwartkop to somewhere in the north of the colony. The NAD refused point-blank, while politely informing the mayor that "the Natives in their habits were not likely to pollute the streams."[19]

Bubonic plague arrived in the city the very next year, and the opinions expressed about it in the press and city propaganda are revelatory. The *Natal Witness* ran almost daily reports to follow outbreaks of the disease in the district, in Durban, and elsewhere in the colonies. As Swanson described, the plague unquestionably raised the profile of public health in the public discourse and expanded the conception of sanitation beyond the hitherto usual narrow confines. That year there were as well regular editorials under the titles "Enteric," "Health," "Public Health," "Our Water Supply," "Some Sanitary Suggestions," and "Inter Alia" (the last a column that was frequently health oriented). One looks in vain in these diatribes and homilies for even between-the-lines allusions to Africans as a source of danger. On the contrary, it is hard to avoid concluding the precise opposite in, for example, a paean to the "good old days" not just of abundant clean water but also rather intimate contact: "when a bath was desired, a kafir was always at hand to dash buckets of water over one" (*Natal Witness*, 24 November 1903). Almost without exception (the main instance being leprosy, brought to the colony by Chinese or Indians immigrants), these articles laid the blame for contagious disease upon irresponsible white property-owners, industry, or the city itself for creating conditions favourable for the spread of infection. The only significant mention of Africans on this topic in the leading daily newspaper in the first big plague year was in an editorial exhortation to white residents and businessmen to protect their employees. They should do this by, first, cleaning up and disinfecting their employees' huts, and secondly, by rigidly excluding "visiting boys" ("The Plague," *Natal Witness*, 3 December 1903). The editorial "Some Sanitary Suggestions" in November also listed underpaid labour among the five priorities citizens needed to address if they hoped to protect the city from health scourges. It recommended the appointment of a full-time professional sanitary inspector whose tasks would be focused primarily on the presumed main proximate threat, "European residential quarters" (27 November 1903).

In terms of external threats, far from constructing the plague as endemic to African reserves or urban quarters, the press and city council unambiguously represented it as coming into the city from outlying commercial farming districts. Maritzburg was thought to be more vulnerable to the disease in that respect than Durban "because it was in the centre [of the colony]" and a transit point for people and rats from all directions ("City Council Debate," *Natal Witness*, 9 December 1903). Indeed, the only suspected cases of bubonic plague among Africans reported in 1903 occurred among employees of the railroad or men passing through the city on the train. For his part, the medical officer of health urged calm and to focus on the known best practices: keeping the city clean so it would not harbour infected rats brought into the city by trains, and killing rats. A rat-extermination squad was set up and a bounty offered to encourage the slaughter. The same approach was reiterated in preparation for subsequent outbreaks. Rather than targeting Africans or Indians as a prophylaxis in the 1912 outbreak, for example, "council wisely decided to take active measures to rid the town of rats; it being generally recognised that rats are the agents for the spread of the disease" (Pietermaritzburg [1904–90] 1912, 14). The city's response to yet another threatened outbreak in 1924 is also noteworthy, given that the Native (Urban Areas) Act had by then explicitly given the city power to target unfit African dwellings for demolition. The medical officer of health pointedly did not propose this action in his strategy to contain the plague. Rather he targeted the train station at the (then still somewhat) fashionable top end of the town grid as the main source of danger. The train station was the entry point to the city for rats hiding in grain shipments coming in from the Orange Free State (1924, 16). The daily paper *Times of Natal*, which was often sharply critical of city officials and liberal sentiments, agreed on this point in its leader "The Plague Menace": "the reservoirs of the disease are not in the towns and villages, but on the veld" (6 March 1925).

This apparently calm, non-scapegoating or non-opportunistic approach to the plague presents a stark contrast to the reaction of Cape Town officials discussed by Swanson (1977), Echenberg (2007), and many historians of Johannesburg's brutal destruction of its inner, mixed-race locations. One possible explanation is that Pietermaritzburg officials purposely downplayed any connection of plague to Africans in order to protect the city's cherished reputation. Maritzburg by 1903 had already entered into a long period of economic and demographic stagnation, with an irreversible decline in political importance relative to Durban. Demonizing the hygiene of the majority of its population would only have added to the woe by undermining one of Maritzburg's strongest claims to attract white immigrants – its otherwise generally healthy ambiance.

Yet it is hard to credit such a conspiracy theory in light of the remarkably vocal and graphic denunciations of sanitary conditions in the press and by city health officials. District Surgeon Charles Ward's annual report of 1896 certainly does not pull punches regarding the apathy of white citizens (and politicians) towards cleanliness and disease: "Filthy streets, sanitary conditions worse than they have ever been, the roads covered with a layer of desiccated excrement, sluits emitting pestilent odours, water supply quite inadequate even in the midst of summer, night soil cars allowed to poison the night air with their emanations, the same story year after year" (Natal [1892–1908] 1896, B88). The state of the Msunduze River as an open sewer, meanwhile, drew a lot of public discussion. The *Natal Witness* described it in 1900 as "a collection of filth that causes the air to positively reek. A little distance ahead and the hospital refuse percolates through a juvenile malarial swamp and discharges itself into the river. Passing along to Alexandra Bridge [in the heart of the city's premier recreation area] and the weir and another collection of odours is met with, followed by the tannery inferno" (29 October 1900). Lead editorials in 1903 poured scorn on the city for actively making the sanitary situation worse by, among other things, discharging sewage from its nascent underground system straight into the river and by the nightly trucking of sewage buckets and animal waste through the streets (18 December and 27 November 1903. In at least two well-publicized cases, the city was compelled to clean up its dumping practices by successful lawsuits from property owners (Pietermaritzburg [1904–90] 1910, 80).

To be sure, deteriorating conditions in Edendale and the emergence of other insanitary peri-urban "mushrooms" on the fringes of the city did cause some alarm in the city about the dangers of contagious diseases. Risk was inherent in a system where poorly paid nannies and cooks worked in European homes but lived in unsavoury conditions elsewhere and where migrant labourers sooner or later would return to their villages in the reserves to introduce the ailments of an urban lifestyle to an ever-widening circle. Such fears reinvigorated public debate about creating a properly managed native village in the 1910s. Inspector of Nuisances James J. Niven, notably, in 1914 cited the prevalence of phtisis (tuberculosis) among urbanized Africans as one compelling justification for such a village (1914, 98, 107). Yet, as with earlier arguments, Niven's largely fell on deaf ears in these years. In part this was because hardly anyone else seemed to believe the danger was serious enough to demand expensive public interventions. To the average citizens, many indicators of sanitation and health in the city actually seemed to be getting better. The water situation had by that time unquestionably and very visibly improved, both on the supply side, with the opening of Henley Dam, and on the disposal side, with the completion of the Southern Intercepting and Outfall Sewer in 1906. As a result of the latter, the Msunduze River where it passed through the city had rapidly recovered from open sewer to "one of the City's most popular attractions," with boating and swimming a short walk from the city centre (1913, 74). The Spanish influenza, as another example, caused only eighty-nine African deaths in the city in 1918, less than the toll on a single day in October that year in Bloemfontein. Judging by mortality rate, Africans in Maritzburg (1 per cent) were far less affected by the epidemic than whites in Bloemfontein (2.7 per cent) (1919, 22; Phillips 1987, 216, 222). As late as 1923, while council admitted that poor housing for Africans was a worry, it still believed the relatively dispersed population protected public health from the worst effects: "The danger arising from these conditions is, however, greatly minimised at present by the fact that the population is small in relation to the area of land occupied by it" (Pietermaritzburg [1904–90] 1923, 56).

All this seemed to vindicate the town council's preferred primary strategy of selective sanitary inspections and moral suasion directed at white property-owners, with prohibitions against Indian "penetration" (that is, purchase of property beyond their existing enclaves). As Mayor P.H. Taylor put it in his 1914 report, in effect undermining his own elsewhere-stated preference for an African location, "many defects in the sanitation of private properties have been discovered and removed. The offending parties, or others responsible for these insanitary conditions, have been interviewed or written to. A friendly word of advice has, where necessary, been given, and, on the whole, much good work has been done" (1914, 13). Indeed, even after a national policy framework had been put in place requiring cities to clean up slum areas and provide housing to Africans, and after Maritzburg council had finally committed itself to building the long-mooted location, city officials still did not cite a health threat to Europeans from African housing as a motive. The mayor's report of 1926 quite unequivocally refuted that notion. Drawing on the report of his medical officer of health, he identified the greatest threat of infectious disease to whites from Africans to be enteric fever (typhoid). This had nothing to do with the spatial distribution of the races within the city. It had to do with milk, and it had to do with soil (where the bacillus lived) contaminated by the urine and excrement of infected animals or humans. The problem was exacerbated by budget-minded white consumers who preferred to send their servants to fetch milk from unregulated dairies in Edendale rather than pay a premium in town. Control was needed to interrupt the connections. "The large part which the Native plays in the handling of milk in South Africa, the prevalence of Enteric Fever amongst the Natives, the residence of considerable numbers of Natives on the dairy farms, and the failure of many farmers to realize their responsibilities when illness occurs amongst these resident Natives renders this control more necessary in South Africa than in countries with only European populations" (1926, 64). What this meant in practical sanitation terms for the local authority was not enforced urban racial segregation but a concerted propaganda campaign to educate consumers, establishing or requiring proper latrines for Africans working in and around town, ridding the city of cow sheds, and closing down the remaining dairies within city boundaries. This process continued well into the 1930s and helped to shape the form that the native village eventually took – that is, urban as opposed to semi-rural, as in the earlier proposals (1926, 72; Brain 1988).

As for Edendale itself, one might have expected the Edendale Trust or Christian missions to cater to the evidently growing needs of the area. That was not the case. The Trust actually opposed even the creation of a malaria committee to address the appearance of that disease in the valley on the grounds that it would impose a burden on the people who were, it disingenuously claimed, "living in perfectly healthy rural surroundings" ("Proposed Incorporation of Sutherland's," *Natal Witness*, 22 March 1933). By the time the disease became an issue to Maritzburgers in the early 1930s, anger at the incompetence or indifference of the Trust and the missions was a strong theme among letter writers. The very first health clinic in the valley was established by Victoria Mthimkhulu, the wife of the Methodist pastor, only in the aftermath of that unwanted attention.[20] But her good work was undermined by a further problem revealed by Selby Msimang when he conducted the first health survey in 1942. Among his shocking discoveries was the extent of sectarian callousness. As Msimang describes, he came upon a widow near starvation in Georgetown. "Conditions were so bad that I drew the attention of the [Methodist] minister whose manse was just about 20 yds away. He said because she did not belong to his church he could do nothing except to call the attention of the minister of the church to which she was attached."[21]

Before concluding this overview, one final concern by government about public health needs mention – the management of conflict between contrasting and competing approaches to health among the city's diverse cultures. Traditional African medicine involved a mix of herbal remedies and psychosocial interventions advised by herbalists (*isinyanga*) and diviners (*isangoma* or *isanusa*). Europeans often sneered at the idioms of witchcraft, rain doctors, and such through which those interventions were expressed. However, they also often observed that the mix of herbs, rituals, and faith tended to have a relatively high success rate both in healing and in maintaining a harmonious social climate. Western biomedicine, by contrast, was capable of some stunning technical or pharmacological health achievements and was an important element in winning a degree of African respect for the colonial project, but it was too costly for the majority of the population. Moreover, stubbornly blind to the social determinants of health or to local cultural sensitivities, Western healing practices were often frustratingly ineffective against common ailments and could even have perverse effects. Laws to suppress "witchcraft," without distinguishing it from poisoning or recognizing its positive psychosocial effects, were widely blamed by Africans for giving rise to a host of new ailments, with some empirical justification. Indians meanwhile added their own eclectic mix of potions, incantations, and healing practices, which sometimes worked in defiance of scientific expectation. People of all races thus tended to shop around, and this trend opened medicine, health, and healing to itinerant charlatans and raised concerns not only about health but about socially or even politically disruptive impacts of healers and healing. As Karen Flint (2008) describes, by the late nineteenth century competition in the health market between the different healing cultures contributed to increasingly racialized and sometimes bitter conflict over appropriate levels of state regulation and professionalization.

Evidence from Maritzburg specifically is scarce, but there is no reason to believe that trends in the city did not conform to those in the colony as a whole. In the early days of colonial rule, the government sought to suppress an expanding circle of "superstitious" practices, including witchcraft and rain making, and the work of herbalists. For pragmatic reasons, however, it gradually began to appreciate the positive health and other stabilizing effects of some of these traditional practices. After 1891 it introduced a system of licencing *izinyanga* (herbalists) to control their numbers, but the government was thereafter tacitly protective of African health practitioners against the exclusionary claims of white professionals. It was tolerant of eclectic health paradigms to the extent that they kept people happy and seemed to work at minimal cost. The provincial government, for example, in the early 1940s supported the right of *izinyanga* to consult and diagnose African patients, in effect recognizing that they possessed skills that were formerly practised by *isangoma* and that white doctors and chemists sought to preserve for themselves (Flint 2008, 142). The state also effectively turned a blind eye to Indian "hawkers" and *muthi* (medicine) shops that catered to a multicultural clientele, implicitly recognizing their therapeutic niche role.

In sum, Dyer's overall conclusion about municipal approaches to infectious diseases over the city's first eight or nine decades appears sound. Public health interventions were often characterized by paternalism, with strong racist and sexist presumptions by white officials about Africans and Asians. They were consistently underfunded and, notwithstanding their typically moralistic tone, they tended to a narrow focus on the biomedical factors of ill health. At the same time, however, government health officials also showed compassion for the victims of ill health and "rigorous" analysis with reference to international best practices (Dyer 2012, 353; see also Wright 2006–07 for progressive voices at the colonial level). In a context where scientifically justified best practices were often not politically tenable or appropriately supported in municipal budgets, public health officials offered an important voice of reason and pragmatism that tempered the gathering sanitation, housing, and health crisis among Africans in and around the city.

≈

The emergence of extensive slum areas in a "black belt" around Pietermaritzburg in the second decade of the twentieth century was similar to what was happening elsewhere in the country. Overcrowded and degraded reserves had been collapsing as economically viable spaces for some time. The eviction of tenants and sharecroppers from white-owned farms after the Land Act of 1913 only added to the flood of dispossessed migrants towards cities in search of jobs. Aggressive enforcement of the Native Beer Act pushed brewers, particularly African women who were in that business, out of the city with no place to go but the pockets of freehold land just outside the five-mile enforcement zone. Maritzburg exacerbated the situation in 1909 by holding a referendum on the future of the town lands. A small plurality of voters decided in favour of turning the bulk of those lands into commercial wattle and gum plantations, effectively closing the door on a relatively easy answer to the question of where to put any future native location. Thereafter, any proposal for a native village had to overcome the obstacles both of the cost of purchasing additional lands (or the loss of plantation revenues) and local citizens' resistance to having an African location placed near their neighbourhood. Indeed, "very strong objections" sank yet another proposal for developing a native location near the city's Fitzsimmons (bachelor) barracks in 1919, despite strong lobbying by the mayor and the expediency offered by the Spanish influenza epidemic (Pietermaritzburg [1904–90] 1919, 19).

It needs to be emphasized that these were politically fraught years nationally. The Industrial and Commercial Workers Union (ICU) had been founded in 1919 with the mission of organizing African labourers to demand better wages and working conditions. It quickly grew in strength and militancy, with strikes and boycotts around the country, which in many cases were met with violent repression. An unprecedented strike of 70,000 African mine workers on the Rand in early 1920 in particular rattled the complacency of the country's main industrial sector. Although that strike was quickly broken, cost-cutting measures in the aftermath sparked further unrest among white workers that ultimately burst into the Rand Revolt of January–March 1922. Its defeat ultimately led to a remarkable political victory for white workers and Afrikaner nationalists, presaged in 1923 by an electoral pact between the National and Labour Parties. They promised – and followed through when elected to power in 1924 – greater racial segregation to protect white workers from black competition, and white "civilization" from the many forces ranged against it.[22]

In the context of competition between cities and different industrial sectors over persistent labour shortages and roiling, often violent, conflict, the ad hoc approach to dealing with African housing and health created a theoretical danger that some cities might gain an advantage over others in the labour market or, by introducing "radical" public health initiatives, might disturb the hard-won status quo of cheap African labour and white supremacy. Bloemfontein's primary response to the influenza epidemic of 1918–19, notably, was to create Batho, a clean and tightly managed location that not only addressed the immediate health crisis in Bloem but made the city an attractive destination for migrant workers coming from Basutoland or Pondoland (Phillips 1987). From the perspective of some sectors in Johannesburg, this was a threat to its own insatiable labour requirements. Johannesburg consequently led the way in lobbying for a national framework to provide legal mechanisms that would harmonize actions between jurisdictions and economic sectors that were competing for African labour. The resulting Natives (Urban Areas) Act of 1923 (NUAA) required that, as a prior step to demolishing slum quarters, municipalities establish publicly owned and managed townships for African families whose male heads worked in the municipality (Parnell 1988, 1993).

Maritzburg council's first response to the NUAA proposal was to say "no," on the grounds that the bill as originally written did not explicitly protect the beer monopoly system (the city's established way of funding native administration). Once government in Pretoria clarified that it would not interfere with local preferences for funding, the city then easily agreed to do its part. After all, the basic idea had been approved in theory in Pietermaritzburg nearly seven decades earlier, and the housing/health crisis was becoming more noticeable as unoccupied spaces in and around the city filled up. Maritzburg had been spared the worst of the Spanish flu pandemic, and an element of self-congratulation in white public opinion was strong. But two slums completely within the city limits had in the meantime blossomed into eyesores and imminent threats to public health: Hathorn's Hill in the east, clearly visible from the town centre, and Camp's Drift/Pentrich in the west. Both were on land mostly owned or rented by Indians but crowded with the poor of all races. The NUAA promised a practical means by which one stone could kill three and maybe four birds at the same time. The new homes it required would enable the city to demolish shacks that presented the most pressing health hazards, without causing undue hardship to the affected people; new homes would attract and stabilize a section of the African labour force (one of the goals of the original village proposal); and, given that the worst slum conditions appeared to exist on land owned by Indians, those demolitions would create space for "the better type of Indian" to erect their own new homes.[23] If all went well, income from rentals and beer would not only cover costs but even generate a profit for the city. Any remaining opposition to the idea (for example, "not in my backyard" or "too risky") could be overridden by pointing out that the mandate came from a higher level of government and the city had no room to negotiate.

The model of township proposed under the NUAA was also not controversial. The original "site and service" proposals in the nineteenth century had been for one-acre plots and a substantial communal kraal to enable residents to keep cattle, as befitting Africans' economic and cultural needs. By the 1920s scientific knowledge about enteric fever meant that cattle were not welcome within the city. The village would therefore comprise standardized homes on a compact grid with room for fenced gardens only. Homes would be rented to qualified African families at below market rates subsidized primarily by beer revenue so as to protect white ratepayers from the expense. As before, the city reserved for itself wide powers to control the behaviour of residents and the access of non-residents so as to prevent the re-emergence of informal structures, illicit shebeens and lodgers, and cattle.

The only foreseeable problem was where exactly to put these homes, to locate the location. Pietermaritzburg still owned a lot of land. Little of it, however, was suitable for the intended purpose, as it was either on the flood plain, hence too vulnerable, or on steep hillsides, much of which the city had by then developed as revenue-generating wattle and pine plantations. Anticipating this problem even before the bill became law, the city discreetly purchased an additional 1,267 acres (512 ha) of private farmland abutting its eastern boundary on Bishopstowe Road. It got a good price in large part because of the farm's proximity to the city's conservancy depot. Declaring its intention to use this land for the new township, council provoked the very first direct intervention by Africans in the debate. In a March 1922 petition to city council, African signatories appealed against the Bishopstowe site, principally on the grounds that it was unsanitary (Peel 1987, 10).

Enough councillors were embarrassed by this image of putting the city's African population out by the dump that the decision was put on hold. Yet pressure on the city intensified, including a stern letter from the secretary of native affairs in 1924 demanding that the city "terminate without delay" the slum conditions at Sutherlands on the west side of the city.[24] Meanwhile, the African petitioners picked up some influential allies to support their case for their own preferred site, also on the west side of town. Mason's Mill was just upstream from Camp's Drift/Pentrich, down from Sutherlands. Supporters included prominent white liberals in the Native Welfare Society and the chief native commissioner of the province (hence the pressure from him on the city to clean up neighbouring Sutherlands in preparation). Objections to Mason's Mill, however, immediately arose from a range of voices. Most obviously, at only forty-four hectares the available site was a tiny fraction of the size of Bishopstowe and would require another large expense to purchase room for expansion. A location at Mason's Mill, much of which lay on shale, would entail more than double the expenses to hook up to sewer and electricity compared to Bishopstowe, with long delays to be expected from the technical difficulties and expropriation of rights-of-way from unhappy white farmers in the area. Mason's Mill also implied that the city take responsibility for the cleanup of Sutherlands, an area outside its jurisdiction that could hardly be left in its current anarchic state were Mason's Mill to be developed, which the city balked at for obvious reasons. Then there was the river itself. Mason's Mill sat at a point where the Msunduze River was "an emulsion of faeces and water, with an admixture of urine,"[25] so polluted that, as the medical officer of health for Pietermaritzburg later described it, discharge from a nearby notoriously smelly leather tannery "would probably have a beneficial effect" upon the quality of the water from which the population drew its drinking supplies (Dr Albertyn, quoted in *Natal Witness*, 28 February 1933).

African opinion was not united on the issue. Notwithstanding the African petitioners and their friends in high places in favour of Mason's Mill, city officials were aware that the place had an unsavourary reputation. In 1923, the city had donated a piece of land there for use by African football clubs. However stingily and belatedly, it was attempting to support the development of a healthy recreational activity that would be convenient for African men on their way to homes in the Edendale direction from work in the city. Almost unanimously, the African clubs rejected the gift on the grounds that it was too close to shebeen central at Sutherlands and thus inimical to the health objective – indeed, Mason's Mill was a "menace" in the reported word of J.J. Magwaza, secretary of the Maritzburg District Bantu Football Association.[26]

The site of the proposed township then became "the shuttlecock question" of municipal politics, going back and forth without resolution for nearly three years. Inability to resolve that question made the city, in the words of the *Natal Witness*, "a farce and a laughing stock" (11 March 1925). In February 1925 council finally decided to break the deadlock though a plebiscite, the announcement of which set off four more months of free-wheeling debate in the press. It also sparked further "emphatic" opposition from concerned Africans and their allies. A mass meeting of an estimated 200 "town natives" in central Pietermaritzburg unanimously resolved to oppose the Bishopstowe site. Their petition, signed by prominent leaders of the Natal Native Congress (including future president of the African National Congress Josiah Gumede), the *amakholwa* chiefs of Edendale (Mini) and the city (Siota), and other prominent African property-owners, drew headline coverage in the press. This included an unprecedented front-page interview with one of the African leaders, W.M. Msimang. In it, Msimang elaborated on the petitioners' view that "natives object to living within the vicinity of filth" ("Natives Say They Will Not Go to Bishopstowe," *Natal Witness*, 10 March 1925). Bishopstowe was also inappropriate, he reasoned, because it was far removed from existing African communities on the west side of town. It would in consequence entail disturbance to whites as Africans commuted across the city to visit their family and friends.

Msimang, and a subsequent petition by African women directly to the NAD, did not dispute the principles either of segregation or of population removals for the purpose of improving sanitary conditions. On the contrary, they were strongly in favour of both. As Msimang put it response to his interviewer's question, "Did he support the idea of a Native Village?" "Yes; we hope to live in better dwellings. At present many of our people live in immoral and unclean surroundings at the back of Indian dwellings. The village will do away with this sorry state of affairs" ("Natives Say They Will Not Go to Bishopstowe," *Natal Witness*, 10 March 1925). What they wanted in such a village was enough land for each family to have two cows and a garden, and "everything like a European town." The women's petition also appealed for a segregated village at Mason's Mill in the name of "Public Health, Social Welfare and Morality." "As mothers and persons intimately in touch with native life and much affected socially by existing disintegrating and unwholesome conditions, we are convinced that the establishment of a Native Village in Pietermaritzburg is the only practical remedy towards checking the many social evils that abound."[27]

Council strongly denied that the conservancy presented a health risk; in any case, it was already committed to closing down the alleged nuisance once a modern, underground sewage system under construction was completed. The original intention had been to finish that job by the following year.[28] This did not imply that council had made up its mind with respect to the site for the village. On the contrary, the advertisements it took out in the press just prior to the plebiscite did not give any explicit guidance or hints of preference on the issue. They were in fact remarkable for the almost complete lack of information that might assist voters to make an informed assessment of the comparative merits of the two sites. The *Times of Natal* (a penny daily, read more widely by working-class whites than the *Natal Witness*) denounced this silence as a "hopeless failure" of leadership by council ("The Native Village," 26 June 1925). It then proceeded to provide the assessment itself. Conceding that the conservancy depot was a "solid objection" to Bishopstowe, the *Times* nonetheless came down heavily against Mason's Mill for a long list of reasons but emphasizing its small size, high cost, and proximity to existing slum areas.

Two days later, just over 1,200 white citizens voted roughly two to one in favour of the Bishopstowe site. By the end of the year, and privately expressing its regret for "running roughshod" over African opinion in a decision that potentially prejudiced the supply of African labour, the province gave its approval for the site as the best-suited one *on health grounds*. The proviso was that the city expedite the removal of the conservancy as promised, which was indeed closed, only slightly behind schedule, in 1927. The remaining slop service was transferred to a sewage farm located a few hundred meters downstream from the village site (Pietermaritzburg [1904–90] 1928, 17). Although "nuisance" (smell and flies mostly) remained a concern, the new site, and its new disposal methods (which involved spreading waste over a wide area as fertilizer within a perimeter of willow trees), was generally regarded as a huge improvement. A year later, the first hundred homes were opened and the demolition of slum housing in Camp's Drift and Hathorn's Hill finally commenced.[29]

Can this decision be interpreted as proof of the "sanitation syndrome," which is to say, of the incorrigible if sneaky racism of the city's white decision-makers? A close look at the evidence and the arguments deployed by the two sides suggests a much more complicated story. In the first place, Africans living in approved accommodation in the city (the majority) were *not* required to move. Second, as the *Times of Natal* had pointed out, and whose points stand the test of time, Bishopstowe was on several levels clearly the more rational choice to satisfy the multiple objectives of a housing project. Mason's Mill was utterly inadequate for accommodating the numbers of people who needed to be housed. It promised to add to the already astronomical burden of faecal pollution that flowed down the Msunduze straight into Maritzburg. Bishopstowe, at over ten times the area, would allow both for immediate needs, including residences, gardens on a clean tributary stream, public lands, churches and recreational facilities, and for future expansion. Bishopstowe was also significantly closer to the city centre than Mason's Mill (four kilometres from centre to centre versus six).

The last point reveals an ironic twist to the debate that was noted by some Mason's Mill critics, if not by subsequent historians. Mason's Mill was *more* segregationist that Bishopstowe. It fit like a puzzle piece between Edendale, the peri-urbanized chieftaincies of the Zwartkop borderlands, the mixed but largely African communities within the city's western boundary (Pentrich, Camp's Drift, and New Scotland), and two of the most notorious slums in the province just beyond (Sutherlands and Macibise). Mason's Mill as an African location would thus consolidate a wide swathe of mostly non-white communities spilling over at least three jurisdictional boundaries. Anyone on the way out of town for the illicit alcohol and sex that were readily available in those places would need to pass through Mason's Mill to get there. Anyone from the "overberg" or eastern Cape seeking access to Pietermaritzburg without legal documents would likely use Sutherlands (and thence, Mason's Mill) as a staging point. Moreover, it would be a challenge to keep cattle from wandering in from the surrounding uncontrolled areas.

By contrast, Bishopstowe's neighbours were mostly white farmers, who would provide a buffer between the village and the raucous liquor and prostitution sites found elsewhere, an essential element of the whole project of attracting a "decent" class of native. It was for this reason that the police favoured Bishopstowe, as did the president of the Native Welfare Society (Councillor Taylor, who broke with his own membership on this issue). White farmers in the area did not appreciate the role chosen for them, and they opposed the Bishopstowe site by appealing to national policy. As those petitioners put it, the Bishopstowe proposal was "in direct conflict with the principle of the Native Land Act" – that is, enhanced segregation.[30] That view did not garner much sympathy, and the apparent alliance between Bishopstowe's white residents and black elites from the other side of town puzzled some councillors. As Councillor Meldrum summed up, Bishopstowe was the better site "both from a health and get-at-able point of view," while fellow councillor Simkins added, "Europeans were content to live there, and why not the natives?" (*Times of Natal*, 11 March 1925).

Why not, indeed, particularly given that the most vocal African petitioners came from Edendale or New Scotland and would not themselves be required to move? Some of the reasons they provided in favour of Mason's Mill certainly struck many as disingenuous. Msimang made the claim, for example, that Mason's Mill was attractive to Africans because the railway from Edendale ran through it. Yet few used the existing passenger train service from Edendale, which, compared to the direct bus link, took a long and circuitous route into town (G.F. Robbins, letter, *Natal Witness*, 10 March 1925).[31] And why were English-speaking, property-owning teetotallers such as Msimang and the members of the Native Women's League advocating to create a decent working-class neighbourhood directly in between a slum area overcrowded with seSotho-speaking rowdies and a Zulu reserve full of polygynists and people who believed in witches, both outside of city jurisdiction? The suspicion grew that *amakholwa* elites were using the plebiscite to rally support for a broader political agenda, with the conservancy as a red herring. That alleged agenda included consolidating a large African population under "radical" leadership that would press for town status for an aggrandized Edendale. It was also well-known that Edendale *amakholwa* wanted freedom of mobility for Africans across jurisdictional borders, among other citizenship rights that would inevitably compromise white domination of the city (if not, in the long run, of the country as a whole). To some voters, therefore, Bishopstowe was thus not just the healthier and cheaper option for the city. To choose it over Mason's Mill was to strike a blow against "agitators" and "negrophiles" who wanted to overthrow the white man in his own country, to use the rhetoric of the time.[32]

Knowing what came later, fears of a power grab by black elites in Edendale may seem to readers today as paranoid to an absurd degree. It would be a mistake, however, to underestimate the political insecurity of the times in a place like Pietermaritzburg and to read the certainty of white domination as achieved in the 1950s and 1960s back to those earlier times in liminal areas like Edendale. As noted above, small but deep pockets of poverty persisted among whites. The Labour Party, with its well-known "red" sympathies, held one of city's two national parliamentary seats and hence a foothold in the National/Labour Pact government in Pretoria. The ICU, which had endorsed the Pact government as broadly supportive of African nationalist aspirations, had recently established an office in Maritzburg. The ICU had not yet split with its overtly pro-Communist element and was consequently widely perceived by whites as a radically disruptive force in Natal. One of its founding members in the province came from Edendale (Selby Msimang, nephew of the anti-Bishopstowe leader cited above). Another of the anti-Bishopstowe leaders, Josiah Gumede, was also a leading figure on the far left of the ANC. Gumede, with the backing of Edendale's Chief Mini, had the year before mounted what Nicholas Cope (1990) has called a "coup" that removed the ANC's conservative leadership and that committed the party to a radical political program of redistribution of wealth, including land. Indeed, soon after the village plebiscite, Gumede toured the Soviet Union and came out with openly pro-Communist views (Van Diemel 2002). As for the Edendale Trust, its later defence against accusations of incompetence is revealing: "They [the Trust] have not, and never have had, any powers of control of the population. And had they attempted to enforce such they would have met with defiance and, possibly, worse trouble" (Rev. L.S.H. Wilkonson, "Edendale and Its Trustees," *Natal Witness*, 17 Septemper 1941).

For white voters who paid attention to such things, it was entirely conceivable that an articulate, socially ambitious, and politically astute African leadership based in Edendale could leverage enough white people's guilty consciences to achieve their broader objectives. The *amakholwa* had appealed above and embarrassed the city before on similar issues. That possibility may well have motivated a higher than expected number of whites to vote in the plebiscite in defence not of their health, or a desire for formal racial segregation on sanitation grounds, but of their power to control municipal structures, rates, wages, and white political supremacy in general.

The eventual "remarkable" turnout for a municipal plebiscite did settle the question of location once and for all (*Times of Natal*, 20 June 1925). Subsequent protests from Africans of Edendale and Pietermaritzburg, and by European landowners in the Bishopstowe area, did not succeed in moving council or the NAD away from respecting the result, and the protests quickly died down. A threatened boycott also fizzled once the village opened in 1928. With electricity, piped water, a mix of types of accommodation at low rents, and strict policing against shebeens, the new village of Sobantu's first hundred units were 92 per cent let after the first year. To be sure, the Msunduze River itself remained heavily polluted and notoriously smelly at that point for decades to come (the Darvill water treatment plant on the opposite bank from the village was opened only in 1957). But on the northeast side of the village, Bayne's Spruit ran more or less straight off the escarpment to provide clean irrigation for market gardens. By the 1950s Sobantu had a population of over four thousand, with a very positive reputation among blacks in South Africa. Reflective of that reputation was the outpouring of opposition to closing it down in 1956. The National Party government in Pretoria banned its expansion and further infrastructural improvements to it in that year on the grounds that it was too close to white-zoned Pietermaritzburg to meet the requirements of the Group Areas Acts (Pietermaritzburg [1904–90] 1956, 31), and it subsequently threatened to evict the entire population. Public protests led by the Liberal Party of South Africa and city council received the tacit endorsement of the national forum of Location Advisory Boards (that is, African township leaders): "[African] delegates were loud in their praise of conditions in Sobantu village," the city reported, which helped it to stave off the community's threatened removal (1956, 31).

≈

Pietermaritzburg did not become formally racially segregated with the opening of the Model Native Village in 1928. Far from it. The majority of

Africans within Pietermaritzburg continued to live in barracks, on private premises, and in the "colony" of New Scotland, albeit subject to tightened residential building codes and harassing restrictions on mobility. Slum clearances throughout the 1920s and 1930s did result in the eviction of hundreds of Africans who could not be accommodated in Sobantu or other approved housing in the city and who consequently found refuge beyond the city boundaries, often in the greater Edendale area. Yet significant numbers of Africans remained within Maritzburg even after enforcement of the Group Areas and Urban Areas Acts commenced in 1963 (Thorrington-Smith et al. 1973, 32). Moreover, just as opponents of the Bishopstowe site had warned, residents of Sobantu did make a regular practice of commuting through the heart of the city to visit friends, family, and shebeens on the Edendale side. Recalling it in the 1940s and 1950s, Sobantu gangster cum musician Godfrey Moloi (1991) describes a seamlessness to African social life extending across the ostensibly white city and suggesting a far less fractured consciousness of place than strict segregation would predict. By the early 1950s, so many African commuters were passing through the central business district to and from their places of residence on opposite sides that council mooted the construction of a rail line that would skirt the inner city in order to connect Sobantu and Edendale directly.[33] Indeed it was not until 1961 that the African population in the city actually began to fall, with the final evictions of several hundred African property owners following the enforcement of the Urban Areas Act in 1963. Even in the darkest days of apartheid, however, the numbers of Africans living legally as tenants within the city remained substantial.[34]

No one would dispute that Maritzburgers shared much in common with whites elsewhere in South Africa in their desire to protect their health, quality of life, and real estate values, and that racial stereotypes were part and parcel of the cultural milieu. In that sense, Pietermaritzburg was indeed a typical segregated city by the late nineteenth century, not particularly different from segregated cities elsewhere in the colonial world and the United States in that period (see Nightingale 2012, for example). However, if Africans' ill health was in fact a key motivating factor behind the creation of formally segregated and geographically distant African townships in South African cities, and if the present shape of South African cities can be explained primarily by the "sanitation syndrome," then Maritzburg clearly was not typical. Other issues were more decisive in eventually overcoming the many objections to creating a managed township, and over where to place it. These included Africans' own concerns for their health and morals, and whites' anxieties about African elites' political ambitions, neither of which was empirically unfounded. Thus, a catchall term like the sanitation syndrome does not capture well the complexity of debates and decision making over many decades or the influence of the distinctive topography and vibrant political culture of Maritzburg.

# FIVE | "Hide as Much as You Can for the Sake of Good Government": Women's Health, Gender, and Local Authority in Edendale, 1930–1958

People commonly cite the legacy of colonialism and apartheid to explain the high rates of HIV/AIDS and gender-based violence in southern Africa.[1] These are among the most debilitating and seemingly intractable aspects of what Elder (2003) termed South Africa's "malevolent geography." That legacy may also account for the prickliness or denialism among some African leaders with respect to contemporary donor interventions around sexual health and sexual rights in Africa that they perceive as racist or colonizing. In this view (very baldly put), racial capitalism combined with a cultural racism that over the decades exoticized, demonized, and transformed many aspects of African gender relations and sexual mores. An effect was to "emasculate" African men, to cut them off from the means of achieving socially recognized/rewarded manhood. Specific sexual traumas within colonial/apartheid institutions (such as prisons and migrant labour camps) sometimes intensified the effect, which over time engendered an unhealthy, heteropatriarchal masculinity. Young men, for example, paid increasingly scant heed to traditional moral economy restraints and obligations relating to women and cohered into gang cultures that specialized in the violent control and sexual exploitation of women (or, in some cases, of young men). Those same processes and structures created space in the burgeoning urban areas for the emergence of an opportunistic "loose" femininity personified in the shebeen queen or wily "wife of the pots." In the context of extreme demographic imbalances in the cities, young men unhinged from traditional restraints and independent women as purveyors of sex and alcohol abetted the spread of sexually transmitted diseases. This hardened many European stereotypes of Africans' promiscuity or essential sexual impulsiveness, stereotypes that through the early decades of the twentieth century were increasingly drawn using the language of science (but without the discipline of the scientific method). Such language justified intrusive, disrespectful campaigns by colonial states to police or to modify Africans' sexual mores in line with normative eurocentric, middle-class values. These campaigns in turn became a focal point for "black anger," which eventually found expression in the African nationalist struggle.[2]

Contemporary initiatives that deny or fudge this history touch a deep core of resentment among some African intellectuals. As the calamitous case of Thabo Mbeki's obfuscation about the link between HIV and AIDS illustrated, that resentment can translate into a reactionary, selective scepticism toward science and a resistance to best practices to promote sexual health as understood from global experience. Economist Nicoli Nattrass estimated over 300,000 additional deaths resulted from Mbeki's stance (Nattrass 2007, 2012). A similar phenomenon may be fuelling the apparent rise of political homophobia across much of Africa since the 1990s, as leaders construe public health initiatives directed at men who have sex with men as an element of Western cultural imperialism, to be resisted in the name of pan-African identity (Nguyen 2010; Epprecht 2013).

Mbeki and his successor, Jacob Zuma, who muddles science to an even greater degree, have been skewered by South African humorists for this prickly streak within their sense of patriotism. Yet mockery, which also abundantly accompanied the Zuma rape trial of 2010 and the great Zuma penis uproar of 2012, sometimes only exacerbates defensiveness or anger.[3] And the stakes are high. The sources of scepticism or reaction against science as applied to gender and sexuality in South Africa thus deserve careful disentanglement from the political, economic, or cultural factors historically in play. They are especially pertinent to consider in a place like Msunduzi where, conservatively, a hundred thousand people have died from a theoretically easy-to-prevent disease since the transition to democracy began.[4] Msunduzi is a place, moreover, where female-headed or female-centric households are now the predominant form in Greater Edendale and where a highly gendered social welfare program has emerged as the state's preferred strategy for managing political expectations of the poor (some would say, "disciplining" the poor – see Goebel and Dodson 2011 and Goebel 2015 on these debates). Masculinity is a big, if unstated, political issue in this context.

My goal in this chapter is not to dispute the broad outlines of arguments about the long-term pernicious legacy of racial capitalism and cultural racism upon African masculinities. Rather, it is to examine a historical case study for the light it may shed on the mixed motives and divergent perspectives relating to women's health – and hence gender relations more broadly – at an important moment in the development of African nationalism in South Africa. The 1940s were a watershed period in many ways but particularly in the context of African nationalist leadership's hardening its position against paternalist developmentalism by the settler state. My focus is upon a place – Edendale – where one arm of that state was attempting a relatively radical experiment in local government and social medicine that placed a high priority on women's health. Limb (2010) and Healy-Clancy (2012) allude to a conservative or even reactionary streak in progressive African politics on these issues, which may help to explain the frustration associated with the social medicine experiment in the 1940s and 1950s and perhaps some of the more contemporary backlash as well. I hope as well to contribute to a related debate about the significance of the "liberal interregnum" between the segregationist era of the United Party in the 1930s and the advent of National Party rule in 1948.[5]

I begin by setting the stage for the crisis that led to the formation of the local health commission (LHC) as the governing body over Edendale. The presumption in the historiography of the crisis is that obstruction, neglect, and overt forms of racism by whites against blacks were the major factors in allowing conditions to deteriorate to such a state. This racism included white fears of "black peril" (that is, black men's supposedly uncontrolled lust for white women), which made agreement on rational and humane solutions to African urbanization impossible (C. Merrett 2009, 81–6). But I also wonder about the role of gender and sexuality among Africans in shaping the debates around governance of the community up to the late 1930s. In the following section I ask, how can we understand the contesting, often ambivalent perspectives on women's health (and science and "native custom") expressed by prominent male African nationalist leaders on the cusp of South Africa's turn to apartheid social engineering? How did African women themselves respond to the intervention of a modernizing, paternalistic state? Not to suggest that any direct lessons for today can be gleaned from it, but this history both complicates the "black peril" explanation for

Edendale's predicament and poses important challenges to today's opponents of sexual and reproductive rights.

≈

Edendale's leading citizens long complained to higher authorities about the harmful impacts that Pietermaritzburg had upon their community, both as a result of city council decisions and through the behaviour of individual Maritzburg citizens and corporations. Concerned citizens wrote petitions to government, retained white lawyers to speak to the local authority, created a breakaway missionary movement (*Unzondelelo* or "Endurance") and a formal lobby group (the *Funamalungelo* or Society of Exempted Natives, established in 1887), and publicized issues through a vigorous independent press, notably *Inkanyiso yase Natal* and *Ipepa la Hlanga*.[6] A common theme in this protest tradition was to rage against racist double standards that the Christians of Edendale frequently encountered from government and from whites in general. The often-heartfelt anger ("under slavery," "treated like dogs," and so forth) evident in these early protests belied a strong guiding sentiment of loyalty to the state. Edendale's leaders hoped that whites would be amenable to passionate but reasoned argument derived in part from their own ballyhooed principles in law, religion, and philosophy. Petitioners were cross but they were not seeking revolution. They were appealing for help against such offences as the dumping of toxic wastes, low pay, the failure of the Edendale trustees to carry out their most basic mandates, petty corruption by the chief, the displacement of shebeens from the city to its border zones after the Native Beer Act of 1909, and, as we saw in the previous chapter, the city's choice of location for its new township in 1925. Almost without exception, these appeals fell on deaf ears, with a wide variety of often specious rationales offered for inaction.

Gender and sexuality emerge as strong underlying points of contention in the Edendale protest tradition, as indeed they were throughout the whole colonial venture in the region. The Shepstone system was predicated upon a constrained "native administration" by chiefs and an inflexible interpretation of customary law that would keep African women subservient to men in the rural areas, notwithstanding their husbands' and fathers' prolonged absences as migrant labourers and the spread of cash and consumerist values. Women found in town without their male guardians' consent could be deported back to their rural place. Incentives were meanwhile put in place to encourage the *gradual* transition of tribal Africans to "civilization," in the language of the day. This included, notably, a hut and a marriage tax, which might cause polygynists to restrain themselves from adding another wife (and hence hut) to the household; laws protecting women from excessive violence; and the prohibition of "repugnant" customs such as witchcraft and child betrothals. However, on central questions of African gender ideology and practices such as *ukulobola/bohali* (in isiZulu and seSotho, respectively, the practice whereby the family of the groom sent cattle to the family of the bride, as discussed in chapter 3), the colonial state did not seek radical change. On the contrary, while settler and Protestant missionary propaganda frequently described *ukulobola* as an "evil" comparable to female enslavement or purchase, and some foresaw the dangers it posed to the long-term ecological health of the reserves, the state not only exempted it from its repugnancy standard but took pains to defend African marriage laws from the influence of missions. The state's conservatism towards a reified "traditional" patriarchy by many accounts actually hardened over the decades, as reflected in its increasingly explicit and reactionary versions of codified Native Law (see, for example, McClendon 2002 and 2010).

One element of traditional African cultures in the region deserves special explication. I do not want to give credence to colonial stereotypes of African women groaning under the yoke of masculine laziness and cruelty, and indeed, we can find ample references to instances of African women's leadership and assertiveness, if not outright insubordination, even in precolonial and subsequent strictly traditional settings.[7] However, traditional culture did enjoin everyone to show *hlonipha/hlonepho*, as "respect" was termed in isiZulu/seSotho. This meant a strict code of etiquette that upheld hierarchical expectations accorded by age, gender, clan, marital status, wealth, and numerous other often quite subtle signifiers of family accomplishment, worth, and dignity as people. Women, and young women in particular, had considerably stricter obligations than men. Public deference to men was an essential component of *hlonipha* for women, as was avoiding any deed or word that could be taken to undermine a man's dignity. For a young wife this meant never pronouncing the name of her father-in-law and inventing substitute names that avoided even the syllables of the real name. Being caught in or admitting to adultery was another extremely grave offence (although there were well-known ways to cover up that particular shame), as were accusations against a husband of impotence or neglect, revelations of family secrets, or any explicit public talk of sexuality.

The contradictions and stresses intrinsic to the Shepstone system quickly proved unmanageable. Despite wide support, enforcement was a losing battle in the face of the widening cash economy (which allowed women and young men to acquire the means to make independent decisions, including to whether and when to marry and have children). Overstocking of cattle in cramped reserves also, as predicted, rapidly hastened the collapse of the communal agricultural economy and hence the viability of the population without recourse to socially corrosive levels of male migrant labour. As factors that undermined Shepstonian ideals, others noted the temptations of city life, the unruly libidinousness of Africans, and African women in particular, the leniency of soft-hearted magistrates, the confusing effects of book learning, and the bad moral example so often set for Africans by whites. A common trope in the settler literature was to wax nostalgic about the days when an African man could expect unquestioning respect and obedience from his wives and children and could beat them without fear of retribution in the rare instance of a "child" speaking back. The progress of bowdlerized Western civilization and the money economy had supposedly upset this fine balance. As an American friend of Natalian colonialism put it, "Children in the native villages are brought up in an atmosphere of happiness, and discord of any kind is almost unknown in Zululand, except where the white man has forced his commercial invasion" (Millward 1909, 291).

The *amakholwa* generally rejected that nostalgic vision of the noble savage and attendant gender relations as well as the idea that Africans could not adapt to the modern world if given a fair chance. They sought to refashion society towards what has elsewhere been termed a "soft patriarchy."[8] Christian ideology considerably lightened the grip of *hlonipha* on African women, but it was not fundamentally opposed to the principle. The Christian husband and father remained the head of the family in this vision, but with close and respectful consultation with his wife in decision making. Companionate monogamous marriage stood in the place of polygynous clan-building as the underlying ethic, while respectability and social standing hung to a large degree upon evincing sexual restraint. As was also very much the case with European Christian women and girls, African Christian women and girls tested their reputations, marriage prospects, material well-being, and even physical safety if they pushed the limits of feminine propriety too far.

There were admittedly many obstacles in the way of achieving the Christian ideals, not least of all economic constraints but also harassment and humiliation from citizens of Pietermaritzburg. In 1875, notably, *amakholwa* petitioners complained of the "deserting us by our women to live in towns and becoming prostitutes to white men," against which offence they demanded government action ("The Administrator and the Christian Natives: A Petition to Sir Garnett Wolseley," *Natal Witness*, 23 August 1875). The first four of the twelve grievances listed in that petition refer to gender issues, and in particular the perceived injustices created by government regulation and taxation of marriages. The petitioners wanted freedom to govern their own affairs under the same laws applied to whites, in order to cultivate the moral, progressive community that had been their vision since the founding of Edendale a quarter of a century earlier.

That argument was not well received even by liberal-minded whites in Maritzburg. The tendency at that time rather was to blame the *amakholwa* for failing to live up to their professed Christian obligations and for clinging or perhaps backsliding to select elements of traditional culture that they were supposed to renounce. The following polemic by the editors of the *Natal Witness* was no doubt overstated and deeply hypocritical, but the Edendale petitioners themselves conceded the point about their continuing preference for *ilobolo* (which they described as "marriage presents"):

> They embrace Christianity, but adhere to polygamy. They are content to pay more than ten head of cattle for their wives, and are still more ready to receive beyond that legal *maximum* when parting with their daughters; but they object to pay the £5 marriage fee to the Government on the acquisition of a fresh jewel for their seraglio … They adhere to the most demoralizing relation between the sexes, and object to their women being prostitutes. No wonder, poor fellows, they do not know what law they are under! ("The administrator and the Christian Natives: A Petition to Sir Garnett Wolseley." *Natal Witness*, 23 August 1875)[9]

Some African women were able to assert a degree of independence from men (and hence show lack of *hlonipha*) that attracted growing concern across race, class, and sectarian divides. Most of that concern focused on the rise of prostitution, adultery, shebeening, and sexually transmitted infections. But it was much more than a moral issue in public discourse. In the context of deepening poverty at the turn of the century, African women's rebelliousness also came to be linked to the deteriorating appearance and general environmental condition of Edendale. On this issue, city officials tended to be more sympathetic towards the men of the community than on *ilobolo*. In his 1904 report, District Surgeon D. Campbell Watt noted the growing untidiness of African villages and respectfully noted how the men explained it: "The men say they have no control over the women now. If they scold them, they go home to their mothers; if they chastise them, they complain to the Magistrate" (Natal [1892–1908] 1904, B22). Unkempt premises meant rats, fleas, and other health threats but also fed into the kind of "demoralization" that fuelled the consumption of alcohol and commercial sex. Like many of his colonial peers, Watt saw the growth of slum areas as a further proof of the declining efficacy of the Shepstone system, which he deeply regretted. Yet as described in the previous chapter, attempts to fix the problems over the next two decades proved utterly ineffectual, if not counterproductive, both in around Maritzburg and in most of the country.

Some African men obviously enjoyed and exploited the new social and sexual liberties town afforded, and Ben Carton (2000) makes a convincing case for generational conflict between men over women as the most visible political fault line in Zulu society in the early twentieth century. While some African patriarchs looked to the colonial state to shore up their masculine prerogatives and dignity, younger men tended to chafe at the costs of buttressing their fathers' authority. The *amakholwa* in Edendale started in the first camp, seeing the state as an ally worth fighting for. Over time, however, harsh experience brought them to view the colonial state as, at best, an unreliable and deceitful ally against the emasculating tendencies of racial capitalism and the venality of individual settlers or white burgesses. The *amakholwa* protest tradition is suffused with a tone of sometimes barely contained rage at those tendencies. Eloquent petitions and, as in the case of the spontaneous battles against white vigilantes during the "black peril" uproar of the 1886, fisticuffs with white racists could restore African men's sense of honour, however briefly.

The black peril scare was an important moment in local political history that has been largely overlooked but deserves attention, at least to the extent that it qualifies claims that the early *amakholwa* protest tradition was too polite and hence ineffectual. In that incident, a black man in Durban had allegedly raped a white woman. This sparked a mass meeting of 700 angry white men in central Maritzburg who demanded that the state act to exclude African men from the city, and *amakholwa* men in particular, through the creation of a strictly policed native location. The Female Protection Society – basically a gang of white youths – was formed to clear "insolent" Africans off the footpaths, as a prelude to getting them out of town altogether. Mobilization against the black peril quickly fizzled, however, in part for want of evidence of a real problem but also because African men fought back. On their response to white vigilantism, the *Natal Witness* reported:

> the natives were noticed congregating together in large numbers at the street corners, when a large amount of talk and gesticulations were indulged in. These harangues reached a culminating point on Sunday evening, when most of the chapel-going Kafirs got together in bands of between twenty and thirty each, and armed themselves with sticks and stones. Their first performance was in routing the so-called "Female Protection Society," and sending them helter-skelter into the out-station for protection. The gang, made bold by their victory over the society, then marched down Longmarket Street into Chapel Street, just as the Wesleyan Chapel had finished worship. Here they pitched on to a party of young men. ("The Pavement Question: The Natives Up in Arms," *Natal Witness*, 14 December 1886)

A further victorious battle with white youth ensued the next day, with "school Kafirs ... threatening to thrash every white person they came across," including the police. With anxiety high among whites about civil war in neighbouring Zululand, the editors urged caution to their readers against aggravating *amakholwa* opinion.

A small victory, perhaps, for the self-respect of African men, but elsewhere the trends looked bad. Land ownership and control over women had by far the greatest symbolic meaning in this regard. Land ownership as a powerful marker of masculine honour had been present from the beginning of the Edendale mission but clearly intensified through the decades of bankruptcies and loss. In line with Wulf Sachs's observations of "black anger" in the 1930s (Sachs 1937), proprietary, protective sentiments by African men over "their" women also intensified as poverty and political frustrations deepened. They were implicit in Chief Mini's lament to the Native Economic Commission in 1931 about the decline in his authority over "children" and in Josiah Gumede's concern that "our womenfolk are getting out of hand" (South Africa 1932, 6772–82, 6825). African men in Edendale could be brutally frank about what needed to be done. As expressed to an enquiry by the city about the sale of "kaffir beer," one Edendale witness at the end of the decade used terms stark enough to surprise even the hardliners in town council. On the question of what the city should do about female brewers in the village, he offered: "Sweep them up. How [one councillor asked]? – Eject them all and send them to their chiefs" (quoted in "Native Women's Bath Water Used for Brewing Beer," *Natal Witness*, 24 September 1941).

The brewers of Macibise did not bequeath an archive of their thoughts and feelings about their lifestyle. We may be tempted to fill the gap by romanticizing them as proto-feminist or otherwise admirably rebellious figures against the patriarchal and the prissy. We need to be careful, however, that this does not blind us to the very real suffering that shebeeners contributed to in their own communities. A contemporary Zulu man who was generally sympathetic to poor women's plight pointedly reminds us that there was substance behind the worries of respectable opinion. Gilbert Coka came from an impoverished rural family himself but rose through education to become a journalist and an important figure in both the Industrial and Commercial Workers Union (ICU) and the South African Communist Party in the early 1930s. He left this description of Malay Camp, an analogous location in Johannesburg:

> People lived in dilapidated buildings, fitter for pigs than for human habitation … [They] bred a type of wild, violent criminal, and generally lived amidst insecurity, squalor, crime, disease and poverty. Stabbings were everyday occurrences. It was a common sight to find women reel in the last stages of drunkenness to fight against men in the same conditions. Houses were small, ill ventilated and full of vermin. Liquor smelled all around. The police frequently made raids and humanity was seen at its worst … Many wives and girls, failing to get jobs, resorted to illicit liquor-traffic. They concocted many intoxicants. Some were poisonous. Snakes were dug out of one of them. Other liquor-brewers washed themselves in it before serving it, so as to attract many customers according to herbalist instructions … The liquor-brewer and her assistants had to humour them [the men] in order to retain their patronage. Love-ties were formed. Friction followed after a clash because the liquor-seller was obliged to have as many sweethearts as she could. (Coka 1939, 308–9)

Coka also lived for a time in Maritzburg, where he worked as an orderly in the city's main hospital (Grey's). The Africans he encountered there made a profoundly negative impression on him and he despaired about the state of African lives in town. Over the course of two pages, he writes of his countrymen and women: "miserable … sickly … ailing … [with] tale of sorrows … stupid … stabbed … lost hope … Destitution, malnutrition, unemployment, degradation, crime, poverty and death claimed countless victims" (315–17). My point here is simply that, when put together with the actual statistics on disease and mortality gathered over the next decade, such commentary is difficult to reconcile with Swanson's description of the social and health crisis in a freehold township in the 1940s as "putative" (1996, 296, on Clermont) or "imagined" by whites (C. Merrett 2009, 89). So too, du Plooy's claim that shebeens around Pietermaritzburg offered "better quality beer" and a place "to relax away from the constant vigilance of European supervisors and away from the stabbings, beatings and pick-pockets of the beer halls" (1988, 143). The beer was not better; the pathology was real. The unhealthy and dangerous conditions of the shebeens may actually explain why men generally failed to support women's protests against the state's alternative to shebeens that broke out in Natal in 1929 (Bradford 1992, 227).

Of course not all African women in Edendale and environs conformed to the loose and dangerous stereotype. On the contrary, besides the reputable wives of male *amakholwa* property-owners and artisans, a class of widows sustained themselves and paid taxes through their own respectable labour, primarily by domestic service, sewing, and taking in laundry. In addition, a landowning class of independent women began to emerge as early as 1884, when Ellen Kunene with her two daughters purchased 54 acres (Meintjes 1988b, 251). More typically, women became landowners through inheritance of their deceased father's property. Mary Mini, the wife of Chief Mini, reportedly owned 66 acres in 1899, and by 1919, at least fifteen women owned more than two acres of land, including Dorothea Mtimkulu (20 acres), Mary Mini (15), and Lucy Rose Dambuza (14.75).[10] Dambuza proved particularly adept over the following decades at turning that asset into a successful and diversified business "empire," acquiring more land for rentals in the Harewood area just outside Georgetown (25 acres), plus a shop and cattle dipping tank in Macibise.[11]

A tiny number of these women sought social advancement by applying for exemption from native law: Meintjes (1988b, 262) records that, by 1895, eighteen Edendale women had successfully applied for exempted status independently of husbands and fathers. More typically, they expressed their social voice through church groups that blossomed throughout the region in the 1920s in the form of *manyanos* or *kopanos. Manyano* women had the reputation of conservative guardians of respectable domesticity (see Gaitskell 1982 or Epprecht 1993, for example). They also helped families survive the travails of the system in practical ways such as pooling cash for funeral expenses and rotating savings clubs known as *stokefeles.* Social welfare, such as it existed, was largely in their hands as volunteers. As noted in the previous chapter, the first health clinic in Edendale was the result of such volunteerism motivated, however belatedly, by Christian charity.

Women's work of this nature fell outside the realm of politics in the popular imagination. Yet as far back as 1885, African women did begin to participate in politics more recognizably. Ellen Kunene joined with nineteen other *amakholwa* signatories in that year to protest the corruption and abusive practices of the local headman.[12] Relative to elsewhere in the country, however, Edendale did not produce any renowned female leaders, and its educated African women had a low profile in such nascent nationalist movements as the Daughters of Africa or the Bantu Women's League. Their first recorded instance of an organized venture into politics as conventionally understood came only in the 1920s. The women at that time showed a decidedly feisty streak. First, they marched publicly in central Maritzburg with placards to promote Mason's Mill as their preferred site for an African location. Then, claiming to speak on behalf of "Responsible Native Opinion," Lydia Msimang and Justina Mdaka of the Edendale Council of Women boldly chastised Maritzburg city council in their petition against the city's proposed alternative site: "Thus Europeans made a choice on a matter in which they were not directly concerned," overruling people like themselves "whose judgement in these matters is worth more consideration than that of the average [white male] Burgess."[13]

This petition was, as usual, unsuccessful. That is not to say, however, that Maritzburg council was completely insensitive to the worries of Mdaka and Msimang. For some time it had been awakening to the fact that its own actions or inactions might be contributing to the social and sanitary problems that were disturbing reputable opinion in Edendale and environs. In 1915, for example, it opened a branch beer hall at Sutherlands, just outside the city limits on the road to Edendale. The hope with this extraterritorial reach was to intercept the city's male African labour force on the way home before the men could patronize local shebeeners and sex workers (Pietermaritzburg [1904–90] 1915, 16). The mayor also began to make exhortations to Maritzburg employers to pay fair wages and to provide decent accommodation for their

African workers. Yet other than moral suasion, city officials felt that responsibility for addressing Edendale's problems lay elsewhere, most probably with the province but, on some issues, with the national government. The single largest employer of Edendale workers, notably, was South African Railways, which was notoriously oblivious to city entreaties to house and pay its African employees properly. Council was further constrained by a budget enfeebled by the city's long-term economic stagnation and then thoroughly pummelled by the onset of the Great Depression. Many ratepayers felt that the city had already gone beyond the call of its technical obligations by the large expense it had laid out for its model native village, Sobantu.

The opening of that village in 1928 unquestionably provided dramatic improvements in the living conditions for, ultimately, hundreds of African families working in Pietermaritzburg. It may partly explain why women's protests against the beerhall system, which led to violence in Durban and northern Natal in 1929, fizzled in Maritzburg. Yet the creation of Sobantu also ushered in one of the most destructive and shameful periods in the city's history. The Natives (Urban Areas) Act (NUAA) of 1923 specified that once insanitary housing was identified, it either had to be brought up to standard by the owner or razed. If the latter, the city was responsible to provide new, up-to-code replacement before demolitions could proceed. Unfortunately (assuming that that was the city's earnest intention), the challenges of adhering to the letter of the law proved insurmountable. As the annual report of 1925 worried, the demolition of a mere six "hovels" in Hathorn's Hill would have created no fewer than eighty-six homeless people, and there were hundreds of such dwellings (Pietermaritzburg [1904–90] 1925, 68). In any case, many of the residents there did not qualify for the new housing, as they had no formal employment or, in the case of the women, could not prove that they were legally married. Building new homes for such people would reward the very behaviour that the city disapproved of. Moreover, many of the legitimate beneficiaries of slum clearance regarded the idea of moving to a formal township with tight regulations (curfew; no visitors for more than three hours a day; a white man's permission needed for this, that, and the other) as something to be avoided at all costs. As a certain Mrs Ndhlovu tersely put it in a slightly later instance of attempted slum clearance, "she would sooner go to gaol than live in a location and requests the Commission to 'refrain from treating us like slaves in our own properties.'"[14]

City council did not wait until such objections could be addressed. Starting as early as 1928, it moved aggressively first to clear squatters living on town lands and then to evict all persons without legal employment from the most blighted ward in the southeast corner of the city. The first demolitions of insanitary housing in the city proper began with the removal of the Hathorn's Hill slum in 1929. The following year the city commissioned a survey on housing and poverty to identify the next priority areas to target. It was no surprise that Camp's Drift came next on the list of urgency – the surveyors found 97 per cent of dwellings there to be "unfit for habitation" (1930, 74). It was a surprise, however, to find that the biggest landlord was the city itself. It collected rent from mostly Indian tenants, but it then waived city building by-laws so that those tenants could sublet to even poorer people in the most overcrowded conditions. This disturbing information was used to justify proclaiming the whole of the city under the terms of the NUAA, notwithstanding protests from less-guilty landowners. In August 1931, town council duly empowered the medical officer of health to proceed with identifying slum dwellings for demolition anywhere in the city and up to five miles outside the city boundaries without first providing replacement homes. People evicted who did not qualify for housing in the new native village by virtue of legitimate employment or marriage would, it was believed (or disingenuously claimed), find lodging in the underutilized male hostel or in buildings specifically licensed as exempt from the act or would return to their rural homes. "With this cleaning up of the city will go an improvement in the health of black and white," the city's preeminent voice of liberal conscience naively opined (*Natal Witness*, 19 August 1931).

The actual results could have been better predicted, not least of all since the removals coincided with the full onset of the Depression, a serious drought, and calamitous declines in agricultural commodity prices. A surge of African men and women desperately seeking employment began to pour out from the distressed reserves and white-owned farms. For those in Zwartkop, this involved a short stroll down the hillside, where hundreds of wattle and daub structures mushroomed, often in people's backyards but also on the Edendale commonage. Basutoland in the early 1930s also permanently disgorged tens of thousands of people from virtual famine conditions. Most went to the Reef or towns in the Free State, but hundreds ended up in Macibise. Some of the first wave of refugees from the Hathorn's Hill slum clearances thus did indeed go to the hostel and expanded native village as planned, and some found rental accommodation in the exempted *amakholwa* "colony" of New Scotland, in neighbouring Pentrich, or in grossly overcrowded rooms in the downtown core. The majority, however, simply moved up the hill from

Hathorn's to a farm called Raisethorpe, or headed west to join the burgeoning camps in Macibise, Sutherlands, or the Georgetown area, just out of reach of the NUAA.

Desperate overcrowding of malnourished people in damp, airless shacks was ideal for the spread of tuberculosis, while the proliferation of privies and rubbish pits on the sodden flood plain facilitated the colonization of the valley by anopheles mosquitoes. As if on cue, malaria appeared for the first time in history in the Maritzburg district in 1930–31, and became an epidemic that claimed ninety-two predominantly African lives in the following year (Pietermaritzburg [1904–90] 1932, 58). Flooding, which had been a recurrent threat, also turned deadly as foliage along the river was stripped for firewood. But gender remained for many, including *amakholwa* leaders, one of the most visible and upsetting aspect of the crisis. The so-called shack or kafir farms that landowners were throwing up to house the displaced populations almost invariably came to have a shebeen as a noisy social focal point.

The mayor's annual reports in these years did not acknowledge the suffering his police were causing. Clearly, however, tension was on the rise in the affected communities, to which African women in Edendale delivered an eye-opener in early 1933. The city had dispatched a crew to oil the water of stagnant ponds in the valley as an anti-malaria measure. A mob of women attacked the team, which beat a hasty retreat when men joined the fray armed with chains and sticks (*Natal Witness*, 9 February 1933). Council interpreted this event as confirmation that the city should stand firm in its refusal to concede any responsibility for the gathering turmoil in Edendale. It consequently boycotted hearings called by the province to investigate the possibility of expanding city borders precisely to avoid having to respond to public pressure. Councillor (later mayor) G.C. Joliffe denounced even the even anti-malaria initiative as "a great mistake … [and] the thin edge of the wedge," meaning a worthy health initiative that would quickly implicate the city in a colonial-style quagmire of unlimited expensive obligations ("City Council to Boycott Boundaris Inquiry," *Natal Witness*, 1 March 1933). When the top health official in the country visited Pietermaritzburg three years later, he reported back to Pretoria how "very disappointed" he was with the city's attitude. Council was not even willing to share (at cost) a portion of its abundant supply of clean water, which passed under Edendale by pipeline from Henley Dam, with the stricken community.[15]

In the absence of leadership from the Edendale Trust or city officials, Edendale's community leaders of all races came together in 1933 to assert their own rudimentary governing body. The Vigilance Committee, or *Isolomuzi* as it came to be known, conformed to the model of a health committee that advised government on where precisely health interventions were needed or what specific local issues had to be considered to make those interventions effective. Edendale's *Isolomuzi* differed from other such committees in the country, however, both in its multiracial make-up (three Europeans, one Indian, and four Africans) and in the fact that it had its own "civic guard" to patrol against dumping, ploughing, depasturing or cutting wood on the commons, and noisy shebeens.[16] The committee, which funded itself through a levy of five pence per house, may have had some success in moderating the worst noise and nuisance. However, the scale of the crisis was far beyond its ability to cope. Someone, likely in the national Union Department of Public Health, came up with the idea of shaming Pietermaritzburg to take direct responsibility as an urgent priority. The aforementioned public hearings, the Maritzburg Environs and Boundaries Inquiry headed by C.F. Hignett, were a key part of that strategy, convened only weeks after the malaria debacle. The province's chief native commissioner came to testify to the moral imperative for the city to act, as did the national assistant health officer, Dr G. Park Ross, who compared the city unfavourably to elsewhere in the union (*Natal Witness*, 2 March 1933). Nor did the *Natal Witness* mince words, denouncing council as "singularly supine" and "reprehensible" for its failure to do the right thing for the people of Edendale (27 April 1933).

The historiography has tended to focus on white ratepayers and councillors in the city for their selfishness or racism in blocking progress on this crisis. These sentiments were undoubtedly factors. Yet this explanation fails to consider the difficulties incorporation of Edendale would have encountered from the people of Edendale itself. It was not just European and Indian landowners who feared that they would be saddled with onerous urban rather than rural rates and restrictions. For the same reason, *amakholwa* in a nearby section of Slangspruit complained that incorporation would require them to abandon a school they had built for themselves, and to uproot from homes with "verandahs, kitchens, pantries, properly installed water tanks, etc." (*Natal Witness*, 10 March 1933). Numerous speakers and petitioners from the area meanwhile denied that the problems merited disturbing the status quo. Indeed, Mr Boola, the biggest Indian landowner, assured the commission that the *Isolomuzi* was already actively cleaning things up and that there thus was no need to incur additional expenses or boundary changes (28 February 1933). And Chief Stephen Mini, as a member of the

Edendale Trust and claiming to speak on behalf of the majority of Edendale property owners, feared that incorporation would open the floodgates to a European and Indian land grab. He demanded protection by the province through formal segregation, as laid out under the terms of the 1913 Land Act. This would have transferred Edendale to the jurisdiction of the Native Affairs Department (NAD) and, presumably, buttressed his own powers over both the trust and unattached women, enabling him to lead an effective campaign to suppress the shebeen trade and other urban indiscipline (11 April 1933). To that end, and pointing to the growing mess of slums within Maritzburg as proof of the city's incompetence on these matters, Mini's right-hand man proposed that a European should guide the people of Edendale under a *bunga*, a parliament of hereditary chiefs modelled on the Transkeian authority in the eastern Cape. "What they wanted was Government [NAD] control only" (Walter Msimande letter, *Natal Witness*, 27 April 1933).

No African women testified before Hignett (which would have been unprecedented for the era), but we can easily imagine that many were listening closely to the debate and quietly making their views known. The implications of both incorporation and a strengthened chieftaincy for the poor majority were obvious, but especially disturbing for small-scale shebeeners and short-term "wives" who could not afford the kinds of bribes expected by Mini's police. The proposals – whether regarding the extension of city authority (threatening to raze rental dwellings and impose rates) or NAD rule (which had recently toughened its definition of customary law) – were also of concern to women with property. The Natal Native Code of 1891 had provided a "rigid" definition of native custom but was extremely difficult to enforce. The revised code of 1932, by contrast, was "blatantly reactionary," in Dubow's terms (1989, 118), and a key part of a NAD initiative to achieve enforceability and uniformity between the provinces in the administration of exclusively African areas. Nowhere in the debates about boundaries was the issue of legal dualism explicitly raised. Yet the humiliations of that system had long been a grievance of the *amakholwa* elite, to which John Kumalo's epigraph at the start of this book pointedly refers. It would be hard to fault the many black Christian women of Edendale and environs for being anxious about Mini's apparent support for extending the reach of customary patriarchal law and traditional polygynist chiefs over them.

Faced with a cacophony of opposition, the boundaries commission effectively recommended the status quo, with better enforcement of existing laws (*Natal Witness*, 17 April 1933; South Africa 1940, 24). Far from scaling back its clearances, Pietermaritzburg council interpreted this decision as a green light to step up its efforts to make the city "one of the healthiest cities in the world" (Pietermaritzburg [1904–90] 1935, 15). At least 1,500 people were rendered homeless by the destruction of slum dwellings in Camp's Drift in 1934 alone, with a further unspecified "large number" following police invasions of Sutherlands (that is, outside the city borders) beginning in late 1935.[17] The province meanwhile sought to broker an Edendale town board but then balked at the behind-the-scenes negotiations. The new body would have been under titular NAD authority but with a staff of full-time health officers on loan from Maritzburg. Ten elected and four nominated members would have been explicitly equipped with all the powers of a normal urban local authority to enforce health, building, residency, and environmental regulations, including the control of animals both in Edendale proper (including the commonage) and neighbouring Plessislaer and Slangspruit. But because there were so few ratepayers relative to the size of the problems to be addressed, and because Maritzburg emphatically would not support additional expenditures, the proposed board would require a subsidy from the provincial administration of approximately £1000 per year. The province refused. The provincial administration simply ignored a subsequent request by the city's Town Clerk in 1936 to revisit the proposal, even as conditions on the ground further deteriorated.[18]

Two factions among Africans in Edendale coalesced around divergent views of what political reforms were needed to address the crisis. The first and initially most vocal faction was led by Chief Mini. He was a descendant of one of the founding *amakholwa* families, and he remained a Christian. Elected chief in 1893, he was not typical of native authority and indeed a common complaint against him was his long absences from Edendale in pursuit of sundry business interests. Mini was also an early active player in the Natal Native Congress (founded in 1900) and then the South African Native National Congress (SANNC, founded in 1912), the predecessor of the African National Congress. These were both self-consciously modern political movements that regarded Shepstonian-style chieftaincies as retrograde cogs in an oppressive system. Mini had initially been an outspoken critic of the Land Act of 1913, and of the exploitation and harassment of African women by the police in towns. Through to the late 1920s, he tended to side with the left or far left of the Congress movement and was a key ally behind "avowed Communist" Josiah Gumede's rise to the presidency of the ANC in 1926 (Limb 2010, 293).

At the local level, Mini's political record was more ambiguous. His integrity as a leader had been forcefully challenged by other *amakholwa* men as far back as 1902, and his position was saved largely because of the strong support he received from the white missionary at Edendale.[19] Mini then experienced personal bankruptcy in 1916 and was forced to sell his prime lands to a European, H.F. Kothe, who turned them into a profitable dairy and citrus farm (letter, *Natal Witness*, 13 April 1933).[20] That experience was traumatic enough, but throughout the 1920s Mini witnessed growing numbers of white and Indian landowners in Edendale work in league with the city to block initiatives that would protect African landowners from further losses. By the time of the Hignett Commission, Mini and his supporters had come to regard formal racial segregation as the only way to promote the development of Edendale (and to regain his own lost lands and diminished influence). He was, he told the commission, "quite satisfied with the [health and environmental] conditions at Edendale"; hence there was no need for a health committee or town board, or to incorporate with Maritzburg. But he was worried that Edendale's historical rights as a "native area" were in danger. "We still insist on the 1913 Act, which provides for the total segregation of Europeans and Indians, but at the moment we are not emphasizing that the full provisions be exercised" (*Natal Witness*, 28 February 1933).

Those claims of historical rights as an exclusively black community and for a hereditary chieftainship were hotly disputed by non-African landowners and the trust. However, after the repeated failures to negotiate a deal with the NAD, the province, and the city, the Union government finally intervened in 1936, more or less taking Mini's side. It declared Edendale to be a "released area" under the Land Act. This simply meant that there would be no evictions of current non-African residents but that non-Africans could not in future buy up any more land that came on the market. If non-Africans chose to sell, it could only be to Africans. The stated intention was that, in due time and without further government interference or expense, Edendale would become all African under a modified form of enlightened NAD administration. Mini seized the initiative with a demand that Edendale be allowed to expand through the purchase of neighbouring farms, a purchase made possible by a provincial grant of £50,000 to cover the cost. Mini in 1937 also imposed a levy of two shillings on his subjects, which was intended to facilitate the repurchase of lands that had been lost to European and Indian encroachment (his own former property presumably a high priority).[21]

The second faction in the community cohered around opposition to Mini and to Pretoria's unwanted interference, and originally included most of the big non-African landowners in the valley. They threw their support behind the idea of turning the existing probably illegal Vigilance Committee into a proper town board and decided to seek professional help to persuade the province to make that happen. Such was the context that led H. Selby Msimang to return to Edendale. Msimang was also a descendent of one of the original *amakholwa* families but had left Edendale as a young man for education at Healdtown and Kilnerton Institutes. He launched his career in Johannesburg as a court interpreter. This exposed him to the gross exploitation of and injustice for African labourers upon which South Africa's industrial revolution depended, and he turned to politics. He was at the founding convention of the SANNC in 1912 and thereafter an activist for workers' rights in Bloemfontein, secretary to the Natal Native Congress, and a key mover in the formation and early successes of the ICU. Over the decades he earned a reputation for being both highly principled and "astute" (see Limb 2010, 226; Mkhize 2015; and Msimang's own modest and moving recollection of his role in these events, Msimang 1970).

Politics ran in the family. Msimang's lawyer brother Richard was also a founding activist in the SANNC, while his uncle, William, had led the campaign to defeat the Bishopstowe site for Pietermaritzburg's proposed native village. Although that campaign had ended in failure, it had generated considerable sympathy among whites in town. Another brother, Oliver, was NAD's preferred candidate to succeed Mini as chief and a member of the Edendale Trust. With such a family reputation for leadership, it was natural that the *Isolomuzi* invited the younger Msimang to assist the community in its struggle for self-determination, a role he began to play in the late 1930s (Msimang n.d.). Selby Msimang ultimately emerged as the leading African spokesperson for Edendale's "moderate" or "traitor" faction, depending on one's point of view. Its demands included the abolition of the trust (and later, of the chieftainship as well) and that the Vigilance Committee be reconstituted as a democratically elected village management board with jurisdiction over a territorially expanded Edendale and without any NAD involvement. Only such a modern form of local authority could deal with "the worst slum that could scarcely be imagined" (Msimang n.d., chap. 3, p. 6).

In early 1938, both factions convened large public meetings to rally support for their rival visions. Tempers frayed. After thugs led by Chief Mini's son invaded an *Isolomuzi* meeting to heckle and threaten violence, Msimang requested police protection. The town clerk phlegmatically remarked, "It would appear that property owners themselves in Edendale are apprehensive of the dangers of allowing the present state of affairs to continue."[22]

Similar conflicts were simmering in the other urban areas of the country, and the governing party at the national level was alert to the danger that violence over local disputes could spill over into much wider conflicts over land, wages, passes, and political rights. In 1938, Prime Minister Jan Smuts struck a commission of enquiry intended to avert that eventuality. The Thornton Committee was to hear evidence and make recommendations for the formulation of a national policy on the question of the "Administration of Areas which are becoming Urbanised but which are not under Local Government Control," to cite the title of the committee's ultimate report. The prospect was greeted with great interest in and around Maritzburg. In the lead-up to Sir E.N. Thornton's visit to the area, the terms of the debate were framed in the starkest language by editorials and letter writers in the press. Not just health, environment, and the appropriate control or emancipation of African women were at stake, but "democratic duty towards the inhabitants of the black belt," as the editors of the *Natal Witness* put it ("Over the Border," 10 May 1939). One of the few regular African correspondents to the paper used the opportunity to insist that "it is high time the native was entrusted with some responsibility in the management of matters pertaining to his welfare. Let him be given a chance to show his worth – let him fight his own battles, and maybe a solution can be found for what is termed the native problem" (William Manyoni, *Natal Witness*, 24 May 1939).

Thornton arrived in Maritzburg in mid-June 1939 as part of his national tour. City council came prepared, knowing well what both the province and Union governments preferred to see happen (that is, that the city administer the clean-up at its own expense). With the divisive debates around the Mason's Mill proposal for a native village in the 1920s still fresh in many people's minds, council was also deeply wary of the "moderate" African proposal. A village management board could provide the template for an autonomous, black majority local authority almost equal in area to the city itself and not much smaller in population. In his submission to Thornton, the mayor thus sought to forestall these options. He first baldly denied that slum clearances from the city had anything to do with Edendale's plight but noted that Maritzburg would generously offer assistance if it were fully compensated for its expenses. The medical officer of health elaborated by proposing a city takeover of Edendale, again provided someone else paid for it. Other stakeholders then presented the familiar arguments that served council's obstructionist strategy, including an impassioned plea by a bemedalled Mini against incorporation with the city: "'Never, never,' he declared, shaking his head vigorously and holding up his hand in protest" (*Natal Witness*, 15 June 1939).

Thornton's only public comment at the time was to point out the "ghastly experience" of mixed-race government in the Cape and Transvaal (15 June 1939). He then moved on to the next city. Pietermaritzburg continued its demolitions of insanitary structures within its area of jurisdiction, including, allegedly, thirty substantial, modern-style brick houses in an African freehold enclave on the city's eastern border (24 November 1939).

When Thornton's report came out the following year, it devoted several paragraphs to the Edendale question. Brevity belied the strength and radicalism of its recommendations: to create a new form of local authority that would be directly accountable to and primarily funded by the provincial administration. A local health *commission* (not committee) would have powers virtually equivalent to those of a normal town, differing only in its non-elected nature and the source of its funding. Paragraph 119 specified that if the city and the new local authority failed to cooperate fully, the province should compel the incorporation of Edendale into Maritzburg under its own conditions (South Africa 1940). That threat finally worked to break the deadlock.

It took another year to establish the exact legal framework and process for a local authority to be established in predominantly but not exclusively African residential areas. As defined under Natal Ordinance 20 of 1941 (the Public Health Areas Act), a public health area (PHA) could be created where people applied for one or where it was recommended by expert advice supported by a scientific inspection. Following receipt of such an application or recommendation, the act required government to convene a public meeting to ascertain local opinion. Only if that were favourable could a PHA be declared and the machinery of governance (the LHC) erected over it. An LHC so created would have powers to levy rates; to take out loans; to enforce environmental, building, and health regulations; to carry out infrastructural development; to expropriate land; to provide public education; to demolish slums; and to remove people to new homes. The mandate of the LHC was to get things done in the expansive spirit of social medicine that was then emerging out of the health centre experience in nearby Pholela reserve and that in 1944 was articulated in the Gluckman Commission on a national health service.[23] That is, it would not simply provide clean water, healthy homes, erosion control, and curative medicine, but also conduct research, surveillance, primary health education, beautification (planting flower gardens, for example), and programs to stimulate household food security and income generation.[24]

The nature of these tasks demanded direct approaches by the state to African women, who were the principal primary caregivers and custodians of the domestic sphere in their own households as well as domestic workers in white households. African women were also, in the discourse of the times, commonly framed as the most direct threat to the health of Maritzburg citizens. The recurrent title for an impassioned exchange of opinions about Edendale in the pages of the *Natal Witness* in 1941 brought that point home to its readers. The crisis in Edendale could no longer be deferred, as it was now literally present "In the Kitchens" in the form of African female domestic workers. Avatars of Edendale's sanitation holocaust were now present in the white homes of Maritzburg, preparing food and looking after babies.

Events moved quickly in late 1941. In September, a raucous public debate took place at city hall following which leading city officials and the *Natal Witness* came out in favour of the Thornton proposal. In October, the provincial administration appointed a triumvirate of white men directly responsible to the executive council to guide the process through the requisite public consultations. These began with the city's medical officer of health in November, followed by the Natal Indian Association, the Edendale Trust, and experts including Dr Sydney Kark of Pholela. No record remains of consultations with African leaders at this stage but, in November, the *Isolomuzi* in Edendale split over the issue, with a new faction opposed to the PHA led by George Mtimkulu in tactical alliance with Mini (more on this below). In December, the province proposed an expansive definition of Edendale's borders, which would include both the released area (the original farm) and the semi-urbanized farms between it and Maritzburg. The Edendale and District Public Health Area (EDPHA) was duly established by Proclamation 9 taking effect on 1 April 1942. This proclamation also obliged Pietermaritzburg city council to partially fund and to fully cooperate with a newly created LHC. *KwaPoyinandi*, as the LHC was known locally, overnight became the third-largest urban local authority in the province. It was an "experiment" on several levels: ecological rehabilitation, social medicine or primary health care, and training in local governance.

Before that experiment could happen, the majority population needed to be won over. Opposition came not just from suspicious village women, as in

5.1 The first Local Health Commission executive, 1941

the anti-malaria incident back in 1933, or from cantankerous and ambitious chiefs wishing to maintain or extend their power. Even respected moderates from the *Isolomuzi* at first refused to cooperate with the PHA proposal. A deputation led by Selby Msimang had presented its demands to the province in June 1942 and had threatened to boycott unless the LHC explicitly limited itself to a probationary period of five years. During that time, the LHC had to concretely demonstrate its commitment to training Africans in the skills of local governance, with the objective of progressively devolving decision-making powers to the majority population. If, after five years, no progress was evident, the commission should be disbanded. Msimang also rejected the original suggestion of creating three distinct advisory boards defined by race "on the grounds that the interests of the community were indivisible and … that all races should be represented on one Board."[25]

Remarkably, and apparently in good faith, the province accepted these conditions. Indeed, Douglas Mitchell, the political superior in the provincial administration and chief backer of Thomas Wadley, the first chair of the LHC, described the initiative to Msimang precisely as an "apprenticeship," the ultimate goal of which was to hand the powers of local authority over to a democratically elected leadership. This was sufficient to bring both the *Isolomuzi* and Chief Mini to make peace with the LHC idea, and so to finalize the required public consultation. There would be a single, multiracial advisory board with elected members and delegated tasks, as requested. Msimang, for his part, demonstrated his own commitment by accepting employment with the new commission as a researcher and taking a seat on the new advisory board. It was, in his memoirs, "the first ever and only multi-racial body in the country" (Msimang n.d., part B), and in Commissioner J.C. Boshoff's words, "a gentleman's agreement" and "a test of your [that is, black] fitness for self-government."[26]

The creation of the Edendale LHC was followed soon after by the declaration of a second PHA in Clermont, just outside of Durban. By 1959 there were more than two dozen other PHAs scattered around Natal. Edendale's, however, remained by far the largest in area and population, the flagship of the province's new approach to the rehabilitation of areas where the overlapping crises in health, gender relations, and environment were most acute. Looking closely at the debates around its establishment helps us to contextualize the strength and sometimes bitterness of feelings towards what briefly flowered as an initiative with strong liberatory potential. In the next section, we look at African women's suspicions about this new arm of the state and how the issue of gender and sexual health brought to the fore some underlying tensions within African nationalist politics.

≈

The LHC charged into Edendale in 1943 like a bull in a china shop. Wadley, its chair, had formerly been mayor of Durban (1924–26) and was a member of Parliament for Point on the liberal-leaning side of the South Africa Party when in 1933 it formed the national government (*Natal Witness*, 19 May 1933, 4).[27] He combined a powerful confidence in technocratic solutions to Edendale's problems with a moral conscience that urged haste and coercion, if necessary. The latter was, in the first instance, directed against Maritzburg town council, which Wadley held to be the main culprit in Edendale's predicament. He urged the secretary of native affairs to "go to the limit" to pressure the city to pay up and, in June 1943, to intensify "all proper means of exerting pressure on 'Maritzburg to meet the needs of the natives employed within its boundaries ... My candid opinion is that Maritzburg should be doing the whole job, but somehow or other they seem to have succeeded in getting someone else to nurse the baby. Edendale is nothing more or less than a huge location for Maritzburg. I find responsible people in Maritzburg most sensitive when I touch on their responsibility, and most aggressive in their revendication."[28]

Moralistic language to shame city officials into action had never been noticeably effective, but in this case Wadley had the backing of the Thornton Committee, a provincial administration that was not afraid to issue threats, and a key liberal ally on city council, future mayor Eleanor Russell. The combination succeeded in persuading council to commit £10,000 to the experiment, a fifth of the initial total budget.[29] Its first project thereafter was a survey of people's state of health and living conditions. This confirmed the worst – no fewer than 94 per cent of the population lived on less than five pounds income a month; infant mortality exceed one in four live births, and so on. But the survey also enabled the state to target its interventions. As a vehicle to start delivering services, the LHC took over the valley's only health clinic. Under the direction of the LHC's first medical officer of health, David Landau, Victoria Mthimkulu's charitable clinic became a health centre on the Pholela model. Its main focus was on antenatal care and education for women to encourage proper breastfeeding, vegetable gardening, and good nutrition.

Shaming and coercion were also intrinsic to one of the LHC's first policy announcements in the new EDPHA: a total ban on the construction of wattle and daub shelters. Wattle (a cheap wood from a tree that grows like a weed around Natal) provided a frame onto which daub (mud) could be plastered for walls. Huts made in that fashion were ubiquitous throughout the PHA – almost anyone could build one for basically no cost. They were also often damp, musty, and vermin infested, and they crumbled quickly. The LHC considered them not only a danger to health but demoralizing to inhabitants and disgraceful to see in a modern jurisdiction. It declared that none would be destroyed before new homes were built to replace them, but that no new ones would be allowed.

This regulation, as much as the imposition of rates on property owners, poured fuel on the already simmering fire of opposition. Wadley, on the appeal of his main African ally in Edendale (Selby Msimang), quickly realized the danger of inflaming tensions and backed down from his initial hardline on "inferior-type housing." New wattle and daub structures would be allowed as temporary, emergency expedients, but only with prior application to and

5.2 A wattle and daub house in Macibise, "improved" with wine bottles

approval by the commission. The LHC further agreed to offer financial assistance to owners to improve their existing wattle and daub structures up to a minimum standard set by the medical officer of health. The commission also accepted the advisory board's request to desist from evictions of squatters on the commonage – as erosion science urgently advised – until alternative housing was available, including "improved" wattle and daub extensions.[30]

These compromises brought some peace, but within the year simmering discontent nearly came to the boil again, this time over intrusive health research that spilled into lurid stories in the press about rampant venereal disease in Edendale.[31] A crisis meeting took place in the boardroom of the LHC office in Pietermaritzburg on 18 July 1944 to address the issues. Present were Wadley and the other two members of the first LHC triumvirate, J.C. (John) Boshoff and Henry C. Lugg. The former was a city magistrate and the latter a career official with the NAD who was on the verge of retirement. Lugg was a fluent isiZulu-speaker who had been seconded to the LHC to provide the voice of experience for dealing with Africans, and perhaps also to mollify conservative critics.[32] Two other top officials of the LHC were there, the secretary, A.H. Norman, and the medical officer of health, Dr David Landau. The meeting was chaired by Mr R. Wronsky, the chief native commissioner of Natal, with invited guests including A.W.G. Champion, the African member of the Native Representative Council, which advised the provincial administration on native affairs, Chief Edgar Mini of Edendale (son of the previous chief Stephen), E.J. Mpanza of the EDPHA advisory board, and "some 20 natives (including five women)." The latter were not named in the minutes but they were likely the leading female representatives from the two main factions that had emerged in Edendale over the previous years of dispute: Lydia Msimang and Justina Mdaka of the Council of Women on the "moderate" or "collaborationist" side, and Mrs T. (Pauline) Nxumalo, Mrs Z. (Joyce) Miya, and Mrs G. Mhlongo of the dissident Edendale Lot-Holders Association on the other, at Champion's invitation.[33]

The occasion of the meeting was a letter that Champion had sent to Wronsky protesting against the LHC with specific allegations of, among other things, scandalous intrusions by the state into the private affairs of African women in Edendale. Champion had learned of these alleged evils at a public meeting he had attended at Chief Mini's invitation in May. The reported crowd of a thousand (Nuttall 1984, 45) represented roughly a quarter of the adult African population of the district and many times more than had ever heeded the words of any Mini in the past. Wronsky had forwarded Champion's letter to Wadley, who now wanted to hear directly from the protesters about their grievances and to persuade Champion not to inflame public opinion in the community against the LHC by propagating false information. Champion, for his part, asserted that his goal was to establish the truth about alleged abuses taking place in Edendale that had been reported to him by the chief and leading women of the community. If the allegations were true, he argued, they justified an impartial judicial enquiry. Such an investigation had the potential to be deeply embarrassing to the provincial administration and to bring the PHA experiment to a standstill.

Protests against the LHC were in fact nothing new. On the contrary, ever since the idea of some form of state intervention in Edendale had first been proposed, it had been a focus of intense, debilitating controversy. The Edendale Lot-Holders' Association (ELHA) in particular – that is, the Mtimkulu faction that had broken from the *Isolomuzi* in the lead up to the declaration of the PHA – stepped up its efforts to oppose the commission after 1942. Some of this reflected the specific class interests of the ELHA's members: whereas the *Isolomuzi* was supported by the old *amakholwa* elite with inherited properties as well as the main non-African landowners, the ELHA spoke primarily for Africans who had purchased their properties more recently. As newcomers, small-scale owners, and mostly non-Wesleyans/Methodists or non-Christians altogether, ELHA members saw themselves as outsiders, socially and economically more vulnerable to the risks that the PHA experiment posed. George Mtimkulu had some bitter experience with those risks, having been a sanitary inspector in Alexandra when a similar experiment in elected, non-racial local authority had spiralled into political crisis linked to a deepening health crisis in the mid-1930s.[34]

A significant element in the ELHA strategy throughout the 1940s focused on the gendered aspect of its members' vulnerability under the new regime. The ELHA brought African women into leadership positions and appealed directly to women to mobilize for mass protests and boycotts. From its inception, for example, the ELHA had one woman on its executive (Pauline Nxumalo), and women were vocal representatives on committees that repeatedly met with NAD officials to protest the LHC and to denounce its African allies as stooges.[35] Demonstrations organized by the ELHA throughout the decade were noteworthy for their high numbers of women. Msimang reported, for example, a rowdy meeting in Georgetown where he was personally accused of selling Edendale out to Pietermaritzburg. "The majority of those present were women who seemed not prepared to listen." Their spokesman, a Peter Kumalo, threatened to lead "all the women in the area" of Edendale into the city to demand an end to alleged LHC oppression and discrimination.[36]

The ELHA's principal leader, George Mtimkulu, was without question a demagogue who frequently employed hyperbolic and violent language. Initially he presented himself as a progressive. After the initial break with the *Isolomuzi*, he had distanced the ELHA from the Mini camp with an appeal to the provincial administration to abolish both the Edendale Trust and the chief's legal powers. Edendale was sufficiently developed, he argued, to "have one law governing Europeans and the enlightened classes, since the Native Code is a hundred year make up, and the times have drifted far beyond the old 'Zulu.'"[37] Yet soon after the multiracial advisory board cohered around more or less that very promise, Mtimkulu changed tack. He called for expropriation of all non-African property and for Edendale to revert to direct control by the NAD – "our father to guide us." Failing that, he argued that the African population should be bought out by the state and removed altogether onto a new reserve that would guarantee racial segregation and where, presumably, the Natal Native Code would apply as the law.[38]

The fact that Mtimkulu was able to muster so many women's support to obstruct the commission's work behind such reactionary proposals speaks to one of the most poignant aspects of the crisis. That is, notwithstanding the crushing poverty that many women in Edendale and environs experienced and the health/environmental/social crisis that so dismayed African intellectuals such as Gilbert Coka, the status quo for African women in Edendale was in many ways still better than elsewhere in the province. In 1919, about 5 per cent of the village's property owners were women (Limb 2010, 293), and we can infer from a petition from the ELHA in 1946 – with fifteen women signatories out of 130 (11.5 per cent) – that the proportion grew steadily over time.[39] In addition to this possibility of legally owning land, and to personally controlling the profits from rents, shops, and dip-tanks or their own vegetable and milk production, women in Edendale had access to employment income in relatively nearby Maritzburg, on the surrounding white farms, and even within Edendale itself. With no influx control, no effective chief or native authority, and no police station, Edendale offered further opportunities for women escaping poverty, neglect, or abuse in tribal areas. The attractions of life in Edendale, however precarious, were manifest in the LHC's first census. It found in 1943 that African women over the age of twelve actually outnumbered male African residents in the community by nearly 10 per cent (4,835 to 4,438), an almost unheard of gender ratio for a significantly urbanized area of that size in South Africa at the time.[40]

Many among that female majority clearly preferred the status quo over what the LHC promised (a gentleman's agreement, a test) or threatened (rates, expensive building standards, intensified police and other surveillance, and competition to make a living off of beer and other petty commodity sales). It is possible that some women shared Mtimkulu's belated confidence in the NAD and customary law, which, however often it had proven ineffective, at least gestured to the control of unruly daughters and sons. More likely, Mtimkulu's frequently stated concern for "poor widows" struck a number of populist notes with women that allowed them to ignore his references to the NAD, including the claim that Indians in Edendale undermined women's ability to eke out a living. He also threw out emotive appeals to family, claiming that Indians seduced African daughters and were giving rise to a growing population of "curly haired bastards, which are neither Muntu nor Indian." He demanded the expropriation of Indian properties by the state to remove this "curse."[41]

Such inflammatory language raised the prospect of mob violence, as indeed was soon to happen in Durban and (to a much lesser extent) Sobantu. Fearing reports that Mtimkulu and a few hundred of his angry supporters were planning to crash the community meeting called to celebrate the establishment of the LHC, the government advised Chief Mini not to attend. Two weeks later, Mtimkulu accused Mini of "*tengisa* [*sic*]" (*thengisa*, "selling out"), for which Zulu culture demanded execution. For his part, Mtimkulu claimed to fear violence if he attended "Toy Advisory Board" meetings, and requested permission to do so accompanied by a bodyguard.[42]

Champion's involvement in this explosive situation came as a worrisome development for the provincial administration. Champion came from an *amakholwa* community north of Durban (Inanda). He had risen to prominence in the 1920s as a leader of the ICU and "minister of labour" within the ANC. The government had long regarded him as an opportunistic agitator and held him to blame for the 1929 beer-hall boycotts that threatened the revenue mainstay of native administration in the province. For this, and for allegedly scheming with the disgraced Zulu king Solomon Dinizulu, he was banned from Natal and Zululand for several years in the mid-1930s. Yet government was also aware of and subsequently came to appreciate Champion's strong anti-Communist views. Indeed, as Shula Marks points out, Champion was a businessman and landowner with deeply ambivalent feelings towards working-class aspirations that might threaten his property rights. He played a significant role in hastening the collapse of the ICU movement in the late 1930s and in 1942 became one of the African members of the Native Representative Council, which had replaced the qualified franchise for Africans in parliamentary elections and was supposed to be a forum for the articulation of moderate, sober African opinion. For his part, Msimang regarded Champion as "difficult, conceited, selfish."[43]

If his politics at the provincial level were somewhat erratic, the LHC had ample reason at the local level to be wary of Champion as a determined, formidable opponent. Among his many business enterprises in the 1930s, he had raised capital for the development of a private, for-profit township – Clermont. Clermont had deteriorated so quickly into a slum, with even worse infant mortality figures than Edendale, that in 1943 it was declared Natal's second Public Health Area. Champion's highly profitable investment had in this way effectively been expropriated by the state. That act may partly explain the statement attributed to him by Wadley, and not disputed by Champion at the opening of the 1944 meeting: "You would fight the Commission whenever and wherever you met it."[44]

Champion's letter to Wronsky did indeed take a scattergun approach to attacking the constitution, motives, and practices of the LHC in Edendale. It listed several formal grievances, which Champion elaborated upon in the meeting and which, to LHC officials, must have sounded like a rehash of Mtimkulu's more outlandish demagoguery. These included high rates for property owners, a conspiracy to dispossess Africans of their land and historic rights, lack of consultation with the people (in this, the LHC was characterized as "dictators"), passes and other shameful impositions upon women, lack of services, incorrect constitution (namely, the LHC's placement under provincial rather than Union government oversight), allowing Indians to run the bus service, and creating "confusion" by promoting civic groups that undermined the authority of the chief. On the last point, Champion specifically mentioned the Council of Women (supporters of the advisory board) as a threat to good governance. He denounced the LHC for "acting communistically" through its subsidies on milk and vegetables that undercut women sellers. Its general maladministration, he claimed, was bringing "untold miseries."

Before addressing these accusations, Wadley testily informed Champion that the constitution of the LHC had been set by an act of Parliament and was not up for discussion. Wadley would respond only to specific accusations of maladministration. As his first and clearly most irritating example of a demonstrably false accusation, he cited Champion's claim that the LHC required African women to have a paid nurse in attendance for all births and that they were being charged by police if they failed to do so. Champion conceded that he may have been misinformed on the issue. But he then revealed a deeper, underlying grievance. Why, he wanted to know, was the commission asking women about miscarriages: "In our Native Code such a thing is not allowed. What good purpose does it serve? … What is at the back of the Government's wanting to know about miscarriages?"[45]

Dr Landau intervened at this point to explain the medical need for such information. Primarily it was to identify the main cause of stillbirths – syphilis – and therefore be able to offer treatment. Landau could have added that research into health, nutrition, and household economy had been the very first project of the LHC, long before any sewers or clinics or houses were built. Its first African employee was Selby Msimang, the principal researcher in that survey. The aim of the research had been to establish the most urgent health needs in the community and where the most cost-effective health interventions could be made. Landau, it should be noted, was one of the pioneers of the social medicine experiment in South Africa that conceived of health in broad, holistic terms. He was a "very close" personal friend of Sydney and Emily Kark, the couple who ran the Pholela clinic (Kark and Kark 2001, 86), and he would have been very aware of the stunning early successes of their approach.[46] In the Karks' conception of social medicine, information on wages, housing conditions, property rights, family composition, diet, household expenditure, and level of education were all equally important to establishing the health needs of a community, as were the epidemiological facts about specific diseases. It was, to be sure, an innovative approach at the time but came to enjoy powerful backing from the Union government until the early 1950s. Judging from Msimang's memoirs, the experience of doing this research strengthened his belief in the LHC approach and brought him to devote over two decades of service to its work.[47]

Champion acknowledged that doctor's explanation as "very good" but then proceeded to reveal yet another underlying grievance. Why was the research about miscarriages, and implicitly the sexual lives of the mothers, being made public?

> This publicity is only given to us because we are very poor.
> It is not done in the case of Europeans.
>
> MR WADLEY: Yes it is.
>
> MR CHAMPION: I have never seen it in the Press.
> We have nobody to bring our complaints to about the publicity given to us. It makes people afraid to go to Edendale.

What, then, were the people of Edendale, as represented by Champion and the women he had invited to the meeting, demanding the commission do? Champion got straight to the point: "I want you to tell these newspapers to stop this information. There are many things which for the good of the country are hidden. Hide as much as you can for the sake of good government."[48]

Wadley was clearly taken aback by this turn in the discussion. The scandal, in Wadley's mind, was not that syphilis was present among the women of Edendale. That was an old story and indeed had provided one of the main political incentives to create the LHC in the first place.[49] Could it be then that he, Landau, and the other LHC representatives were simply oblivious to the nerve they were poking? Indeed, that they were blind even to the fact that suspicion of white people asking questions about Africans' intimate lives was also "one of the most powerful themes" in reactions to recent similar health survey in Alexandra (Bonner and Nieftagodien 2008, 102)? This possibility speaks to the limitations of white medical liberalism at the time and is the subject of considerable recent debate among historians. The documentary record from Edendale does not yet settle the debate, but we can assume that paternalism of the type revealed by historians such as Diana Wylie, Shula Marks, and Mary Caesar would have unambiguously damned the project in the eyes of a man such as Champion.[50] In this case, Champion allowed Wadley to defend himself by turning the tables from blaming Africans to blaming the comfortable white people in neighbouring Pietermaritzburg. Wadley explained that the extent of syphilis, miscarriages, and the many diseases of poverty in Edendale was merely symptomatic of the real scandal, viz., the exploitation and neglect of Edendale's population by their white neighbours. Somehow the citizens of Maritzburg had to be made to realize the consequences of their complacency about the health crisis in African communities. The publication of hard evidence of the health costs of the migrant labour system and low wages for African workers would go farther towards that goal, and towards building the political will for broader change, than moralistic generalizations or tut-tuttings about African women's looseness: "'Maritzburg people will have to meet their obligations one day with better wages for its employees … Anything that can make Pietermaritzburg sit up we should not discourage … I cannot agree that we should hide things. I think the more publicity that is given to Edendale the better."[51] Indeed, the LHC report on ill health in Edendale did spark an extended, very lively debate in the *Natal Witness* under the recurrent heading "Speed the Drains" (a play on wartime propaganda to "speed the planes" to destroy fascism). Much of this was sympathetic to Edendale's population and couched in shaming language against "the ruling class" to support increased salaries for African workers, higher rates, and the work of the LHC (see, e.g., 1 and 11 March, 23–28 October, and 6 November 1944).

Champion let the topic drop at this point. However, from hints he gives elsewhere, it is clear that he was not convinced that the scandal of revealing sexual secrets to certain very specific audiences was worth the theoretical political and health benefits. He asked, pointedly, "Suppose a native came from Bulwer. How would you know that he really came from there?" and "What do the people of Inanda, etc., think of Edendale?" Bulwer? Inanda? The people of the latter, his home town, were *amakholwa*, respectable Christians whose identity and political aspirations to a large extent still hinged on the image of propriety, and feminine domesticity, and who had been chafing at white paternalism in Natal since at least the 1860s. Natives from Bulwer actually meant Pholela and the surrounding tribal reserves including the upper reaches of Zwartkop. They were, in other words, supposedly simple folk who might get the wrong ideas about their social superiors if secrets about life in town were made known. In making these coded references, Champion was underscoring his concern that frank health propaganda profoundly disturbed the people of his own class. Wadley conceded, "I had no idea that there was such a revulsion of feeling, however. We will note this."[52]

Regrettably, the women in attendance had no chance to express their views (or perhaps the minutes taker simply saw no value in recording them). This may in fact have suited Champion's interests. He had reportedly ridden on the coat-tails of women's protests before – the boycotts and demonstrations against oppressive liquor laws that roiled much of Natal in 1928–29 served to boost membership in his breakaway Natal faction of the ICU (Bradford 1992, 212). It is hard not to escape the suspicion that Champion regarded women as either a means or an obstacle to strengthening the position of "my friends" (that is, the chief and his entourage). Indeed, it is telling that neither Champion nor any of the other complainants thought to enquire of the one African woman who had direct experience with women's health (Victoria Mthimkhulu, founder of the first health clinic) or of the women who ran the Edendale Child Welfare Society. The latter had been founded in 1940 and was led by white women from Pietermaritzburg. Its views were pertinent in that it had the full support of the commission and, privately, was acutely conscious of the negative health impacts of the crisis of governance. In a somewhat demoralized-sounding report directly to the LHC a few years after this event, the society specifically identified the primary obstacles to women's health in the community as customary patriarchy, the Natal Native Code, and the "native police" (that is, Mini's rudimentary force, which allegedly entrapped, harassed, and sexually exploited vulnerable women). They called for, among other things, "the emancipation of suitable [African] women who are deemed able to care for their children without the aid of a male guardian."[53]

That Champion was not interested in those women's perspective is noteworthy. It would seem to fit with the pattern of "condescension" towards women that Peter Limb (2010, 158) attributed to the ANC leadership in general terms during this era and of the conservative (exclusionary, fraternal, sexist) notions of "racial respectability" that Healy-Clancy (2012) describes among the men of the New African movement. That said, by the end of the meeting Champion appeared to have been appeased. He departed promising to cooperate with the LHC in the development and governance of Edendale and presumably all the other predominantly African PHAs. Indeed, Champion subsequently sat on the advisory board of Lamontville, Durban's municipal township, and apparently became an ally of government in trying to contain more radical political movements in the townships (Torr 1996).

≈

In Edendale itself, Chief Mini and the ELHA were less easily put off and they continued to obstruct commission initiatives with some notable "successes." These forced the LHC to compromise if not abandon its scientific priorities. Women and gender were rarely mentioned explicitly but they are implicit in many of these compromises, most critically the keeping and slaughtering of livestock. Donkeys were a particularly destructive grazer that contributed to the "fantastic" erosion that heightened flood risks and washed away the remaining agricultural potential. Pigs, often kept in sties close to people's living spaces, not only produced a concentrated foul effluent but their measly meat was a major health risk. Typically slaughtered under extremely unhygienic conditions in people's homes, the carcasses were commonly disposed of by dumping in the most convenient stream or donga. Cattle were an even bigger problem, from the LHC point of view. It estimated the carrying capacity of the area at 200 beasts, while the actual population was close to 2,000. They were a scraggly herd, a repository of ill health and parasites whose milk and urine transmitted the bacillus for enteric fever. With fences around most of Edendale having long since been torn down, they were also a threat to herds on the neighbouring farms and Zwartkop, and hence of concern to the NAD's own betterment initiatives there. One of the commission's early top priorities was to cull all of these animals to sustainable populations.

The LHC did not effectively communicate the scientific imperative to assert control over livestock in an urban area to the people of Edendale and soon came up against an unanticipated level of intransigence. From the point of view of the majority of Edendale residents, even Christians, cattle were prized as a visible measure of a man's wealth, essential for negotiating marriages and for honoring family in other feasts. They were the means by which women acquired social standing. Pigs ("women's cattle" among the Basotho) were an important economic asset for the very poor, and donkeys a hardy, self-sustaining means of transport also widely used by women. Talk of culling them, or of prohibiting backyard slaughtering, thus brought stiff resistance, including from the community's "moderates." African health assistants employed by the LHC expressed fear of violence when approaching butcher shops in private houses, and the message quickly percolated upwards.[54] Indeed, the commission's reluctance to press too hard on the pig issue eventually came to the notice of the province. The deputy chief health officer upbraided the local medical officer of health R.P. Seymour in 1949 for allowing the slaughter of pigs for consumption in Edendale to continue. In his defence, Seymour admitted he was under pressure to relax standards, and that, as a political expedient, "we have already sacrificed a large proportion of our basic hygiene principles."[55] Regulations against chicken runs were allowed to "lie fallow" until 1956, when the law was finally enforced. In the case of cattle, the LHC revised its target population upwards to 500 (that is, two and a half times the carrying capacity). Even so, it made virtually no progress towards reducing the size of the herd until the 1960s.[56]

Women's opposition to the suppression of home-brewed liquor was another major frustration for the LHC, as it had been for city police for decades before and virtually everywhere in the country. The health costs of alcoholism and poisonous intoxicants were very high and often borne by African women. Yet many African women depended on brewing and selling those intoxicants, and perhaps on some sex work, to survive. Evasion of the police was a required skill for the business, and police tactics to catch the women fuelled disenchantment with the new regime. It likely also fuelled disbelief in other health propaganda or initiatives that seemingly had nothing to do with beer. One of the biggest targets of concerted and successful opposition was the Georgetown market hall. The LHC invested considerable funds into constructing what it hoped would be a social, economic, and health hub for a revitalized village. The new hall would offer women in particular a secure, spacious, clean place to sell the vegetables they would be inspired to grow, protected from the weather and the tough marketing practices of Indian competition. That never happened. Rather, the market became the focus of all kinds of rumours with racialist undertones, including that black women were not allowed to enter through the front door (*Ilanga lase Natal*, 21 June 1952). The anti-LHC faction called for a boycott, which proved so effective that in 1951 the exasperated LHC Secretary D.R. Donaldson frankly described the market as a "white elephant." He first agreed to open it other uses (cinema, dances), and eventually the commission sold the building to the YMCA.[57]

Setbacks like this contributed to discrediting the social medicine approach, not just at the political level nationally (which after 1948 was positively hostile on ideological grounds) but on pragmatic grounds locally. Landau's untimely death in 1947 eased the transition away from social medicine in favour of less intrusive, curative medicine. One can clearly discern a shift in tone from his successors and superiors in the years thereafter. The director of provincial medical and health services, for example, in 1949 revealed a touch of cynicism in explaining his own disappointment with Landau's approach: "One got the impression that the local Bantu were saying something to the effect that – 'These White people talk about Baby feeding, Privies and Sickness. They give injections which they say prevent sickness. They write about peoples' private affairs in books and on cards, but they don't give any medicines. These Doctors cannot be much good'" (Natal 1949, 2). Unstated, but obviously understood by his audience, was that it was critical not just to perform curative rituals but also not to alienate male African allies of the LHC by speaking directly to their "minors" (wives) about family secrets.[58]

Landau's immediate successor as medical officer of health was R.P. Seymour, who, over a decade in office, played a key role in the local demise of social medicine. Seymour unquestionably understood the close relationship between ill health and political economy and that primary health care was a cost-effective way to ameliorate the health burden of capitalism as practised in South Africa. The scourge of tuberculosis, for example, drew from him this sarcastic comment while he was making his case for urgent government intervention: "Riding as they are on the crest of a gloriously profitable era, it is unlikely that either Commerce or Industry will take time to see the danger which this disease threatens their business interests" (LHC [1943–57] 1951, 9–10). And yet Seymour was obviously uncomfortable with his predecessor's emphasis on social medicine, which in 1953 he called "fashionable at the time."[59] It may have been fine in theory, he explained, but in practice was not workable. Among other things, a huge amount of staff time was consumed filling out forms rather than delivering what the people of Edendale could clearly recognize and appreciate as effective medical interventions. They wanted pills and injections, not humiliating questions about taboo topics.

The timing of Seymour's declaration against social medicine is of interest in that it followed closely upon not only the departure of Dr George Gale from the Union Department of Public Health and consequent collapse of support for the model at that level, but also a pharmacological fix that seemed, dramatically, to fill the gap. As Seymour wrote in his report for 1953: "It is with an immense feeling of gratification that it is possible to record for the first time a ray of hope in the treatment of Tuberculosis amongst the non-Europeans … The drugs available to the Tuberculosis clinician today justify the feeling that we are at the beginning of the end of the war against Tuberculosis" (LHC [1943–57]).

Compromises with respect to the scientific or social medicine ideals also came in tandem with some major technical successes that incrementally built support for the technocratic approach. First, the LHC secured piped water from Henley Dam, purchased from Maritzburg at market value with certain supply restrictions. Second, a mass immunization drive resulted in a "spectacular" drop in cases of enteric fever, from sixty-six deaths in 1944 to a single death in 1948 (1948). And third, the LHC purchased Ashdown, an enclave of white farmland between the PHA and Zwartkop. Here it began construction on its first, flagship native location, employing local men in its construction. The first phase, completed in 1946, then enabled the LHC to fulfil one of its central promises to the province, city, and the *amakholwa* of Edendale. In 1947, it razed the notorious slum Schoonplaas. Inhabitants were moved into homes that compared in size (425 square feet) and quality to those of Sobantu ("the best in Natal," according to the newly formed Ashdown PHA advisory board).[60]

Health statistics in the years that followed confirmed that sense of cautious optimism. The medical officer of health report on Ashdown in 1953, for example, showed the death rate at 16.98 people per 1,000 as opposed to 60.38 in the rest of the PHA; the difference in infant mortality was similarly dramatic, at 82.35 versus 266.75.[61] Over the whole district, mortality from pulmonary TB fell from over 6 per 1,000 to under 1 for all population groups (University of Natal 1951, 38; Dyer 2012, 234). Falling rates of mortality from infectious diseases did not obviate the critique of poverty, ignorance, and migrant labour, but they did support the view that improving the infrastructure was the most cost-effective way to improve health (University of Natal 1951, 238). They also tended to confirm the wisdom of the province's decision to concentrate its health budget on hospital care. An out-patient extension of the city's main hospital had been opened in 1950 on the road out of town towards Edendale. Injections and stitches could be administered there, and the site proved so popular that soon beds were spilling out into tents in the yard. The following year construction started on the Edendale

5.3 The Edendale hospital at dawn

Hospital, which, when it opened in 1954, was even more powerfully symbolic of the shift from primary to therapeutic medicine. Situated at the entrance to the PHA along the main highway from town, the new hospital towered as a monument to the benevolent state. Indeed, it could be argued that this institution ultimately contributed more to winning the tacit support of the women of Edendale for the LHC than any previous initiatives. In addition to the obvious health benefits of a huge hospital in such close proximity to the area's greatest health needs, it created immediate job opportunities for roughly three hundred black women. Over the next two decades, nearly two thousand African and Indian nurses and midwives received specialized training there. Use of the facility burgeoned, along with the expansion of pioneering services – the first intensive care unit for non-whites in the country, for example, and a highly successful transition from majority white to almost totally black nursing staff, including the first non-white senior matrons in South Africa (Cosnett 1975, 1490). The hospital was not owned or administered by the LHC, but it almost certainly helped to engender a relatively wide base of people willing and able to listen to the advice of the commission's professional health care workers, nutritionists, erosion control advisers, and engineers.[62]

Much of the good will so earned would soon be squandered through the Group Areas Act and other apartheid injustices. A mass resignation of non-white staff at the hospital in 1957 over racially discriminatory wages, and subsequent administrative encouragement of racial exclusivity, left a pall of bitterness that soured relations between African and Indian nurses (Marks 1994, 174). It is nonetheless suggestive, and was certainly noticed at the time, that Africans in Edendale did not lash out at their Indian neighbours when anti-Indian violence erupted in Durban and Sobantu in January 1949. Ten years later, the women of Edendale also did not join their sisters in Sobantu when the women there rioted to protest against rents and living conditions. Nor did Edendale experience anything like the gang violence described in other freehold areas like Alex (Bonner and Nieftagodien 2008, 115–23). From the barely contained civil conflict of the late 1930s and early 1940s, Edendale by the late 1950s appears to have been "meliorated" to a place of relative social and political calm.

≈

African women's resistance to, and evasion of, the state were important factors in frustrating the early LHC vision for primary health care. As I hope to show in the next chapter, we are probably safe to say that that vision was unlikely to have been attained in any case, given the wider context of South African politics and the economic constraints that the province and the city of Pietermaritzburg were under. However, the fact that the LHC could not rely on African women to consistently support projects that it regarded as primarily in their best interests made for slow progress. Women's fears about the intrusiveness of the state into their intimate lives and livelihoods made them susceptible to exploitation or manipulation by African men who opposed the state for this and but also many other reasons. African women's bodies, for these men, were at least partially a means to an end and an evocative symbol for political mobilization.

The LHC meeting of 18 July 1944 in Pietermaritzburg brings our attention to this neglected aspect of the history of political struggle, and to women and gender as factors that complicated the modernist, developmental project in the immediate pre-apartheid era and that have echoes down to the present. I do not mean to suggest that a direct line connects the prickliness of Champion then to that of Mbeki in the first decade of the twenty-first century around perceived slights to African men's dignity by prurient and/or controlling whites. I simply want to emphasize the historical depth of that prickliness, including, on the one side, the preference for secrecy among African leaders about connections between gender, sexuality, and health and, on the other, the evident naivety of white liberals on the issue. Whites' paternalism and/or racism were thus not the only factors in play. A marked tendency towards the suppression of evidence considered to be shameful as well as the opportunism and the condescension towards African women by some of the leading African men in the affected community were also significant elements constraining Natal's brief social medicine era. These made it politically easier, even if ultimately less effective, for the state to opt for expensive but dramatic performances of therapeutic medicine. In the short run, at least, these appeared to be effective for the achievement of both health and political goals. In the next chapter, however, I consider the longer-term failure of the model, which set the stage for intensely violent conflict in the 1980s and early 1990s.

# SIX | *KwaPoyinandi* and the Racialization of Space to the 1980s

The word "experiment" was first applied to the settlement on Edendale farm by Bishop J.W. Colenso following his visit there in 1854 (1855, 51). Colenso was referring to freehold land tenure underpinning the newly established community of black Christians, the *amakholwa*. The famously liberal Colenso regarded the experiment as a success. Indeed, engaged in market-oriented peasant agriculture and artisanal crafts under the tutorship of the Wesleyan mission, the *amakholwa* soon came to be widely regarded as a vindication of the wisdom of British imperial rule. They not only freed themselves from debts to the mission but established other communities on farms purchased throughout Natal (Meintjes 1988b, Etherington 1978). To Sir Garnett Wolsley, Edendale "was the finest object lesson in Christian Missions he had even seen" (cited in Msimang, n.d., chap. 4), and to Theophilus Shepstone, "the most healthily advanced" of all the African communities in Natal (Cape Colony 1883, 2:68). As a sign of their own self-confidence after their much-lauded contribution to the Anglo-Zulu war, *amakholwa* leaders in 1880 drew up one of South Africa's first non-racial "constitutions." The *Imiteto yonze ifunekayo* ("strict laws required for the station") called for an autonomous, self-financing, and (for the times) democratically accountable municipal government with powers to enforce environmental regulations. In beautiful handwritten script, the bilingual document includes one of the first

African-authored articulations of democratic local governance in the whole region. In English it reads:

1. The chair of the Station is to be respected
2. To reject the elders of the station is looked upon as a case of guilt
3. The rules of the station to be observed
4. No man is free to do as he pleases
5. The streets to be kept in order
6. The law is law to all people
7. The judge has power over those who brack [*sic*] the law to punish them
8. A man to be chosen by the people to inspect all matters of the Station
9. And a Station purse is required therefore every man according to his land value must pay a penny in the pound a year, and that money shall be to pay all costs[1]

The *Imiteto* of 1880 were referred to the attorney general for discussion of their legal merits and, notwithstanding the public backing of Shepstone himself (Cape Colony 1883, 2:66), they were found wanting. Item 6 in particular was simply not feasible under the coalescing system of indirect rule. Edendale's population already included a significant number of migrants who fell under the legal authority of their chiefs in Zwartkop, Table Mountain, or other reserves (hence the Natal Native Code held sway over them). There was also a small number of whites in Edendale, who were expected to grow to become the majority through the normal play of the real estate market – "the natives will, for the most part, go" as Shepstone foresaw it (57). A large number of white settlers elsewhere in the colony would also have bristled at *imiteto* 1, 2, 7, and 8. In the years following this rejection, those settlers grew steadily more powerful. Increasingly repressive laws and racialist policies of the Natal state, plus the cumulating impacts of the "colonial ecological revolution" and epizootics, further undermined the *amakholwa* economy and community integrity, and sped the alienation of land to non-Africans. By the mid-1920s, white and Indian farmers claimed to own a third or more of the original farm area outside the commonage. White farmers in the valley expressed confidence that the area would evolve inexorably in their favour in the name of improvement.[2]

In the 1940s, a second experiment at Edendale aimed to deal with the environmental and health crises discussed in the previous chapter. This time, the term referred to an innovative form of local authority whose mandate, shaped by the concept of social medicine, was to rehabilitate the community with the promise of eventual self-government on non-racialist lines. Through the late 1950s, many impressive developments were achieved under *kwaPoyinandi*, as the Local Health Commission (LHC) was known to isiZulu-speakers.[3] These included model housing tracts, tarred roads and bridges, a protected water supply, a subsidized milk program for children, environmental protection/rehabilitation projects, and an aggressive primary health care program, among others (Caesar 2015). Together with a huge new hospital separately administered by the province, widespread vaccinations, and the introduction of effective medicine for tuberculosis, the late 1950s witnessed a dramatic reduction of infectious diseases and infant mortality. By this time there had also been progress on the development of a responsible, non-racial advisory board, with some noteworthy successes. Although no woman of any race was ever appointed to the executive of the LHC, several African women served on the advisory board of the Edendale and District Public Health Area (EDPHA) over the years, beginning with Lydia Msimang and Mrs E.J. Mpanza in 1954.[4] In that way, Edendale for a short time bucked the national apartheid trend. Indeed, notwithstanding a freeze on the sale of land to non-Africans and the looming threat of the Group Areas Act, the heterogeneity of the population continued to increase through the 1950s. Europeans reached a peak population of just over 400 in 1957, along with nearly 5,500 Indians and 550 coloureds scattered throughout variously mixed neighbourhoods alongside roughly 15,000 Africans; thus, over a third of the total enumerated population was non-African (figures are from Seymour 1958, referring to the 1957 census).[5]

The medical officer of health did not specifically mention such diversity as a manifestation of the community's social health. He did, however, note that people of all racial groups paid rates, adhered to building standards, and paid fines for environmental offences. He boasted of the LHC's achievements and sought to defend the experiment against the logic of separate development:

> There can be no doubt as to the success of the experiment, or that this is no longer an experiment so far as the Commission is

> concerned. The system of control has worked and is working with remarkable smoothness. The Commission, like all public bodies, does have its critics but there is no doubt that the bulk of the population is well satisfied with the control imposed. If the area is not again to become a health and social menace to itself and its neighbours, it is essential that everything possible must be done to maintain the high standard which the Commission has set. (Seymour 1958, 32)

In retrospect we can see that these claims went too far. Later that very year, cases of child starvation (*kwashiorkor*) reported in the international press brought sharp recriminations locally (*Natal Witness*, 15 and 26 November 1958, 30 May and 11 December 1959). The promises of political advance, non-racialism, and social medicine had already stalled by 1958, and the writing was on the wall for their definitive closure. The non-African population began to decline that year, and precipitously so after 1963, when the city embarked on construction of its giant, racially zoned Imbali and Northdale/Raisethorpe complexes. Pietermaritzburg began to shift (some would have said "dump") Africans from other parts of the city into Greater Edendale while pressuring the Indian and coloured populations to move out to their new townships on the other side of the city. The deterioration of the environment and living conditions in broad sections of co-terminus parts of Zwartkop (today's Vulindlela) had meanwhile become so "alarming" by the early 1970s that the city of Pietermaritzburg offered to rebuild the whole area using modern, urban township engineering standards. Edendale and environs under that plan would have been swallowed up in "Greater Imbali."[6] While that never happened, the LHC definitively conceded defeat soon after by transferring all its assets to the Native Affairs Department's assertively pro-apartheid successor, the Department of Bantu Administration and Development (BAD). With no consultation with the people of Edendale or even colleagues in the Maritzburg administration, it announced the liquidation of more than three decades of work and relationships on a Thursday to take effect that Sunday.[7] It was widely feared that this presaged the transfer of jurisdiction to the KwaZulu Territorial Assembly (KTA, the planned Zulu "Bantustan"). This also never did take place. But the course was set for a calamitous deterioration through the 1980s into political frustration, a virtual civil war, and the near-total collapse of service delivery and environmental protections.

Why did the second experiment at Edendale fail? The easy answer is "1948" – that is, the victory of the National Party at the national level and the introduction of formal, institutionalized racism – apartheid. Progressively intense pressure from Pretoria on local authorities to conform to its vision of separate development reduced to virtually zero the political space available for promoting social medicine, multiracial communities, and democratic local governance. The material guts needed to sustain an experiment in social medicine were meanwhile eviscerated by economic malaise driven largely by Pretoria's obsessions. Others have suggested that Pretoria offered a cover for local pro-apartheid sentiments or cultural blindspots. Christopher Merrett (2009), for example, argues that Pietermaritzburg citizens and officials often hid behind national legislation to enjoy the short-term benefits of racial zoning while maintaining their self-image of liberal opposition to apartheid. This builds on the arguments of Maynard Swanson, who hardly makes even that distinction in characterizing the LHC as simply another in a long line of "paternalistic and exploitative controls." Its demise was rooted in its "inexorable drive" to limit the development and democratic aspirations of westernized African leaders (Swanson 1996, 296–7). The critique has been extended recently by Anne Digby (2012, 187), when she claims the social medicine experiment nationally had "*already* failed," due largely to bureaucratic and jurisdictional wrangling, even before the National Party began to dismantle it, and by Bill Freund (2012, 175), who dismisses the idealism of the social medicine moment for being imbued by the "unctuous, Panglossian belief in modernity and progress."

Yet "1948," "Pretoria," and "apartheid" do not capture the slowness of Edendale's suffocation. There were still over a hundred whites and nearly three thousand Indians living there in 1970 (Thorrington-Smith et al. 1973, 26), while enrolment at the multiracial Federal Theological Seminary (Fedsem) continued to rise, peaking in the mid-1980s (Denis and Duncan 2011, 144).[8] Revealingly, twenty years after the end of apartheid, former residents had to correct their interviewers' assumptions about the dates. As Muntakabongi Hlophe, a prominent African businessman, told his interviewer when erroneously prompted about racial zoning: "The 60's? No it was in the 80's. It was in the late '70's and early 80's before my father passed away. With the advent of apartheid, Indians were being moved out of Edendale and lots of businesses and buildings were left standing empty. Besides making business, he became involved in property syndicating. He bought a lot of properties from different people … 'If they come selling, I'll buy.' So he started buying. Most of the property he bought in Edendale was bought in the late 70's" (SC, interview by Mabongi Mtshali, Pietermaritzburg 23 December 2009).[9] Prominent Indian property owners still resided in remnants of their historic enclaves until the 1990s. They finally left not because of racial zoning but to flee criminal violence (Diesel 2007 30).

Nor does "inexorable" capture the agency of the people of Edendale *and* Pietermaritzburg, and of many officials of the LHC and Natal government who worked to frustrate Pretoria's illogical and cruel pronouncements, however ineffective they may have been in the long run when seen in hindsight. Indeed, far from obstructing the more radical articulations of social medicine, as Digby suggests, provincial administrators in Natal's case applied intense pressure upon Pietermaritzburg to support the model, at least in the beginning. The use of "1948" as an explanation also hints that, if the National Party had not won the elections that year, perhaps the experiment in Edendale might have survived to be emulated in other parts of the country, so to have sped the normalization of South African cities in line with the vision of the Liberal Party of South Africa. It is a wistful thought, and perhaps not very helpful all these decades later. Nonetheless, it is a fact that more Africans in Natal than anywhere else in the country shared that liberal vision well into the 1970s, not least of all the elder statesman of Edendale, H. Selby Msimang.

Given that a third experiment at Edendale is fitfully underway (GEDI, the Greater Edendale Development Initiative; GEVDI when we include neighbouring Vulindlela), the role of local actors and factors in the earlier development scheme bears reconsidering. The original GEDI aimed to do this through a commitment to public history, and much of the work of the Sinomlando Oral History Project for the GEDI promised sophisticated, and gender-sensitive, perspectives. What may actually be in store with the state's interest in local history for development, however, is captured in Jeff Radebe's description of Edendale as "a seat of power of one of the most powerful Kingdoms in the world – the Zulu Kingdom" and "a birthplace of heroes and a production line for revolutionaries" (Radebe 2009). My worry, also abundantly raised by the essays in Carton, Laband, and Sithole (2009), is that the type of narrative offered by Radebe is not just ludicrously hyperbolic but also erases the fluidity and contested nature of racialized or ethnic identities, the pluralism of Edendale's population even through the years of apartheid, and the complexity of the apartheid state. Oversimplying or grossly exaggerating may not be as development-friendly as politicians seem to assume.

My focus in this chapter will be on the years of transition from a flexible, patronage-based system that was able to stave off crisis with a mix of concrete achievements and vague but plausible promises of liberal-minded, "fiddling welfarism" (as Bozzoli 2004, 53–53 described an analogous experiment in Alexandra), to a rigidly bureaucratized system whose ideological dogmatism, along with venal and arrogant police, profoundly alienated the local population. Why did the promise of a self-governing, multiracial experiment fail or, to put the question somewhat more provocatively, how exactly did Edendale "become black"? There are two aspects to this: the cultural process by which today's black Africans came to see themselves and Edendale in those terms, and the political process whereby decisions were made that incrementally sabotaged the non-racialist promise. Let me start with the cultural question. I do so because, first, it is so counter-intuitive to many young people today who have been raised on patriotic narratives and reified racial categories (Maré 2014), and, second, because the cultural shift helped to ease the political transition to formalizing the race-based urban geography.

## BECOMING BLACK

Probably 98 per cent of the current population of Greater Edendale would identify as black or African, and a large majority of those as Zulu. It would probably surprise them to learn that the founders of Edendale and, subsequently, large minorities within the valley explicitly rejected the contemporary equivalents of those identities. One of the greatest grievances of the *amakholwa* leaders in the early period was precisely that the state treated them as if they were "Native," against their eloquently stated wishes. "Johannes" spoke passionately to the issue about a decade after the founding of Georgetown. Apparently delivered in English, his speech to other community leaders was reported with admiration in the colonial press:

> Who has ever heard in Natal the cry of "kill the witch?" Who has had his cattle slaughtered because his chief was hungry since we came under the shadow of Victoria's wing? Yet the law is bad. Let me put a case. I quarrel with a black man. I go to law. What do I find? An *Induna*. I speak my case to him. His answer is a stroke of the chin. I am told to return tomorrow, and this goes on for a month. Have I not left my plough? Have I not walked ten miles

a-day for a month? Who pays me for my time lost, my money spent, and my heart hardened? Let us represent this evil to our superiors. Let us tell them we have left the black race, and belong to them. Will they send us back to barbarity? They may send our bodies, but our spirits they cannot send. Our spirits belong to civilization, though our bodies are the colour of the earth. ("Meeting of Civilised Native at the Edendale Mission Station," *Natal Witness*, 27 March 1863)

Racial, tribal, and ethnic identities remained in flux as the demographic make-up of the community changed and as people attached different meanings to different markers. Skin colour was obviously one such marker, and few dispute that those with power in Natal could rarely see beyond it. Indeed, the cruelties, prejudices, and stupidities that flowed from cumulatively entrenched racialist ideologies over the course of the nineteenth century are abundantly evident in the various commissions of enquiry about "Kafirs" and "Natives" and in much of the literature discussed in chapter 2, and these sources have been ably analyzed by scholars such as Saul Dubow (1989 and 2000). But even during apartheid, which imposed formal bureaucratic definitions based on skin colour, possibilities existed to "pass" and even to legally change assigned racial category. In a place like Edendale, where people of diverse hues and ethnic backgrounds lived relatively closely together, that area of liminality was as great as anywhere else in the country. Skin colour was in fact sometimes seen as less significant than language, accent, ancestry, faith, marriage and funeral practices, and so forth, as a marker of difference. Without in any way questioning the pain caused by imposed racial identities, it is important to remember this history of liminality, ambiguity, and debate over how people set themselves and each other apart.[10]

The people of Edendale from the beginning saw themselves as a distinct, new, non-tribal people who deserved, and expected, equal treatment in the law and society with whites. In this, the *amakholwa* or *oNonhlevu* had to fight against a powerful, simplifying colonial narrative. Theophilus Shepstone, notably, who was usually a stickler for exactitude on tribal identities, continued to refer to the people of Edendale not as *amakholwa* but as "Amaswazies" as late as 1881 (Cape Colony 1883, 66). Yet the founders of Edendale commonly identified themselves by diverse ancestral affiliations: "your Petitioners are of different tribes and consist of Swazies, Basutos, Griquas, and Zulus," as one letter the secretary of native affairs explained in 1902.[11] Leading citizen

Mark S. Radebe, for another example, identified himself as Zulu in 1904 but of "Sesutu extraction" (South African Native Affairs Commission 1905, 521). In both cases, his evident modernity of style and fluency in English demonstrated that the cultural content of those connections was moribund. In his testimony to the Native Affairs Commission he seemed perplexed, and perhaps irritated, that his white interviewers thought he might understand "the Natives" or "the heathens" by virtue of his skin colour.

That diversity gradually faded after the exodus of many of the Afrikaans-speaking Griqua to a new colony carved out for them in East Griqualand in the 1870s, and as isiZulu became the lingua franca. By the time of the 1917 Commission of Enquiry on Native Affairs, both Chief Stephen Mini (Swazi) and Josiah Gumede (of Ngwane/Swazi ancestry displaced to the Eastern Cape) identified themselves by "tribe" as belonging to the "Christian people at Edendale" but by "race" as Zulu.[12] Scepticism towards the larger affinity with Zuluness nonetheless remained strong. Mini and Gumede were subsequently leaders of the faction of the ANC that in 1924 drove John Dube from the presidency of the party in part on the grounds that he was getting too close to the Zulu king and all the retrograde, tribal connotations that that implied (Cope 1990). Indeed, while Gumede took pride in his "Zulu" ancestry, he also became a pioneer of the pan-ethnic identity "black" through his writings in the newspaper *Iliso Lesizwe Esimnyama* (Eye of the Black Nation) (van Diemel 2002). Early colonial terms for Africans like "Kafir" (isiZulu-ized as *kafula*), which originally had a neutral or generic connotation, meanwhile began to acquire a derogatory meaning and to pass out of polite conversation as early as the 1880s. So too did "native" became progressively more demeaning in intent. By 1923 it had acquired enough political liability to be dropped from the name of the main African nationalist political movement, the South African Native National Congress (SANNC), which became the African National Congress that year.

Magema Fuze was probably the most articulate local *kholwa* intellectual to wrestle with this identity question in the early twentieth century. He claimed that "brown" (*ntsundu*) was the most accurate descriptive term, and that other terms in common usage were misleading when applied to the general population. "Red" (*bomvu*), for example, literally meant light-skinned in isiZulu but metaphorically suggested a weak character, while "black" (*mnyama*) meant possessing gravitas or commanding respect as an individual, irrespective of physical colour. Fuze mocked isiZulu-speakers who were confused about how to name their colour or who unwittingly used terms like *kafula* imported from isiXhosa or other languages. Yet ironically, despite his stated preference for the brown identity (*abantu abantsundu*), Fuze helped give blackness greater currency by entitling his 1922 book *Abantu Abamnyama Lapa Bavela Ngakona* – Black People and Whence They Came (as translated and discussed by Mokoena 2011, 211–17). Fuze also promoted an expansive definition of Zulu to include people who would more likely have asserted an independent identity (the Hlubi, notably). Paul la Hausse de Lalouviere, (2009) has termed Fuze's perspective "cosmopolitan Zulu," a proto-nationalist ideology that sought to offset the petty, reactionary tribalism fostered under the Shepstone system. Yet for all his critique of colonial injustices, Fuze's promotion of such an expansive Zulu-ness spreading seamlessly throughout the whole of the colony actually suited the colonial project, which, as it matured in the 1920s, sowed the seeds of future ethnic chauvinism and conflict.

Becoming Zulu, African, or black (and so on), it should be stressed, was often not a choice. Identities were commonly imposed by a bureaucracy or other authority, in many cases with hurtful or repressive intentions but also with casual, unthinking prejudice. Yet to project a uniformity of intent, as well as identity, back in time seriously misrepresents the process. In many cases, becoming Zulu, black, and so on was a positive choice stemming from various motivations. Mini and Gumede seem to have chosen to become Zulu at least in part for political expedience. Did Fuze eventually opt for *abamnyama* to sell more books? For women, meanwhile, marriage to an African or Zulu man was the main route to travel from a non-African or non-Zulu identity. Love, and gendered notions of respect (*hlonipha*), could ease the transition with a rapidity that was sometimes startling, even to an African man:

> You see in an African manner, men have not built any families. Families are built through women ... In fact a very interesting observation of my wife: She got married at 19, and it was a Saturday when we got married and on Monday we were in the business working ... The first thing I noticed was that my mother was teaching her to bank and she showed her where to sign. And she didn't sign Nxumalo, but signed PJ Hlophe. What? Just after a day or two we were married, her attitude had already changed. She was no longer an Nxumalo – not that she forsook her family. I just don't know, maybe it's what God has instilled in a woman, for them it's very easy to change. (SC, Muntakabongi Hlophe, interviewed by Mabongi Mtshali, Pietermaritzburg, 23 December 2009)

At the risk of being misinterpreted, let me also note that there could even be some humour in the process – or "cultural intimacy," to use Herzfeld's (1997) term – meaning shared mockery about the absurdity of stereotypes, labels, and other naming practices. Here, I reproduce a part of an interview with Moosajee Allie, who brought the issue up twice with reference to the 1940s, to the apparent confusion or scepticism of his young, black interviewer. Allie identified himself as a descendent of freed slaves from East Africa brought to Durban in the late 1870s, which is to say one of the "Zanzibaris." These people were mostly Makua from northern Mozambique plus some Yao from Malawi and they practised Islam (Oosthuizen 1992). Several hundred were settled in Natal under the protection of the same official in charge of immigrants from India and, by virtue of their faith, they came to be socially close to the Indian Muslim community. Presumably for that reason, the apartheid regime eventually classified them as "other Asiatics" and had them removed from a mostly white working-class area to Indian or coloured areas (Sicard 1981; Clark 2004). Allie, who elsewhere in his interview was scathing in his anger at racialists and racism, had this to say about earlier attempts to make him fit into a knowable category:

> I told them [white employers] I needed a vehicle, a gas stove, pots and pans because I eat my own food as I don't eat pork, and don't eat meat bought from a white man's butchery or anywhere else except meat bought from a Muslim butchery. Mr Shuter said "You black Arab" [Laughter]. In fact, he's the one who gave me the name Moses. I'm not Moses. My real name is Moosajee. You see it's long. He said, "You black Arab, your name is too long, I'm not going to call you Moosajee, I'll call you Moses." I accepted it and the name stuck. [In my follow-up, second interview I asked Mr Allie whether he has any regrets about the way his name was changed, he said, "Oh, no, he didn't change my name, he only tried to make it short for his convenience. It was not an insult. He was joking. Jokes must be accepted."] (SC, Moses Allie, interviewed by Mabongi Mtshali, Nhlazatshe, Edendale, 29 March and 8 April 2010)

The term "Bantu" followed similarly twisting trajectories. *Bantu* literally means "human beings" in isiZulu. But it implies people from Africa speaking related languages as their mother tongue. In the early part of the century, the term was used positively by African nationalists, as in the slogan "Vukani Bantu" or the socialist-leaning paper *Abantu-Batho* (la Hausse de Lalouviere 2009, 266). It acquired an increasingly pejorative meaning early in the apartheid years. Victor Bheka Mshini Madisha, for example, explained the difference between two local football associations to a Sinomlando interviewer: "The term African was used for those who stayed in the townships. Those people were regarded as more civilized than those who stayed in rural areas. The term Bantu was mostly used for the rural people. We were using the Bantu term because we were staying here in Edendale away from the city. Those who were living in the townships such as Sobantu were using the African term. As time goes on we told the Association that we do not like the word 'Bantu.' It must be removed from our Association" (SC, interview by Nicholas Ziqubu, Edendale, 19 October 2009). For its part, the local authority in Edendale first used the term *African* as a courtesy to the majority population in 1945, formally adopting it at the advisory Bbard's request in 1957. Yet even then, some African leaders worried about moving away from the native or Bantu designation on the grounds that "the uneducated type of African would not know what was meant by the term 'African.'"[13]

Even greater confusion was to follow with the terms non-European and non-White. Today, one can sense something of a generation gap in discussions of this question in the Sinomlando interview of Sana Moodley by Lekhanooe Solomon Mathaha (SC, Northdale, 24 July 2010):

LM: What do you mean by non-Whites?
SM: Non-Whites are Indians, Coloureds and Africans, we were non-Whites.
LM: Hmm ...
SM: Yah.

And again, in an interview of Charles Donnelly:

CD: It was very free at that time very free because everybody seemed to accept everybody ... That time we used to be non-Whites yeah! But now ... but now it's Coloured, Indians, Africans, Whites.
LM: In Edendale?
CD: Ya. At that time when I knew, it was non-Whites, we were non-Whites, Coloureds, Indians and Africans ... and we were non-Whites.

LM: So you were …
CD: We lived together.
LM: You just allowed … ok you were living together.
CD: [Until the 1980s] when that fighting was coming on we were already coloured people and Indian people. (SC, interview by Lekhanooe Solomon Mathaha, Pelham, 16 May 2010)

The term *black* developed over this time as a political project to unify all people who experienced discrimination and oppression on account of their ascribed identities. The Rev. Enos Sikhakhane alluded to the shift in his interview with Dee Shirley Deane. In founding the Edendale Lay Ecumenical Centre in 1965, he seemed to be using the term synonymously with African. "Our motivation was the historic fact that Blacks had always been the followers, the dependents. We were turning this around. Of course, the Whites couldn't get over the shock. At our official opening, not one White minister from my own church showed up. I suppose that says something" (quoted in Deane 1978, 188–9). But his centre operated as a place where people of all races from around the country could stay. It developed over time to offer a wide range of community services, including workshops that broached sensitive health topics, language lessons (especially for non-Zulu to learn isiZulu), and adult education. In the 1970s, the centre created its own director of agricultural programs to provide extension learning to African farmers and gardeners in the valley, and it provided a temporary home for Fedsem, the country's premier non-racial, interdenominational educational institute. At Fedsem, which subsequently moved to a permanent campus in Imbali, from 90 to 123 students a year learned about ecumenism, multiracialism, intellectual creativity, and freedom within a Black Consciousness or black liberation theology paradigm.

Fedsem was "the most involved" of all South Africa's seminaries in seeking to apply Steve Biko's vision of liberation theology in the anti-apartheid student movement (Denis and Duncan 2011, 85). Black Consciousness and its political arm, the Azanian People's Organization (AZAPO), are only barely alluded to as having been present in Edendale – a single mention in Laband and Haswell (1988, 211), for example, and a nod to the student arm of the movement in Radebe (2009). Yet from Denis and Duncan (2011) and Gennrich, Inglis, and Kromberg (2009) we learn that the idea of black consciousness had a clear political presence in Greater Edendale that both preceded the United Democratic Front and had strong ties to the predominantly white liberal activist movement. Fedsem hosted the inaugural meeting of the Azanian Students Organization (AZASO) in 1979, held public prayers calling for the downfall of apartheid in 1980, organized liberation hymns in central Maritzburg to protest against Republic Day in 1981 (for which forty students and three lecturers were arrested), and in 1983 called for a boycott of events associated with the KTA. The seminary also had links to perhaps the most significant local successor to the Liberal Party, the Pietermaritzburg Agency for Christian Social Awareness (PACSA). PACSA was established in 1979 following long discussions among liberal, white Christians on how to respond to Biko's murder, the banning of the ecumenical Christian Institute, the intensifying repression of the apartheid state, and the evident failure of liberal politics as conventionally understood. Following the logic of Biko's philosophy, they accepted that black racial pride and exclusivity was a means to an end that would eventually erase racial distinctions in a liberated society. The primary role of white liberals in the short run was thus to make other whites aware of the horrors of the system rather than to presume that they could lead blacks to liberation. PACSA launched itself into that role with a series of fact-finding missions and newsletters that exposed the grotesque inequalities and injustices of South African society to a presumed predominantly white audience. Led by Peter and Joan Kerchoff, it comprised a small circle that extended to influence the Association for Rural Advancement, the Black Sash, and the End Conscription Campaign, among other mostly white anti-apartheid groups (Gennrich et al. 2009).

This conceptualization of race ran hard into a growing Zulu nationalist movement spearheaded by Chief Mangosuthu Buthelezi, leader of the Inkatha yeNkululeko yeSizwe, a "cultural liberation movement" that was the predecessor of the Inkatha Freedom Party (IFP). Inkatha has a bad reputation in South African historiography as a conservative or reactionary movement, taking much of the blame for perpetrating the "black-on-black" violence in Natal and around the migrant labour hostels in the Johannesburg area (South Africa 1999, notably). That conflict is beyond the scope of this study. I do, however, want to note that Inkatha's appeals to tribal identity and local racial exclusivity found support not only in the supposedly conservative rural areas like Vulindlela but also in Greater Edendale, for various reasons. An independent poll taken in 1980, for example, indicated that a clear majority of Imbali residents welcomed their proposed transfer to the Inkatha-administered KwaZulu (Butler and Harley with Aitchison 1993, 12). There was also an early progressive wing to the movement, which was subsequently marginalized. The Inkatha Women's Brigade, notably, was "at the forefront of development strategies in South Africa, introducing notions such as participatory development, women's empowerment and community development" (Hassim 1993, 2). It drew inspiration from the hugely popular Christian *manyanos* and its leadership from *amakholwa* women. Indeed, many Edendale/Imbali elites with a history of progressive politics became members of Inkatha in the mid-late 1970s, including the aforementioned Sikhakhane. The pioneering advocate of organic urban agriculture Robert Mazibuko and jazz musician/promoter Solomon Mkhize were also Inkatha members. Selby Msimang for his part joined Inkatha in the belief that it was a disciplined movement that would restrain the power of reactionary rural chiefs. It thus promised to open the way for a "bootstraps" kind of African entrepreneurship, or self-improvement within a market economy and an eventual non-racial democracy. In Msimang's words, "the government uses the chiefs as informers. However, under Inkatha, the chief is bound to the will of the community" (quoted in Deane 1978, 118).

To project a uniform Africanness, blackness, or Zuluness on Edendale identity back in time requires us to forget some of these complications and "hybridities" (Carton, Laband, and Sithole 2009). It also requires us to forget four distinct non-Zulu communities within Edendale, which have now mostly disappeared – Basotho, Indians, coloureds, and whites. The latter three, which at their peak constituted up to one-third of the population, were mostly removed or induced to move in the 1960s–1970s, with stragglers remaining until the intensification of violence in the early 1990s. The Basotho, however, are intriguing not only because many of their descendants still remain in Edendale today but also because they were at one time an explicit source of concern to the local authority. Shepstone in 1881 testified that there were "very few indeed" Basotho anywhere in Natal, while SeSotho-speakers among the original *oNonhlevu* seem quickly to have adopted a Zulu identity. A wave of immigration then came sometime between 1910 and the early 1930s, when floods in Mount Frere (present-day Matatiele) created an exodus from that district, then part of Eastern Cape.[14] The Basotho of Matatiele were relatively recent migrants from Basutoland proper, but had a long-standing reputation of intransigence or rebelliousness to modern methods of native administration. It appears they brought that reputation with them when they settled in Macibise/Schoonplaas, becoming some of the principal purveyors of alcohol and sex to the African workers of Maritzburg. Ashdown, which was opened in 1947 primarily to house people cleared from

Schoonplaas, was thus primarily Basotho by ethnicity in its early years. This created a stubborn problem of legitimacy for the government. Whereas the advisory board consisted of Christian, Zulu-identified *amakholwa* who desired to develop Ashdown as a respectable, middle-class community, the majority of its inhabitants were not so inclined, and indeed their behaviour often seemed intended to subvert the goal of respectability. As the LHC secretary put it in a confidential report on slum clearances and overcrowding in Ashdown, "it is an accepted fact that most of the Basotho families in the Location are no asset, to say the least, but a distinct liability."[15]

The above comment is noteworthy not just for its ethnic stereotyping but also, in the context of the LHC archive, for its rarity. If I may sum up the thousands of pages of *KwaPoyinandi* documents it would be: basic conformity to wider bureaucratic and cultural norms. LHC officials may have been slightly ahead of the curve in making their terminology respectful to African leaders, and they certainly did not promote racial exclusivity. But they also showed no interest in the nuances of African ethnic or other identity lines. The word *kholwa* was never employed in official documents, for example, outside rare acknowledgement of the official, archaic title of the chieftaincy inherited from NAD,[16] and references to migrants from Pondoland or elsewhere are as scarce as the admission of control issues with the Basotho. Rather, as time went on, such distinctions were almost always subsumed within the homogenizing categories of Native, African, and Bantu, with a strong assumption that the population was mostly and unproblematically Zulu.

Like Africans, "Asiatics" and "Indians" in Greater Edendale were divided by ethnicity and class, which often aligned. They were also deeply divided by religion, colour, caste, and language. The first reference to Indian settlement in the valley is to 1887, the year Harden Singh bought land on Plessislaer farm (*Natal Witness*, 3 March 1933). He and his family were among the lighter-skinned Muslims mostly from Punjab and Gujarat who came to South Africa as fare-paying passengers and tended towards the professions and business. By the 1940s, these people formed a significant middle-class colony in the Plessislaer area, owning shops and rental properties throughout the valley, as well as the Chetty and Maharaj bus companies that connected Georgetown to Pietermaritzburg. In contrast, generally darker-skinned Hindu descendants of mostly Tamil indentured labourers had earlier settled on the east side of the city and its lower fringes. They began to settle in the greater Edendale area as renters by the turn of the century. While often very poor, they were also among the most productive small farmers in the valley, mostly growing

6.1 Old and new above Dambuza: a formerly Indian farmhouse built in the Cape Dutch style sits among post-1994 neighbours

rice and fruit and generally reliant on their own extended families for labour. Farther up the Wilgefontein valley (KwaPata today), overgrown orchards among otherwise barren tracts of RDP housing are most of what remains to attest to Indians' historical presence there.

Let us be wary of the nostalgia in the following interviews conducted with Indians who grew up in those neighbourhoods. It is well to recall, too, that Indian political associations in the 1920s to the 1940s often deployed very explicitly anti-African, segregationist discourses (Bhana 1997). But let us also respect the very powerful unifying memory that shines through Alleyn Diesel's (2007) compiled oral histories of women of Indian descent now living in Pietermaritzburg. Bunny Bhoola was born in 1955:

> I think back to how we used to take our bicycles and ride all over Edendale, or take the bus and go to the movies in Edendale or in town, after seven at night … It was a very free society in which one could feel very comfortable and safe walking down the road, even

> at ten at night. We were very privileged to be brought up in such a society … Edendale was an environment that was very conducive to people growing up happily there; there was no difference between Black families and Indian families. Everyone was a child of the community, and there was an interest in each person's child. People shared the joy and the pain of life. (29)

Vidya Satgoor, who was born in 1936 and grew up in Plessislaer, echoed these sentiments, remembering her home as "a very beautiful place with green hills, and a very united, happy community in those days, with Indians and Blacks living close together in harmony, and all the children playing together. We all spoke Zulu fluently, and the doors of our house were always open as there was no fear of crime. We always moved about perfectly freely and safely over the whole district" (37). Durga Bundhoo was born in 1921, grew up in Maritzburg, and moved to Plessislaer in 1947. Like the other two women, her memories were positive: "a flourishing and vibrant Indian community lived in close and friendly contact with their Black neighbours. If anyone was in trouble, the whole community got together to help. It was 'one big family and no-one could starve there.' Nobody had cars in the early days, so everybody walked wherever they wanted to go. It was so safe that there was no need to lock one's doors" (169).

Similar memories were expressed in the following two interviews, conducted for the Sinomlando project by Lekhanooe Solomon Mathaha and Zama Mnguni, respectively:

> SM [SANA MOODLEY, BORN CA 1935]: Well my experience personally, it [Georgetown] was a beautiful place, it was a lovely place … We could go anywhere in Edendale the people respected us and wanted us.
>
> LM: Were you staying together … your neighbour could be coloured and white …
>
> SM: Oh no, well our neighbours were Africans …
>
> LM: What language were they using there?
>
> SM: Well they … they speak Tamil, they speak Hindi.
>
> LM: Then what was the medium of communication in Edendale?
>
> SM: Eh … some of them [Africans] they [were] so used [to being] in an Indian home … they learnt the Indian language yah, they learnt the Indian language because speaking … they could even

cook like our Indians like they were … so that's how I'm telling how close they were … they came into our homes and our parents showed them everything … this our life style.

LM: What about Indians could they speak IsiZulu?

SM: Yes fluent Zulu, fluent Zulu they could speak yah fluent my uncle, my father, my mother all fluent. (SC, Sana Moodley interviewed by Lekhanooe Solomon Mathaha, Northdale, 24 July 2010)

JM [JAY MAHABEER, BORN 1948]: My childhood was a very joyful one in the Edendale area. I enjoyed living in that area because we were very free; there were no rules and regulations. We walked and went anywhere with no crime. There were no fences around our house. We played; the Blacks, Coloureds and Indians children together … My neighbour was Mr Dlamini, my coloured neighbour was Mr Kusheen and my white neighbour was Mr James. Mr James was a farmer, my grandfather when he came from India he came to work in his farm … Ya I used to picture the farm. We use to pick cow dung and use it to keep our floors neat using our hands (*ukusinda*).

ZM: You were able to do all that?

JM: Ya we were taught to do all that and also to put cow dung in a tin once we are done and spread it on the yard so that the yard looks clean and neat, interesting is it?

ZM: Of course … (SC, Jay Mahabeer, interviewed by Sama Mnguni, Pietermaritzburg, 2 June 2010)

Little has been documented about the relatively small coloured community specifically in Edendale.[17] In the wider Pietermaritzburg area, we do know it was not a community at all but highly diverse groups of people who were only subsequently lumped together in this identity category. There were Griqua among the original *oNonhlevu*, these being descendants of Afrikaner men and their Khoisan or Malay slaves/servants or of later unions between Griqua families and Xhosa, Basotho, and other people on the Cape frontier. There were the descendants of John Dunn, the white trader and elephant hunter who became an important Zulu chief and major polygynist (forty-eight wives) in the mid-nineteenth century. A group of highly creolized immigrants from the island of St Helena arrived in 1873, and others from

Mauritius. They were generally regarded at the time as respectable "white" people and, indeed, were sometimes (as in East Griqualand) treated by colonial policy as "place holders" for hoped-for white settlement (Bardsley 1982). As for *whites* as later defined, they included the original missionary families, Afrikaner farmers, and, the biggest landowner and entrepreneur in Edendale, the German immigrant H.F. Kothe.

Whiteness is often assumed to be self-evident. Yet, as Robert Morrell (2001) and Christopher Merrett (2009), among others, have shown, cultural whiteness has to be learned, performed, and policed. White Natalian masculinities, for example, with all the cliquishness and arrogance for which they became notorious, were achieved and demonstrated through the military, sports like hunting and rugby, and the cultivated silencing of untoward emotions and social "misfits." This was undoubtedly the case with much of Edendale's relatively small white population, who for the most part seem to have kept aloof from their neighbours outside of teacher, employer, or other commercial relations. They had their own schools and churches, and in the early decades of the Union a vocal minority even pressed for the expulsion of all non-whites from the valley.[18] Meanwhile, whites who engineered the liberal experiment in Edendale in the 1940s and 1950s – employees of the LHC – did not actually live there. But it is worth noting that the majority of whites who did live in Edendale did not support racial exclusivity. Through the 1920s they lobbied vigorously to prevent the Land Act from applying to Edendale, which would have imperiled their investments there.[19] At key moments in the debates over the form of Edendale's government, such as the Hignett Commission, local whites often defended both the multiracial status quo and the reputation of their darker-skinned neighbours. J. Herbert Plumpton, for example, who claimed in 1933 to live only 200 yards from the reputed worst slum, wrote against incorporation to the city on the grounds that he had never suffered ill health, could only praise the good air and scenery of the Schoonplaas area, and saw rowdiness as merely a weekend phenomenon that simple policing could contain (*Natal Witness*, 29 March 1933).

In theory, interracial sexual mixing was certainly frowned upon by most people, but in practice opportunities for such mixing sprang up in the border zones between the different neighbourhoods. This did give rise to some tensions and the derogatory racial category "coolie," referring to the children of African women and Indian men. But who knows what secretive adventures and romances may have flowered there in defiance of the tides of exploitation or fear. One African reporter found "social integration" in that respect to be somewhat advanced compared to Durban, with Indian men dancing with African women to an Indian band at the Georgetown "Y."[20] The Hotel DuSalle in Plessislaer was also renowned as a place where, in the 1960s and 1970s, white, coloured, and Indian men from Pietermaritzburg could groove to African jazz with African women.

> CD: Ya ... no, no ... you ...you would never ... you would never allow black people to ... at that time black people were not allowed to go up to that to ... to that section [of the hotel], there was a section just for ... if you say black people as black males (Laughs ...) black males were not allowed to go up.
>
> SM: Women were the ones ...
>
> CD: Women would be allowed like, they would be allowed to go inside but the males ... not and it was very secret ... secret because ... [no explanation]. (SC, Charles Donnelly, interviewed by Lekhanooe Solomon Mathaha, Pelham, 16 May 2010)

The conflation of ethnic or racial identity with political affiliation is another aspect of this tangled issue. It becomes apparent when looking at the ways that isiZulu words for Indian (*amaIndiya* and *amaKula* – "coolies") changed meaning in the late 1980s. By that time most people of Indian descent had moved out of Greater Edendale, and National Party efforts to court Indian and coloured support for apartheid fuelled old resentments against "coolies" as profiteers off of African suffering. In 1988, prominent Inkatha leaders began labelling African supporters of UDF/ANC as *amaKula* on the supposed grounds that they had abandoned their presumed Zulu culture (no chiefs, among other things). "Real" Zulu were incited to cleanse the valley of such foreigners: "They said that we were Coolies," one Zulu woman reported. "They would say, 'Here is a Coolie, it smells of a Coolie here.' You would hear them saying, 'Mff, mff (sniffing sound) there is a Coolie, it smells here'" (Denis, Ntsimane, and Cannell 2010, 60).

The responses of UDF/ANC comrades also, unwittingly, drew on a long-standing ethnic stereotype: they referred to their enemies as "Russians." The informants in Denis, Ntsimane, and Cannell do not exactly understand the connection between Russia and Inkatha but it seems to be a reference to the MaRashea, the much-feared, pseudo-traditionalist Basotho criminal gang that had terrorized the townships on the Rand since at least the 1920s and may have made in-roads among Basotho in Schoonplaas as well. Their reputation for violence lingered beyond those years. The term Russians, according to one informant referring to the 1980s and 1990s, "originated because of politics. It was a name that already existed although I do not know where it came from ... The Russians are people who like to sing during the night and do Zulu dance and carry traditional weapons. They go to these Zulu parties, doing stick-fights, you see, all those things" (Mthoqotho Man, quoted in Denis, Ntsimane, and Cannell 2010, 60).

As a final point, I will observe that any attempt to build a sense of community pride based on self-evidently shared blackness will need to address the fact that only a small minority of today's residents have any historical connection to Greater Edendale extending back to before the 1960s or, more commonly, the 1980s, when civil war and the breakdown of influx controls fuelled massive in-migration. Over those troubled decades, the Edendale valley became in effect a dumping ground or refuge for dislocated people whose emotive home remained elsewhere: working-class families from over-crowded Sobantu; "tribal" refugees from violence-wracked enclaves of KwaZulu, Lesotho, or Pondoland; and evictees from so-called black spots who came from Durban and elsewhere in the country. Africans from the rest of the continent have also moved in in number since the first decade of this century, although Zimbabweans and Somalis – the main targets of xenophobic attacks in recent years – have tended to settle in the inner city and nearby places like Jika Joe settlement (Peters 2014). These people – the *amakwerekwere*, in the commonplace highly derogatory term – are only likely to increase in number as economic migrants and political refugees flee from their own troubled countries. Where will they fit in the public history of uniformly shared black or African identity?

## THE PROMISE OF SELF-GOVERNMENT

While debates, jokes, stereotypes, slanders, and laws related to identity and who could dance with whom were ongoing, changes were taking place to the political economy architecture upon which the fate of the *kwaPoyinandi* experiment would eventually hang. Let us next consider the political record of the LHC in two periods: promise (1942–53) and betrayal (1954–74). The aim is not to elevate the first to a liberal Camelot but to discern how it achieved or failed to achieve core goals in ways that attracted the loyalty of an impressive cohort of African leaders, and then, in the second period, what factors contributed to the collapse of the earlier optimism. While avoiding the kind of overgeneralized critique that lumps all "township" authorities and advisory boards together as inherently failed, I draw attention to some concerning continuities between these periods. To an even more concerning extent, some of these issues linger in present-day "growth" discourse, as I will discuss in the final chapter.

Enemies of the LHC experiment abounded in all of the different communities in Edendale and at all levels of government, and they found ample cause to focus their attacks. As discussed in the previous chapter, sometimes insensitive intrusions by the state into Africans' private lives in the name of health initially alienated key potential supporters. Overambitious approaches to social medicine, animal control, and inferior housing all had to be tempered in the face of sometimes organized opposition, sometimes blatant disregard. The aforementioned and evidently formidable Mrs Dambuza provides a case in point. As a major landowner and proprietor of a key service (a cattle dip-tank in Macibise), Dambuza had some influence in the community. It is not apparent whether she was a supporter of the Edendale Lot-Holders Association (as we have seen, the Mtimkulu faction in Edendale politics), or where she stood with the chief's faction, but she was definitely not one to alienate. When she raised her dipping fees from one penny to three, the advisory board requested that the commission build another tank to compete with or replace hers. Instead, the LHC secretary approached her to conciliate the two sides, asking her to reduce her new charge to two pence. She refused, and over the next few months she also refused even to meet with LHC officials beyond shouting at them from a distance. The secretary finally had to admit that her position was in fact "fair and reasonable" and that the LHC could do nothing.[21]

Another early infrastructure project also attracted much criticism. Ashdown was to be the centrepiece of the initiative of rehousing people from slums to quality homes that reduced ill health and promoted human dignity. The long-term benefits eventually became evident, and homes in Ashdown today remain desired properties. When Ashdown first opened, however, it had no sewers or pit toilets, hence necessitating the undignified bucket system of household sanitation. One of the LHC's most ardent opponents among the white population held up Ashdown as symbolic of all that was wrong with the new authority, fulminating in particular against its high costs and the erosion allegedly caused by cutting roads and building sites into the hillside. Meanwhile, the ELHA spoke for many property owners in denouncing what it saw as an attack on their right to earn a living from rentals. The LHC, by becoming a landlord offering "sub-economic" rents and requiring the demolition of substandard dwellings as its own properties were developed, was in effect driving less well-capitalized landlords out of business. Even the NAD intervened against LHC plans to develop "economic" housing in Ashdown, presumably to defend hard-working natives by vetoing a potential petty bourgeois enclave right up against the border of their tribal jurisdiction.[22]

As with housing and public health, political development under the LHC also proceeded in a fitful manner that attracted copious criticism. The provincial administration had accepted the idea of a single, multiracial advisory board against its original preference for three racially distinct boards. But it held firm that representation on the board and elections to it would be conducted on a mostly racial basis, with a quota of seats for each racial group. Although it allowed for two categories of Africans (tenants and landowners), quotas were set in a way that left the African majority with a minority vote on the board – initially a total of 9 out of 22 seats, later 9 out of 18. This approach left the board wide open to attack from those who saw Africans as window dressing for white domination. Sometimes violent rhetoric against the board, and repeated boycotts by the Indian community (which feared its permanent minority vote spelled trouble), then made people reluctant to serve on the board. The LHC described that reluctance as "apathy," a constant frustration that meant it frequently had to co-opt members for acclamation without any rival candidates. It became increasingly difficult to find enough candidates even to run uncontested. By September 1959, with ten seats to fill, only two nominations were received.[23]

There is no doubt that the founders of the LHC had little appetite or inclination to proceed too quickly on the political front. Sir Edward N. Thornton, the intellectual founder of the LHC, himself had indicated a strong aversion to non-racial governance in his hearings for the peri-urban committee in Maritzburg in 1939 (*Natal Witness*, 15 June 1939), while one of the province's chief promoters of the PHA concept was a well-known segregationist (and frank white supremacist), Senator George Heaton Nicholls.[24] No hint of mentoring, let alone actively promoting Africans in self-government, exists in the original (confidential) terms of reference for the LHC superintendents' duties. And the first superintendent (J.C. Boshoff) did not consider it necessary to include Africans among the various local experts and community members he thought were important to consult prior to the proclamation of the PHA (as was required by law). The first minuted meeting between the LHC and any African leaders came three months after the fait accompli.[25] At a subsequent meeting, Boshoff informed them of the nature of their important task: "guide the people wisely and cautiously" – that is, support the LHC in population management and control, rather than represent the will of the people to the commission.[26] A decided lack of enthusiasm towards political development may also be surmised from Henry Lugg's memoirs (1970, 125), in which his five years of work for the LHC do not merit a single comment to differentiate them from his long career with the NAD. Even Thomas Wadley expressed strong scepticism about the commission's political potential at the time of its establishment. At a special meeting requested by the new advisory board in 1944, he accepted the importance of training Africans in government "as long as it doesn't consume too much time."[27] And indeed Wadley was sometimes terse in his impatience with the discrepancy between Africans' political demands and their perceived lack of readiness or temperament for modern governance.[28]

That said, the accusation that the LHC regarded its advisory board as a lapdog underestimates the opportunities it presented to Africans to articulate sometimes quite strong political critiques. In April 1945, for example, B. Nomvete and Selby Msimang tabled their "Report on Social and Economic Conditions at Edendale and District." They explicitly identified exploitative wages as the root cause of the deplorable health situation. They called for a minimum wage, collective farming with a marketing board, and government support for local manufacturing industry. "It is of no use at all for the Commission to preach self-help through better ways of family upkeep, cleanliness and balanced diet; to people so oppressed, so underpaid, so ignorant and so overworked as the people of Edendale." They further called for translation of all directives from English into the main spoken languages of the district and for ward committees to facilitate communication between the people and the advisory board. The latter would be directly elected without property or educational qualifications: "as long as a man is sensible and has balanced views and a thorough knowledge of his locality he should be eligible for election."[29] In the same month, Mr F. Mazibuko tabled a motion to approach the provincial administration to create two new members of the LHC executive with the title "assessor." An African and an Indian would be appointed to this role in the same manner as the white officials.[30]

6.2 Ashdown today, seen from "Peace Valley." The LHC's original township (1947) has stood the test of time, while in the foreground the wattle and daub homes on the flood plain probably date from the 1980s

The advisory board directly asserted its ambition for political autonomy in the next meeting by politely insisting that Commissioner Boshoff step down as chair in favour of the board's own elected president. Boshoff would in future attend board meetings to present items for discussion and provide information as requested, rather than to run them. Boshoff seemed deeply offended at this request, and he defended himself with the claim that the LHC adhered to the board's advice in 95 per cent of cases.[31] Yet he acceded, and over the next decade the board elected its own chairmen from among all the racial groups represented on it. This included F.J. Mazibuko, who, as deputy in 1948, may have been the first African to chair a multiracial jurisdictional entity in Natal.[32] The LHC further accepted the board's request to create subcommittees for the different departments as they emerged – including health, engineering, forests, and finance. The intention was that these would be places for Africans to observe and learn the modern art of urban governance but that, eventually, as committee members acquired expertise, they would take on substantive oversight responsibilities. According to Msimang, Wadley actually agreed to the latter in 1950 in the case of the financial sub-committee, which would have been headed by Msimang. That move was vetoed by Wadley's bosses in the provincial administration.[33] Wadley none-theless continued to press the issue. In its annual report for 1953, the LHC went on public record to advocate greater independence for the advisory board and speedier replacement of European by non-European technical staff (LHC [1943–57] 1953, 55).

The debate on the political efficacy of advisory boards in the early apartheid years has not been settled. One extreme of the spectrum, which was the dominant view from the ANC and other liberation movements at the time and subsequently in the struggle historiography, tends to dismiss them as "Useless Boy's Clubs," stooges, gulls, and, in Nuttall's rather loaded terms, "collaborationists" or "evolutionists" (1984, passim). Other historians have been a bit kinder. Baines (1994), for example, found the New Brighton (Port Elizabeth) advisory board was able to advance a leftist nationalist discourse despite the intentions of the local authority. Bozzoli also identified an early left- or liberal-leaning tendency in the Alexandra Health Committee, which only gradually gave way to "power hungry mad puppets" by the 1970s (2004, 57). Goodhew (2004) recasts the aspirations to social respectability by Sophiatown's advisory board members as a form of resistance to racism (tenants were thought to be more easily co-opted by the state than owners were), at least until the mid-1950s. Most pertinent for this study, Mary Caesar found that the Edendale and Ashdown advisory boards, at least through the 1950s, were constantly "testing the (liberal) boundaries of the LHC" (2015, 181). In a context where there were almost no other legitimate structures through which Africans could express political opinions, standing up for seemingly mundane local issues reflected considerable integrity and tenacity.

My reading of the LHC documents inclines me to this latter interpretation, with the advisory board functioning as a relatively effective loyal opposition that not only advised (usually some form of "go slow") but also queried commission hiring and spending decisions, pro-actively proposed new projects, and protested in sometimes sharp language when it perceived discrimination or hardship arising from too-dogmatic an application of policy. Msimang's first offer to resign from the advisory board is telling in that respect. This was in 1946, after two years of service, not in protest over any of the controversies up to that point but so that he could advocate for the LHC model in the Native Representative Council, the body that advised national government.

"My purpose is to get our national movement to recognise the need for an institution such as the L-H-C in order that pressure may be brought to bear on the N-A-D from outside."[34]

Movement towards non-racial governance was of course anathema to the National Party, which came to power nationally in 1948. LHC officials do not appear to have taken Pretoria's racialist ideology as a serious concern and, like many in Natal, probably regarded the party's first win as a fluke.[35] The LHC did eventually begin to talk to Maritzburg officials about coordinating their approaches to conform to the Group Areas Act (GAA), as mandated by law in 1950. However, no particular sense of urgency or desire to change the status quo imbued the process. In 1952, for example, Wadley informed the advisory boards of Edendale, Ashdown, and Clermont that LHC policy was to "preserve the present situation" and to "recommend no change" to settlement patterns.[36] Consultations dragged on for a decade, during which time the LHC gradually accepted that change would have to happen while still clinging to its belief that the Edendale PHA could be organized into group areas without either coercion or even stating racial zoning as a goal. That is, areas that historically were predominantly Indian or white could be preserved within the wider African-majority area simply by more clearly demarcating their "natural borders" (streams and steep hillsides, which needed protection in any case). Racially mixed areas could be subtly encouraged to become exclusive to the dominant group through the existing building-permit system over a period of twenty years. The first provisional draft town-planning scheme, presented in 1955, for example, set aside about 5 per cent of the total area of the PHA as "specified." This meant that non-African inhabitants, while encouraged to move to new housing being built elsewhere in town, would be protected from eviction. One of a series of planning maps shows an enclave of Plessislaer preserved for whites and the western part of Slangspruit and Wilgefontein/Mount Partridge for Indians. As late as 1958, the LHC continued to give permission for Indians to build in the latter (present day KwaPata), while in private correspondence the secretary maintained that it opposed any restriction on residential qualification. His job was to control unhealthy buildings, not occupiers of those buildings: the LHC "felt that the black man of Edendale was comparable with the white man of Pietermaritzburg and he should, therefore, have no interference with his liberty."[37] The engineer for EDPHA (S. Newmark) the following year also privately advocated that the LHC "wash its hands" of the GAA planning process with respect to Wilgefontein/Mount Partridge. Given the unsuitability of its topography and soil for a dense African township, he argued, that specific section should be allowed to develop "naturally," meaning to continue on its present trajectory as a semi-rural Indian community.[38]

The LHC approach to the chieftaincy is also suggestive of its relatively liberal attitudes. Long-standing chief Stephen Mini died in 1939. Despite his son's announced distrust of elections, Edgar Mini had accepted the requirement to run for office rather than simply inherit the position, even though he felt that the latter was more appropriate for an African chief. On two occasions he duly defeated the NAD's and LHC's preferred candidate (Trustee Oliver Msimang), only to have the NAD and LHC reject the results on technicalities. After a third electoral victory in 1943, the LHC begrudgingly accepted Mini as chief in his deceased father's place.[39] Many of the commission's key African allies nonetheless repeatedly pressed for the complete abolition of the chieftaincy, which was, in one submission from the advisory board, characterized as a "parasite and a beggar," an "incubus" "hampering the progress of an industrialised and virtually detribalised community."[40] Wadley agreed with the critique and allowed it to be publicly aired (University of Natal 1951, 27, for example). Out of fear of inciting the kinds of clashes witnessed in the 1930s, however, and in line with NAD preferences, he abjured from pressing for abolition.

The most immediate external threat to the progressive vision of the LHC through the 1950s was not from a retrograde chief or even Pretoria, but rather from Maritzburg. The record of negotiations with the city shines a light on a sometimes difficult relationship where scientific (health) and political (human decency) principles could founder over strikingly petty financial quibbling. Wadley, a Durbanite, may have brought some cynicism towards the city when he first took up the commission and had to employ strong-arm methods to get the city to contribute to the project. Negotiations around council parochialism over the following years could be terse. Wadley sounded understandably sharp, for example, when the city disputed a charge of less than six shillings per day for tuberculosis medicine for a man who lived in Maritzburg five days of the week (but whose bicycle licence said he belonged to the LHC).[41] Another early point of contention came in 1946. The Natal administration had wartime assets to dispose of, including its military hospital in Oribi, not far from the border with Slangspruit – that is, proximate to the southeast side of the Edendale District PHA. It offered to give the property, valued at one million pounds, for free to the LHC to use as a tuberculosis hospital, a truly urgent need. But the city balked at the idea of an enclave of sick Africans with a stream of visiting relatives within its borders and it effectively vetoed the deal by inserting a racial clause on the allowed uses of the property. It was instead transferred to the city, which used it first as a hostel for single African women (mostly domestic servants) and then as a housing estate for working-class whites.[42]

Perhaps the most substantive issue over the whole period arose from the city's monopoly over the supply of potable water. Council finally agreed to sell its water in 1944, after two years of discussions, and it placed tight restrictions on amounts in order to protect its own supply against drought or other eventualities. On those grounds, it refused in 1951 to increase its sales to the LHC to meet growing demand. In 1952, against LHC protests, council tripled the rates it charged to supply the PHA. When the LHC proposed to build its own water treatment plant in 1953 on the east side of town, the city sought to obstruct it, and succeeded in doing so by suddenly offering to increase its supply at rates the market could bear. By 1957 this meant that the LHC was paying two shillings ninepence per 1,000 gallons compared to Durban's charges for Pinetown of 1 shilling sixpence, and two shillings plus a metered amount for Maritzburg residents. The LHC estimated that Edendale was paying double what a 10 per cent annual profit for the city would allow, but this protest too fell on deaf ears.[43] No wonder the Maritzburg City Engineer later defended its water sales to the LHC as "based on sound economics" (Pietermaritzburg [1904–90] 1961, 44).

The city had no direct jurisdiction over the LHC, but it had invested a large sum to get the LHC off the ground and it stood to lose these funds if the experiment failed. On that basis, and notwithstanding its ongoing miserliness, it often felt it had a right to make polite requests to the LHC about how to run the show. A significant theme running through LHC correspondence in the early years was polite rejection of those requests. Probably the most important of these rejections came in October 1950, when the native commissioner for the city approached Wadley on behalf of town council. Mr R. Ashton suggested that Edendale should be included within the city's jurisdiction for the purpose of keeping out undesirable transients, or "strays," in his language, including those he presently anticipated as refugees from Durban's more aggressive racial zoning and slum clearances. In short, Ashton sought to extend the city's power to assert its pass laws over anyone entering Edendale, implying the ability to arrest and deport all those found without proper documents (such as family visiting from up the hill in Zwartkop). This request led to what may have been the highlight of the advisory board's development as a political body. Wadley declined to use his executive power to make a decision. Rather, he directed Ashton to make the request directly to the representatives of Edendale's residents and to defend his case before the advisory board. When the board unanimously rejected the proposal on the grounds that it feared the city would usurp administrative control from the LHC, would impose passes on women, and would inhibit the community's progress to responsible local government, the LHC took that as the final word.[44] As Msimang later recalled, "Whenever I think of this matter I always feel a deep sense of regard and admiration for the late Mr Wadley and Mr Boshoff, the first members of the LHC. If this suggestion had been made to the 1970 members of the LHC Edendale would be no better than a concentration camp."[45]

Not long after this rebuff to Pietermaritzburg, the LHC rejected ("on principle") another request from the Union Department of Public Health acting on behest of the city. The central government wanted the commission to donate land in Edendale to house destitute non-European children from Maritzburg, a transparent attempt to use a health argument to effect racial ordering.[46] The LHC also sided with the advisory board in its ongoing battles with the city over commuter transport. The city operated a heavily subsidized bus service to Edendale and wanted to reduce its losses by hiking fares. The commission advised the board to approach the city as an independent entity to protest.[47] The board did exactly that, arguing that Africans could not afford the hike, that the city benefited from the exploitation of cheap labour and hence could not plead economic necessity, and that the city incurred a higher than necessary wage bill by insisting on white drivers. If it sincerely wanted to cut costs, it would employ people who, because of inherited discriminatory laws, earned less. Threatened boycotts of the bus service throughout 1952–55, at least tacitly condoned by the LHC, eventually clarified matters for the council. It backed down on the fare hike and scrapped the race bar against African drivers (*Natal Witness*, 10 and 13 October 1955).[48]

To be sure, the commission did not always take the advisory board's side. Frustration at the slow pace of devolving political responsibility to Africans combined with instances of casual disrespect for Africans' feelings to spark sometimes tense exchanges between advisory board members and the commission. A stubborn refusal by the LHC to introduce a mortgage scheme for "economic" (that is, private, improved, middle-class) housing was a particular irritant to Msimang. This was a pet project of Msimang's within his grander vision of "industrial democracy." Subsidized loans from government would allow property owners to build extensions and outbuildings that were up to code and hence to increase their property values, their ability to attract a higher quality of renter, and their ability to pay rates to subsidize other development or social welfare initiatives. The LHC opposed this proposal on the grounds that it would be too dangerous for Africans to incur such levels of household debt, although it did allow such a scheme for Indians. Msimang could not see any reason, other than the racism that permeated South African political culture, to block this self-improvement scheme.[49]

The issue of vegetables provided another spark for angry accusations by advisory board members against the LHC. As early as 1946, the medical officer of health reported in worried terms that Africans in Edendale ate almost no fruit and only "minute quantities" of vegetables. This was a cause for malnutrition but, as the Georgetown market project floundered, the situation also provided the opportunity to blame African culture for a political miscalculation. "I am of the opinion," Seymour wrote, "that the African is not fond of vegetables," which explained why the market lacked customers. The advisory board responded to correct him: "Africans would not be human if they are not fond of vegetables. The truth is, Africans are great vegetable eaters, but are handicapped by the existing economic system which compels them to make both ends meet on a sub-economic rate of pay or below breadline."[50]

Advisory board complaints took place against a background of widespread apathy, non-cooperation, and opposition from people such as George Mtimkulu and Chief Mini, with, as we saw in the previous chapter, an implicit threat to boil over into violence. There was also a fairly sophisticated, ongoing resistance through petitions, letters to the press, and even lawsuits by private individuals. A certain Vumisa was among the most successful of the latter, taking his case against the LHC right to the supreme court of the province. The court ruled in October 1950 that the commission (after more than six years of conducting slum clearances) had no authority to order the demolition of Vumisa's (and, by implication, any other) inferior housing.[51] Correspondence from 1956 is also suggestive of the limitations of the LHC's ability to carry out its stated intentions. Five women near the Georgetown area – whom the commission secretary described as "shack farming" on twenty-five acres of property each – petitioned for infrastructural improvements to serve their properties. The commission rejected their appeal on the principle of fairness (the rich should not jump the queue over the poor for services). Yet it then offered to consider building a road for them if they donated some land and funds.[52]

The necessity to negotiate and sometimes compromise with African landowners may have been regretted by some LHC officials but, in the short run, such compromises did much to undermine early opposition to the commission. By 1950, the LHC boasted 150 employees and had won sufficient confidence to have its mandate renewed in Edendale and replicated in a growing number of PHA's throughout the province, including several small enclaves around Pietermaritzburg. We can rightly be sceptical of its self-interested assertions of success and growing local acceptance or of Msimang's many instances of praise.[53] But even *Ilanga lase Natal*'s "Rolling Stone" (H.E.I. Dhlomo), among the most critical voices in the African media, in 1952 allowed himself to wonder why the "away with Poyinandi" attitude was so entrenched in African nationalist politics, and to hint in favour of greater African participation in board elections as a promising strategy to improve Africans' lives (*Ilanga lase Natal*, 28 June 1952).

## BETRAYAL

What, then, happened to sink the promise?

First, it must be stressed that the seeds of eventual disappointment were planted from the beginning and were manifest even in the Wadley/Boshoff years. We saw in chapter 4, notably, that the idea of racial segregation long predated the LHC and had significant purchase among both white liberals and African elites. Selby Msimang was a strong proponent of non-racialism at the political and macro level, but at the level of local housing, even he at one point advocated for population removals and de facto racial zoning. In 1944 he was hired by the LHC as an investigator and social sorker to scope out the situation in Hollingwood, an enclave of freehold land on the east side of town. There, he found a community mostly of widowed African women who survived in deep poverty and poor health by doing laundry in one of the most polluted sections of the Msunduze River. His recommendation was to relocate them to lands in Edendale bought out from non-Africans, who would in turn be moved to Hollingwood, something that the women themselves implicitly desired.[54] Similarly, Chief Mini, the Edendale Lot-Holders Association, and other prominent African leaders in Edendale and environs actively promoted population transfers and changes to land tenure in order to make Edendale an exclusively African community.

Aside from this significant obstacle to the non-racial vision, the goal of self-government was also often compromised by the paternalistic attitude of LHC officials, who sometimes seemed to believe that their good intentions and technical achievements were sufficient, that African apathy or hostility towards their work was Africans' own fault, and that Africans did not read. The second medical officer of health, R.P. Seymour, could be especially impolitic in his language, stating in his published annual report that African women's "ignorance" and "stupidity" explained the high infant mortality rate (LHC [1943–57] 1948/49, 43–4, quoted in University of Natal 1951, 237). Africans who served on the advisory board expressed frustration at the lack of support they received from the commission to counter such insulting opinions and to sell the LHC idea to the people. It was not until 1949 that the LHC invested in a monthly newsletter, *Ikhwezi*, directed at a popular audience and intended to offset the perception of callousness or disregard for local sentiments. *Ikhwezi* was reported to be a great success after its initial trial run, with articles and letters in four languages (isiZulu, Hindi, Afrikaans, and English). It eventually had a print run of 750, which implied a much wider readership. Indeed, by 1956 it was so popular that the advisory board appealed explicitly to the LHC to maintain its publication, an appeal the commission rejected on the grounds of cost.[55]

Seymour's lack of diplomacy was offset to an important extent by the evident good will of Commissioners Wadley and Boshoff and their willingness to compromise. However, in 1953 the original leadership of the LHC underwent an almost complete changeover, and that year marked an important turning point in relations between the commission and the advisory board. Wadley stepped down as commissioner due to failing health and then both his successor as acting chair, R.A. Short, and stalwart J.C. Boshoff died in quick succession. Belief that 1948 had produced a political fluke also died in April with the decisive re-election of the National Party. A period of administrative disarray and demoralization followed, which Bunty Biggs described in an interview for the Alan Paton Centre and Struggle Archives. Biggs was a social worker who had arrived in Edendale in 1954 hoping to revitalize the social medicine experiment. She recalled "a sort of residue of what I would call English speaking officials who seemed to be slightly embarrassed at times about the kind of problems that I was coming [to them] with … They were a bit ashamed but they were all – I can only think of two I think – they were getting towards retirement age and they were obviously not wanting to get too heavily involved in any confrontations with Pretoria." The two ashamed ones did indeed retire soon after, to be replaced by Afrikaners "without a kind of warmth or humanity."[56]

Reflective of the shift, the tone of advisory board communications with the commission became decidedly more confrontational. Already in July 1953 the board expressed its "genuine fear … that with the present change in the personnel of the Commission, the present permissible character of the Board may be reduced to impotence unless it is protected."[57] Indian members resigned en masse the following year in protest at what they thought was a shift in the LHC towards acceptance of eventual zoning for Africans only. As another illustration of the shift, annual reports had until that point been thorough and substantive, comparable in form and professionalism to those produced by Pietermaritzburg city council. Starting in 1955, however, the annual reports shrank dramatically in size and deteriorated in quality to cursory statements with minimal detail and printed on the cheapest cyclostyle. Translation of the annual reports from English was offered for the first time but not into any of the other main languages spoken in Edendale, as the board had requested. The translation was into Afrikaans, which also began to be imposed soon after as the language of instruction in African schools.[58]

In 1956, Medical Officer of Health Seymour precipitated the first real crisis in commission–advisory board relations. It began with him ordering the demolition, without notice, of three illegal fowl runs. The board demanded an apology and compensation for what it described as "irresponsible action … tending deliberately to harass the residents and to foment opposition." For its part, the commission took grave exception to the accusation and its tone, and demanded a retraction and apology from the board.[59] The commissioners got their apology soon enough, but they also got a slap in the face when Edendale residents joined other PHA residents elsewhere in the province in a rent and rates boycott. A drastic (85 per cent) reduction in the LHC non-beer revenues ensued, with consequences for the delivery of new homes and a proper sewage system. A body ostensibly devoted to public health was in the discomfiting position of relying for a dominant share of its revenue upon men's consumption of alcohol in its Edendale Beer Hall. Although rates and rental revenues in Edendale returned to almost normal the following year, confidence in the LHC was seriously undermined. The province compounded the commission's problems by cutting its budget by 10 per cent in 1958.[60]

Pretoria was meanwhile stepping up its pressure on Natal and Maritzburg in a manner that not only promoted "separate development" in morally positive terms but also created persuasive new facts on the ground. Sobantu became a source for the latter and indirectly dealt a serious blow to the LHC. In late 1954 Pretoria blocked the city's proposal to expand Sobantu on the grounds that the township was too close to the town and to white suburbs. As an emergency response to this freeze, the city relaxed its rule on the number of inhabitants allowed per house, enabling it to squeeze an extra 1,500 people into the existing structures. This, Pretoria informed the city in 1956, was unacceptable, and the minister responsible (Hendrik Verwoerd) threatened to use his powers to have the community declared a coloured area if the city did not do so itself. Under the GAA provisions, this meant that the entire existing population of Sobantu would have to be evicted.[61]

To its credit, the city stood firm and refused to carry out the order, which Verwoerd also quietly backed away from. But the freeze on new housing in Sobantu remained, even as the problem of extreme shortage of housing for Africans in the city and environs worsened. Edendale's African population, estimated at roughly 15,000 in 1957, began to surge, and by 1959 had nearly doubled.[62] A pro-growth faction in Maritzburg politics increasingly saw the LHC as an obstacle to resolving the crisis. Indeed, the city's pre-eminent voice of liberalism, the *Natal Witness*, also threw its weight against the LHC around this time. Already in 1957 the paper began to lament how the city "lags behind" in attracting industry, to criticize the LHC as ineffective, and to feature business leaders who denounced council's dithering on the delivery of rail, water, and other subsidized services that they demanded.[63] In 1958, notably, the Chamber of Industries attacked city council's hesitancy to embrace an all-out growth strategy as "downright stupid" (*Natal Witness*, 2 May 1958). Only rapid industrial expansion, it argued, could provide employment and relieve the misery and ill health among African residents of Edendale. Such expansion would require stepping up the pace of state intervention, even if it meant running roughshod over the LHC. As a sign of the times in that regard, the city announced it would proceed unilaterally with plans to build its own new township on Slangspruit farm, against the commission's "urgent" requests not to do so.[64]

Riots in Sobantu in April 1959 made the need for haste more obviously a political issue and, shortly thereafter, municipal elections returned a pro-growth mayor who promised that the creation of jobs for Africans in Maritzburg would accomplish what the local authority in Edendale had not. The new mayor, C.B. Downes, found a crucial ally for his industrialization agenda in Pretoria, and LHC resistance to racial zoning soon collapsed in the face of multiple assaults in quick succession. A direct order came from

Pretoria for the LHC to cease giving permission for additions or improvements on non-African property in its area. Even applications by Africans were forbidden if they were for non-residential purposes. The provincial administration joined in the assault by demanding that the LHC amend its planning scheme in order "to preclude business and industrial development" along the Edendale highway, that is, to ensure that Edendale could not develop except as a black dormitory community.[65] The commission was also stripped of its remaining medical health capacities, including public health education (*Natal Witness*, 2 May 1958). Shortly thereafter, the commission's acceded to a joint GAA plan with Maritzburg that zoned the entire district for the "Bantu group." As part of the ostensible rationalization of the administration of formal African townships, the LHC handed jurisdiction of its showcase Ashdown village over to Maritzburg and began to buy up non-African properties that spilled onto the market in response to the Bantu zoning (Butler and Harley with Aitchison 1993, 15).

One of the new mayor's first acts was to establish a "City Committee for Industry," which included the mayor himself and three men who subsequently became mayors.[66] Their lobbying efforts paid off in 1963, when Pretoria announced a range of incentives to spur rapid industrial expansion in areas that bordered impoverished reserves, now termed homelands. With Zwartkop only a few kilometres away, Pietermaritzburg easily qualified as a "border industrial zone" that could receive central state subsidies for transportation, electricity, relocation, and wages. Pressure (and shaming) intensified on the city from the central government to ensure the marginalization of critics. As the national minister of Bantu administration and development put it when turning the sod on the first phase of Pietermaritzburg's new planned township of Imbali, the aggressive promotion of industrialization was nothing less than the city's "Christian duty."[67]

Such moralistic language makes the subsequent disregard for the health impacts of barely regulated industrial growth, and for the scale and cruelty of population movements carried out by the city to facilitate such growth, all the more shocking. One of the first big clearances to be carried out (circa 1963) was of the multiracial area of Pentrich along the Msunduze River. According to the Surplus People Project, Africans cleared from white farms around Maritzburg in 1967 were sent to an emergency camp created just outside of Edendale on Polltax Farm (New Politique). No water source was provided until 1969 (Surplus People Project 1983, 66).[68] Another clearance in 1969 moved roughly 400 Africans out of Hollingwood on the east side of town

6.3 Borderlands, looking from Imbali to Mason's Mill. Through the smog, one can discern industry, gum trees, old apartheid-era matchbox homes, a newer middle-class neighbourhood, an even newer informal settlement, and rubbish spilling down the steep slopes

for future expansion of the Willowton industrial estate, while M'kondeni on the southern approach to the city was also finally cleared of a "black spot" to prepare for that industrial zone.[69]

The LHC was frozen out of all of these decisions. By its own concession as early as 1964, it was being forced to set aside its development mandate in order to deal with the emergency humanitarian consequences of "what one might term 'residential displacement.'"[70] The growing irrelevance of the LHC to substantive discussions about development is also suggested by city correspondence directly with the Department of Bantu Administration and Development from the late 1960s to 1973, which included proposing to takeover Edendale as part of "Greater Imbali."[71] The development of Imbali itself meanwhile proceeded rapidly, with generous loans subsidies from all levels of government. Higher levels of services (including tarred and lit roads, homes with toilets connected to the sewer system, and subsidized rent), and the right to travel and work in the city for Imbali residents, consolidated the impression that Edendale proper was being left behind if not actively discriminated against.

The people of greater Edendale did take important initiatives to keep the non-racialist vision of development alive, albeit with rapidly declining effectiveness as police repression intensified. Notably, the All-In Africa conference in 1961 saw over a thousand people of all races come together in the Arya Samaj Hall, the Indian community centre that is in what is now Imbali and is dedicated to a non-sectarian form of Hinduism. The fugitive Nelson Mandela appeared, gave a fiery speech committing the ANC to non-racialism, among other things, and was arrested shortly afterwards. The Liberal Party of South Africa (LPSA) escaped the immediate nationwide crackdown on political dissent that followed, and it remained locally active through leaders such as Peter Brown and Selby Msimang. But Brown was eventually arrested in 1964. Whatever spine the LHC had possessed earlier, its will to stand up to higher levels of government seemed to completely evaporate. According to Msimang, the commission abruptly terminated him from the advisory board in 1965 with secret prior information from Pretoria about his imminent arrest for LPSA activities. The Liberal Party itself was shut down in 1968.[72]

In addition to direct political repression and population removals, the tightening of control over people within existing townships precipitated community alienation and the violence that was to come. As Iain Edwards (1996) has shown in the Durban context, such controls over African families fuelled a gathering anger that was to wrack that city in the 1970s and 1980s. A similar process was deepening resentment against "improved housing" in and around Maritzburg. The area secretary of Ashdown, for example, turned down advisory board pleas to stop the expulsion of widows from their deceased husbands' homes, began publicizing rent defaulters' names to shame people into payment, and intensified its surveillance for liquor and tenant violations.[73] We can also get a sense of what *uPoyinandi* was becoming in these years from interviews collected from a nearby township (Mphophomeni, just up the escarpment from Edendale). Michael Xaba and Victoria Shibase recall one LHC official from the late 1960s ("uBhodlendlini," from *bhodlela* – "to be insolent towards, to belch"). uBhodlendlini was so-named "because out of nowhere you would just hear him talking inside the house. We would be sitting like this and you would just hear him, 'Sir, I will arrive tomorrow. Remove all your belongings. I am coming to pick up everything.'"

MX: Every Thursday the inspector would come and check where you throw your rubbish, do you have a toilet, is the house clean? They would come inside the house and check everything. He was going with the nurses and would come inside and even look under the bed. They would go into every house, him and four nurses. They would check outside if the yard is clean, is there a toilet, is there a place to throw your rubbish?

VS: ... that white man was insane. He used to come to our homes at night around 12 to 1 a.am to wake us up.

RN: Why was he waking you up?

VS: He was counting everyone in the house to make sure that right people were occupying homes. If ever he found out that someone was in the house and you had not reported that person, you would get arrested.[74]

The freehold part of old Edendale initially escaped such intrusions. But after 1974, when administration was transferred to the absurdly named Drakensberg Board, the dogmatic application of rules was also imposed there. Tellingly, the first act of the new board was to "unearth" and enforce – with many arrests and fines – an existing regulation that had never been enforced by the LHC before. This required all property owners *and* tenants to get permits from the board to rent.[75] Petitions and lawsuits succeeded after some months to stop the madness. But over the longer term, this type of tightened control had a cancerous effect. Indeed, rage at the state did not at first manifest in the political or legal spheres as much as in the domestic and criminal. Worries about youth gangs and isolated reports of sexual predators and gender-based violence had long been aired to the LHC and in the press. The fact is, though, that notwithstanding its extreme poverty and political frustration, "good old Edendale" and even the new townships like Ashdown were remarkably peaceful through the 1950s and 1960s. The murder rate in Edendale was reportedly about half of that in supposedly genteel Pietermaritzburg (Dyer 2013, 158). This began to change dramatically across the whole of the urban area. The sociology of crime is very complex, and statistics cannot be fully trusted. It is nonetheless difficult not to be struck by the contrast between 1962 (when there was only one homicide or suicide among Africans in Pietermaritzburg (Pietermaritzburg [1904–90] 1962, 5) and the mid-1970s: the toll from violence climbed steadily from

15 deaths in 1975 to 39 the following year and 72 in 1977 – murder becoming by then the leading cause of death among Africans (1977, 6). Recalling his own youth from those days, Muntakabongi Hlophe put it this way:

> Socially apartheid did a very bad thing to us. We used to join gangsters, because to us it was more of a form of recreation than crime. Guys who used to pick pocket and steal would come back to the township to enjoy the fruits of stealing from town. So you would end up not looking at crime as a bad thing, but as a sort of a Robin-Hood attitude where you say, "Oh these guys have gone to town and have come back with something and would share and enjoy the loot." So you grow up with that attitude that it is not a crime, but you're taking from the rich and bringing it back to the poor – that was the power of apartheid. (SC, Muntakabongi Hlophe, interviewed by Mabongi Mtshali, Pietermaritzburg, 23 December 2009)[76]

Youth criminality, in the context of the repression of other means of political expression, attracted many middle-class Africans to an alternative vision of development in the mid-1970s. Inkatha yeNkululeko yeSize reconstituted itself in 1975 as a "national cultural liberation movement" that also promoted African empowerment through small-scale entrepreneurship, property ownership, and female-led rural enterprise. Inkatha promised as well to re-instill traditional *hlonipha* (respect) consonant with the Christian ethics of the founding fathers and mothers of Edendale (Hassim 1993). We know in retrospect that Inkatha later colluded with some of the most notorious branches of the apartheid state and played a disproportionate role in escalating violence, not just in greater Edendale but throughout Natal and the Johannesburg area. However, in its early years, many of Edendale's leaders, including, as noted earlier, Selby Msimang and Rev. Enos Sikhakhane, saw it as a force for progressive change.

Unscrupulous political leaders ultimately made the decisions that resulted in the descent into violence. All levels of government contributed to setting the stage under the rubric of development and growth. The province, notably, used its responsibility for education to justify clearance of the historic, mostly Indian community of Slangspruit in 1971–73 to make room for a community college intended to serve rapidly expanding Imbali. The city, as the original engineer and manager of Imbali, created a local advisory board (later called a town council and with a mayor) that Inkatha supporters dominated from the beginning. In 1973, the city facilitated their ability to dispense patronage, by handing responsibility for the township over to one of the proliferating apartheid bureaucracies and providing a disproportionately generous subsidy towards Imbali services and infrastructure compared to its historic township, Sobantu. Eager to cover the costs, the city built a brewery in Imbali that by 1973 was churning out two million litres of "Bantu beer" and earning over a million and a half dollars in profit per year for the city (Butler and Harley with Aitchison 1993, 11). In Edendale, meanwhile, an attempt to negotiate a meaningful local authority to succeed the LHC fizzled after nine years of state refusal to guarantee private property rights. The last vestige of democratic input into Edendale's governance then collapsed in 1984 with the resignation of all remaining members of the advisory board (*Echo*, 20 January 1983; *Natal Witness*, 30 September 1984).

Perhaps the most devastating move from a health *and* political perspective, however, was the transfer by the province of Edendale Hospital to the central government in 1970, which in turn used that prestigious institution to shore up its creation, the KwaZulu Territorial Assembly (KTA). The KTA's leader, Chief Mangosuthu Buthelezi, also subsequently leader of the Inkatha movement, was thought to be a bulwark against socialist or other radical African nationalist movements. The KTA reputedly exerted political interference in health management and used the budget for patronage purposes in sometimes crude ways, displaying increasingly brazen corruption. Employees of the hospital were required to swear an oath of allegiance to the KTA constitution, meaning in effect to Buthelezi and Inkatha. Many simply quit instead, worked to rule, or destroyed documents as a gesture of protest (Robbins and Hartley 1985, 27–35; Robbins 1988, 191).

From this politicization of institutions flowed a rapid deterioration of basic services and waste management, abetting growing political tensions between factions in the various communities. By the late-1970s the Imbali advisory board (later, town council) had begun to actively agitate for Imbali's full excision from Pietermaritzburg and transfer to the KTA. The writing was on the wall by early 1982, when the local councillors and the police colluded in the creation of armed, pro-Inkatha vigilante groups to "keep the peace."[77] In this climate, notwithstanding expensive projects like new housing estates and a sewage-treatment plant, the delivery of rudimentary services stalled or abruptly declined. To give a sense of the challenges, in 1984 there was reportedly a single ambulance for the estimated 100,000 people living in the valley (*Natal Witness*, 9 February 1989), and hospital waste routinely spilled directly into the Msunduze River. By the time of the outbreak of overt war between the Inkatha/"Third Force" and the United Democratic Front at the end of the decade – accompanied by an influx into the valley of up to 30,000 traumatized refugees from other war zones in the district – victims of Inkatha attacks reportedly preferred to risk their lives to get to hospitals on the far side of the city than go for treatment at Inkatha-controlled Edendale Hospital (Truluck 1990). An imposed levy by the KTA on nurses' salaries sparked another strike at the Edendale Hospital in September 1990, a "nightmare" that quickly spread throughout the province (Marks 1994, 204, and 275n38).

The long-term significance for public health of such a breakdown in governance can hardly be overstated. Perhaps above all it fuelled a perfect storm for sexually transmitted and other infections, notably HIV/AIDS and resurgent tuberculosis.[78] Impoverished women dislocated from their homes and families; men disempowered by unemployment, political humiliation, and high levels of alcohol abuse; children sent to dysfunctional and violent schools (if at all); the collapse of primary health care; and profound cynicism towards, indifference to, and/or mistrust of anything coming from government, almost especially public health messages – all of these factors played out against the backdrop of pre-existing high levels of migrant labour and an informal liquor and sex industry, with virtually zero condom use, no male circumcision, and a prevalent culture of secrecy and shaming around sexuality. The city's annual report from before any cases of HIV/AIDS had actually been reported in Maritzburg is poignant in retrospect: "The devastating disease of AIDS looms on the horizon. Fortunately, as yet it does not present a serious problem in South Africa. This, however, is certain to change in the not too distant future and the State Health services are preparing to meet this challenge" (Pietermaritzburg [1904–90] 1987, 337).

≈

The LHC finally succumbed to death by a thousand cuts in April 1974. Not surprisingly, the LHC is only vaguely recalled in Edendale today. The names of prominent commissioners are occasionally memorialized – for example, in the names of Wadley Stadium and Landauville suburb and in a memorial in the Edendale Hospital to J.C. Boshoff – but with virtually no public explanation of or reflection on these officials' significance. Among academics, in my observation, there is little awareness of the difference between health *committees* and the health *commission*. If remembered at all, *uPoyinandi* is assumed to be the Natal equivalent of Transvaal's hated Peri-Urban Areas Health Board, which is to say little more than a means to facilitate the extension of racialist controls over Africans.

All this is understandable, given where the LHC ended up, but it nonetheless strikes me as somewhat unfair. For the first decade and a half, *uPoyinandi* officials and advisory board members of all races had a great deal of optimism that they could buck the national trend through the promotion of primary health care, negotiated solutions to environmental challenges, and non-racial local governance. Achievements were impressive enough to sustain the commitment of a strong cohort of African leaders well into the apartheid era.

Hope in the second Edendale experiment died incrementally thereafter, with different actors often unintentionally contributing to its demise for diverse reasons. With their brusque and unsympathetic language, LHC officials were sometimes their own worst enemy, while the city was generally ambivalent and opportunistic (profiting off water sales and beer, notably). There is no question, however, that the National Party government in Pretoria deserves the lion's share of opprobrium for this failure. It is easy to imagine that things might have been much worse (for example, the eradication of Sobantu or of freehold land tenure in Edendale) had it not been for stiff resistance and principled non-cooperation by leaders in both Edendale and Maritzburg.

In reconstructing this narrative, it must be conceded that the people of Edendale and Maritzburg did not necessarily make the rehabilitation of the area easy. In addition to evasion or outright resistance to specific projects and initiatives in the early years, there was the more widespread, insidious problem of apathy. Strong voices within Edendale meanwhile agitated in favour of racial segregation or transfer to the Native Affairs Department and its successors, while pro-growth industrialists in Pietermaritzburg made it politically easier for Pretoria to impose its will against the non-racialist vision. It is a complex story that resists easy black/white, good/bad dichotomies, but that we should bear in mind as (or if) the next experiment at Edendale unfolds.

# SEVEN | From "Demented" to "Democracy": Continuities and Conflict in the Growth Model of Development from the 1970s to the Present

Much has been written about the traumatic events of the late apartheid era and the bloody but ultimately successful transition to democracy. I have little to add to the political narrative, which appropriately emphasizes the criminality, social corrosiveness, and environmental destructiveness of apartheid, plus the heroism of local activists and international solidary movements against it.[1] I do note, however, a strong tendency to portray 1994 as such an unambiguously positive historical watershed that certain continuities between the apartheid era and democracy tend to get overlooked. In the spirit of decolonization, and while still honouring the inspiring achievements of the freedom struggle, I feel that that line needs to be blurred a bit. My focus will be on a powerful continuity in thinking about economic development through "growth." Far from steadily progressing under democracy, critical thought about the contradictions of the growth model may in fact have eroded in current planning. As a small example, the first comprehensive plan for Greater Edendale called for the development of four regional services centres in the areas of greatest need, two of them along the main transportation corridor but two placed off the beaten track in the densely populated historic villages of KwaPata and Wilgefontein. These were connected to the city by a new double-track commuter train line. The present urban renewal plan for Edendale and environs envisions five service nodes but has put them all on the main highway, with no train

7.1 The river still needs work. The low-level bridge here offers an approach to Sobantu not far from the spot Ian Player wrote about in disgust at the start of his first river journey

in sight, as befitting the municipality's broader enthralment to the private automobile. The demolition of one of Greater Edendale's few sports facilities (Qokololo stadium) to make room for a gated mall (already showing signs of blight) and a proposed new private hospital to bring "quality health care for those who can afford it"[2] also speak powerfully to the possibility that the growth model itself is an important part of the problem. The river still needs work.

To voice doubts about growth in South Africa today is to risk inviting accusations of ultra-leftism, utopianism, and even racism (whites were enabled to get rich and consume – why can't blacks?). In this chapter I want to brave that risk by assessing the limits of the economic growth model as it has unfolded over the period of "transformation." No reasonable person disputes that political freedom from the former system is a monumental improvement for the vast majority of South Africans. However, following Low and Lonsdale's concept of the "second colonial occupation" (1975), which sought to transform Africa in the 1950s into a capitalism-friendly place, it strikes me that a third, more sophisticated, more subtle, and yet more insistent "colonial" occupation is now well under way. Edendale is being redeveloped, and symbolically reterritorialized, as a place to shop and borrow money in order to achieve fulfillment through cars, bling, electronics, sexy times, and more loans. The Wadley Stadium is now the FNB (First National Bank) Wadley Stadium, just to remind people where the loan repayments should go. The conundrum is that this is what the people themselves apparently desire: 73 per cent of votes cast in KwaZulu-Natal in the 2014 national election went to the party that has facilitated the occupation. The ANC plurality in Greater Edendale was likely closer to 90 per cent. This is the same party whose two decades of economic management have been glowingly endorsed by the multinational investment firm Goldman-Sachs (Coleman 2013), and whose evident mismanagement at the local level has sparked nationwide violent protests.

The chapter begins by picking up the environmental history narrative begun in chapter 3, with a primary focus on the rush to industrialize in the late apartheid era. I then enumerate achievements in environmental justice since 1990, followed by a critical assessment of growth under a democratic system, using the feminist political ecology/degrowth analytic lens discussed in chapter 2. In this way I hope to contribute to the often pointed debates about growth/jobs versus community/justice, neoliberalism versus *uBuntu*, and brown versus green environmental strategies.

## "DEMENTED" INDUSTRIALIZATION

As we saw in the previous chapter, one of the keys to the relative success of the Local Health Commission (LHC) in its early years was that it did not dogmatically impose its regulations but, rather, preferred to let certain regulations "lie fallow" and to "bend the rules" if necessary for the sake of peace. It hoped to insinuate scientific principles through a consultative process with the local population as represented by the advisory board and voluntary associations. That process involved a significant amount of negotiation and compromise. It took more than a decade to get community leaders to agree to the wording of new regulations restricting the cattle population, albeit with little apparent impact even then. As long-time resident Eric Matshatsha recalled, as late as 1957 "good old Edendale [was still] a sort of free society where they could plough and rear cattle and everything in the area."[3]

That attitude frustrated many in the city and province who, for various reasons, urged rapid industrial expansion in sometimes moralistic if not apocalyptic language. How and where to achieve such growth was a long-standing debate in Maritzburg politics, often tied to the fate of Edendale. At the time of the creation of the LHC, the "anti-industrialization" faction was led by Eleanor Russell. Since becoming a city councillor in 1931, Russell had played a key role in the development of Sobantu and, as deputy mayor, in winning council support for the creation of the Edendale and District Public Health Area (EDPHA). As mayor of the city in 1943–47, she consolidated her reputation as a liberal, a friend of the native, and an ally of conservationists. During the Second World War, Russell oversaw the city's participation in the national "Wealth from Waste" campaign to support the war effort through a municipal Anti-Waste Department that exhorted people to reduce consumption, to recycle and re-use household materials, and to practise home gardening (*Natal Witness*, 14 January 1941).[4] She and her allies stood firm against persistent calls from industrialists to freeze or decrease rates in order to attract new industry. It was not that Russell opposed industry in principle – the Alcan aluminum smelter, the city's first heavy industrial enterprise was established under her watch on the Edendale side of town in 1947. However, she and her supporters wanted "the right kind" of industry – that is, development that would provide employment without either disturbing the tranquility of the historic parts of the city or unduly exploiting African workers. In 1949, for example, council imposed the condition that no new industries would be allowed to open unless they included housing for their workers

("Maritzburg Industrial Sites Decision: House Non-European Workers or No Land," *Natal Witness*, 13 April 1949).[5] For years it blocked a proposed new industrial estate on the south side of the city (M'kondeni), while in 1953 it established an industrial effluent tariff by which companies were required to pay for cleaning up the added burden that they imposed on the water supply (Moodley 1997, 85).

The pro-growth faction in city politics perceived such moves as unfriendly and, while not necessarily convinced of the merits of apartheid ideologically, they began to see it as a way to press their vision for industrial development. This faction attained political power in the city in 1959 and, with Pretoria's support, enacted a wide range of tax and other concessions designed to attract industrial investment. New industrial zones were laid out around the city strategically, both to protect the historic core from pollution but also to be close to the main sources of cheap labour. Indeed, the strategy went hand in hand with racial residential zoning, buffer zones, and architectural modernization of the city, with the razing of certain historical neighbourhoods and buildings in favour of open spaces and concrete/steel structures.

It was not long before people began to notice unintended environmental impacts of this growth strategy. As early as 1962, the city engineer complained that the car was "pretty well dictating" the development of relatively cheap land distant from the city (Pietermaritzburg [1904–90] 1962, 43). Not only did sprawl result in a huge imposition of time and cost on domestic servants, but it also led to a major increase in vehicular pollution and an inflated bill for the city to provide and maintain an overstretched infrastructure. The burgeoning commuter problem was also complicated by the state's need to win African allies to its cause. In the absence of more efficient public transportation, and with a mind to co-opting an element of the African middle class, the city in 1974 licensed African-owned "taxis" or kombis to ferry people through the city centre to their various destinations, with predictable increases in congestion, accidents, air pollution, and litter.

Racial zoning had another deleterious impact. Jurisdictional uncertainty and dispute over environmental issues was an old problem that had contributed to Edendale's underdevelopment in the nineteenth century, and it plagued relations between the LHC and the city well into the 1950s. LHC officials believed, for example, that the city was obstructing their proposal to develop an independent source of potable water downstream of the city in part because the municipal government did not want the LHC to discover how polluted the river was at that point (which is to say, how negligent the city was in treating its effluent).[6] But apartheid hugely amplified the problem. By 1980, no fewer than ten distinct local bodies administered the present area of Msunduzi, which the city engineer of the time denounced as an "over-sized and greatly varied 'Dagwood' sandwich."[7] Ample evidence can be found of direct obstructionism by different authorities worried about encroachments on their turf, including, notably, blatant disinformation and resistance by the Department of Bantu Administration and Development (BAD) to city initiatives to regulate urban sprawl in Zwartkop in the late 1960s and early 1970s.[8] This disjointed system, exacerbated by chronic shortages of qualified professionals in the civil service throughout the apartheid years, resulted in what with hindsight can only be described as stunning ignorance and callousness or cynicism by responsible authorities. We saw some tragic examples of NIMBYism in previous chapters. One that took place on the cusp of the transition to democracy, however, is revealing. The *Natal Witness* reported on 23 June 1989 that a sewer pipe had once again broken at Edendale Hospital and was spewing ninety litres of faecal matter, blood, and other detritus per second into the Msunduze River. Although this was in the heart of an area that had been relatively densely populated by Africans and Indians for at least five decades, and was scarcely five kilometres upstream from the city's premier recreational park and central Pietermaritzburg, the director of the parastatal responsible for the city water supply calmed white citizens' fears. As Bernard Chamberlin of Umgeni Water explained "Only those in rural areas who take water from the river will be affected."[9]

The sorry tale of Azalea in these years reveals both the dangers of fractured jurisdictions and how population removals were not necessarily as conclusive as intended or announced. As the city developed Imbali in the 1970s, an unknown number of Africans were removed from land on its eastern border that was supposed to be a white farm but had sprouted all kinds of informal settlements. The intention was to return it to productive agriculture and have it act as a buffer zone between Imbali and Maritzburg. It soon proved impossible to keep squatters out from the strategically located spot, however, and so the owner privately sold the land to KwaZulu Finance and Investment Corporation, a parastatal of the Kwa-Zulu Territorial Assembly (KTA). This body promised to formalize the settlement of squatters in a proper township linked into Imbali and thus still appropriately buffered from the city. For this task, the corporation contracted another company, Azalea, whose details of ownership were murky. Azalea followed up in the mid-1980s with homes built in large numbers but of notoriously poor quality.

The population boomed. By 1993, Koekie Maphanga estimated 20,000–25,000 structures housed 150,000 people without electricity. These numbers may have been overstated, but systematic deforestation around "Unit S" in particular was not. The area bordering KwaPata stream was largely denuded and thus vulnerable to extreme flooding, as indeed happened on Christmas Day 1995, leaving hundreds of people homeless (Maphanga 1997, 12).

But, to go back to 1979, it had become apparent by then to many that the so-called border industry strategy was not just extremely expensive, was worsening pollution and desecrating the appearance of the city. It was also not generating nearly a sufficient number of jobs to absorb the burgeoning oversupply of labour. Part of the problem lay in the costs of servicing Maritzburg's complicated topography, but the perceived isolation of the area, compared to the major centres on the Rand and in Durban, was also an issue. Critical shortages of staff to carry out the expanding engineering agenda were also noted (with some alarm) as early as 1970 (Pietermaritzburg [1904–90] 1970, 46). These shortages picked up as anti-apartheid protests and the nationwide economic malaise of the late 1970s sparked emigration of white professionals.[10] The global oil shocks starting in 1974 added their own specific stresses, including the decision by South African Railways to stop plans to phase out its coal-burning locomotives. Throughout the remainder of the decade, black smoke from trains chugging up the escarpment constituted a "serious menace" that undermined initiatives to mitigate pollution in the city centre (1974, 98).

The supply of new homes for Africans meanwhile never remotely met the demand created by the racial re-ordering of the city. So, too, the new industries attracted by state incentives never remotely kept pace with the demand for jobs. Not surprisingly, this period witnessed the emergence of heightened levels of criminal violence in Edendale and, significantly, "deliberate arson" targeting state afforestation initiatives around Henley Dam in Vulindlela (1971, 60). Hopelessness and anomie also began to take their toll, in the form of alcohol-related disease, injury, and suicide. As a result, the spectre of chaos and ill health in Greater Edendale as a threat to Maritzburg – invoked in the 1930s and 1940s as a spur to create the LHC – was now conscripted once again. In 1979 an even more aggressive pro-growth faction of industrialists called the Coordinating Committee for the Development of the Pietermaritzburg Region began to agitate for further concessions from the city to facilitate profitable industrial investments (Seethal 1993, 66, 79). Again Pretoria came forward as an ally to this cabal by declaring Maritzburg to be a "deconcentration zone" and providing generous subsidies for industries to relocate there. The city sweetened the offer by providing free industrial waste disposal, apparently a first in South Africa. In a single three-week period in early 1983, ratepayers picked up the cost of dumping over a million gallons of industrial effluent and 521 truckloads of refuse (*Natal Witness*, 17 February 1983).

Deconcentration ushered in what the mayor herself called "increasingly demented" planning and a profound lack of respect by pro-growth leaders for democratic or even technocratic oversight (*Natal Witness*, 10 October 1983).[11] In the short run, however, it seemed to work. While international investment in South Africa as a whole slowed in the mid-1980s, Pietermaritzburg managed to attract an unprecedented 199 new industries in 1985–91, more than doubling the number established in the three previous decades (Seethal 1993, 33). The footwear industry in the district in particular ballooned, to account for nearly 70 per cent of the country's production (Harrison, Futter, and Meth 1997, 15).

Four especially notorious polluters set up production in the city on partially serviced land beside Sobantu. Many of these were relocating from elsewhere in South Africa where the enforcement of anti-pollution legislation constrained cheap disposal of effluent. As Siva Chetty and "Mungo" of Durban Metro Pollution Control put it with respect to Capital and Willowton Industries, "They moved from Durban thankfully, twenty years ago – Chased them out twenty years ago" (Pole 2002, 88). Did Maritzburg city council secretly invite dirty industries with a promise of lax enforcement of the law? Did it turn a blind eye to growing evidence of blatant violations (pouring tallow and other toxins directly into watercourses, illegally dumping truckloads of refuse in surrounding communities, burning toxins under cover of nightfall, and so on), or were the infrastructure and monitoring systems simply overwhelmed? Whatever the case may be, Adrian Pole's informants in Sobantu, just downstream on the Bayne's Spruit from the four edible oil–processing plants, specifically cite 1981–84 as the time when fish disappeared from the stream and skin lesions began to appear on children who swam in it (Kwanzi Mngadi and Mazwi Glenn Sithole, for example, cited in Pole 2002, 64). Industry by then was estimated to be paying only 3 per cent of the total operating costs of the Darvill water purification plant while contributing nearly a quarter of the load (Moodley 1997, 93).

As it had been at the turn of the century, the river became a powerful symbol both of the mess the city was making of itself and of its problematic responses to that mess. The canalization of the Msunduze at Camp's Drift in the late 1980s was one example of a very expensive, disappointing, and ultimately dangerous intervention. Camp's Drift, as readers may recall, was originally settled by small-scale Indian farmers but over time became a mixed-race slum area that the city aggressively targeted for clearance on health grounds in the 1930s. Much of it subsequently developed as industrial, although the flood-prone sections remained wasteland until 1986. In that year, the city began to straighten the river out, with an eye both to flood control and to creating an attractive water sports amenity. The project created a whole new burden on the city budget – constant dredging – while "Dusi guts" (dysentery) tainted the water sports potential. The one event that became a success was the annual Dusi canoe race, which was launched from this spot. The race developed as an important part of the Maritzburg brand, which drew a steadily increasing number of adventure tourists to the city. From another perspective downstream, however, the race became emblematic of the green/brown divide, as rich, white, "thoughtless" tourists invited growing hostility from the poor, black, unconsulted communities they passed through (Macleod 2007, 169). Canalization upstream meanwhile intensified the flow of water downstream at times of peak volume. A tragic outcome of flood protection for a Maritzburg green space was to deliver drowned bodies from upstream of the city to Sobantu in the aftermath of the catastrophic flood of Christmas 1995 (Great Edendale Environmental Network 2002, 13).

Such developmental "dementia" gave rise to growing resistance, in some cases by the city itself (which introduced controls over smoke in the early 1970s) and in others by civic associations inspired in part by local activists such Ian Player (see, for example, Player 1972). Eventually, these associations became linked to the gathering anti-apartheid movement or other self-consciously development-oriented initiatives. Conflict along the Dusi, notably, sparked the formation of the Valley Assistance Fund, which levied a "paddle tax" that went towards building toilets, clinics, and classrooms along the canoe route (Macleod 2007, 174).

Sobantu, the former "model native village," provided another success story – indeed, it emerged at the forefront of the brown environmental struggle not just in the Maritzburg area but in the country. Pretoria had failed to have it rezoned in the 1950s but had succeeded in preventing expansion, improvements, and even regular maintenance. Desperate for new housing, the city turned a blind eye to illegal, jerry-built extensions to rental properties in the village. But grievances against the calculated neglect, combined with the proliferation of dirty industry on its doorstep, fuelled civic activism. In this, Sobantu had a considerable advantage over Edendale or the newer African townships, having been a relatively stable community for decades with a strongly unified political orientation towards the ANC and the United Democratic Front (UDF). When Pretoria in 1984 commanded a new form of local authority that implied future transfer to the KTA, Sobantu leaders formed a Committee of Twelve to organize resistance. In the stalemate that followed, the committee won allies in the city and buttressed its own legitimacy both with its "fairly representative form of direct democracy" and development initiatives such as a market garden (Napier and Mtimkulu 1989, 13). Remarkably, in 1986 city council voted by a large majority (11–3) to accept the committee's demand for Sobantu to be re-incorporated into the city, with direct elected representation on the council. This proposal fell through, ostensibly over financial questions but likely also because it was just too radical for the province or Pretoria to countenance. It nonetheless demonstrated the power of popular, non-violent protest to make the case that radical changes to city governance were urgently needed.

Green environmentalists also scored some early successes in these years, notably in the 1984 city council decision to block a proposal that would have paved over a whole new industrial zone on the floodplain of the city's most biologically diverse area (Willowton IV, on the former Hollingwood PHA). That ruling not only preserved a wide area of green space and a bird sanctuary on the opposite riverbank from Sobantu but also mobilized public opinion against, in Mayor Reid's words, the "impertinent, irregular and improper" attitude of pro-growth advocates towards council (*Natal Witness*, 30 November 1983). Thereafter the Metropolitan Open Space System Committee began identifying and promoting other recreational green spaces throughout the city. With weak legislation, overstretched resources, and pushback from the industrial lobby, council was mostly powerless to act in this spirit. But in 1990 it secured its first conviction against a factory (Willowton Oil and Cake Mills) for the flagrant violation of anti-dumping bylaws (Pole 2002, 44). That same year, while Mayor Patricia Rainer hosted the newly freed Nelson Mandela for tea and scones, another powerfully symbolic victory for civil society forced the national government to change policy: orchestrated by Earthlife Africa in solidarity with unions, community groups, and global activists, public pressure convinced the central government to ban the importation of toxic chemicals as were then being processed and unsafely stored at nearby Thor Chemicals (Hallowes and Munnik 2008, 38). Alongside the conventionally conceived but highly active Keep Pietermaritzburg Clean Association, two pioneering black civil society groups emerged soon after to offset the stereotype of green concerns being a white agenda: the Greater Edendale Environmental Network (GREEN) and the Sobantu Environmental Desk.

The transition period between the lifting of the ban on opposition parties in 1990 and the first democratic elections in 1994 saw a whirlwind of change in thinking about urban governance, in civic association activism, and in violent conflict. I will end this section by describing what was widely regarded at the time as a major success for environmental justice arising out of that flurry. Several dozen refugees from the UDF/Inkatha fighting in Table Mountain found a place to shelter on a vacant plot of city-owned land in 1991. They called their little collection of shacks, which nestled beside the railway tracks on a steep slope but was conveniently accessible to the town centre, Happy Valley, or Nthutukoville. The city, then dealing with ballooning informal settlements on all sides, wanted to consolidate such refugees in a camp proximate to Imbali. It attempted to have Nthutukoville removed several times, including by armed force. But with the help of a new, Durban-based civic organization called the Built Environment Support Group (BESG), residents not only won the right to stay in Nthutukoville, but they were also eventually given funds to upgrade their shacks to properly serviced homes. An important component of this transition was that the majority of the inhabitants were – and remain – single women. They were "empowered" through specific training in the skills necessary for community organizing and to effect and maintain their own upgrades, including in plumbing and carpentry (Seethal 1996; Ndinda 2003; BESG 2009; Goebel 2015).

## DEMOCRACY AND SUSTAINABLE DEVELOPMENT

South Africa's first democratic elections produced an ANC-dominated Government of National Unity at the national level. The ANC also dominated the vote in cities and towns in KwaZulu-Natal. However, based primarily on strong voter support in the rural areas and select townships like Imbali, the IFP formed the provincial government. Reflective of the fraught dualism of the new province, KwaZulu-Natal gave itself two capital cities, Pietermaritzburg and Ulundi. Among the many daunting challenges officials in both places faced was to bring peace to the warring sides (and factions within sides) while shepherding the transition of local authorities to encompass hugely expanded jurisdictional boundaries. The aim of such expansion was to override the crippling, irrational, and highly politicized borders of the apartheid era. Local authorities were then tasked to begin delivering a wide range of services to historically neglected populations across those historical fault-lines. Towards that goal, Pietermaritzburg council commissioned an independent investigation into the environmental issues that would need most urgently to be addressed. That report, by the Institute of Natural Resources (INR), proved sobering. It found that apartheid and ill-considered growth had combined to leave a legacy that included "high levels of industrial and domestic air and water pollution, land degradation and land pollution, problems associated with waste disposal and management, and increased noise (traffic) pollution" (Mander 1995, iv). In four out of six months in the first year of political transition, Pietermaritzburg had the highest levels of ambient lead in the entire country. In another study conducted soon after, the INR found that at least 10,000 people in Greater Edendale were at direct risk of catastrophic flooding. Informal settlements in particular were exacerbating the severely degraded riverine environment due to "poverty, lack of choice, ignorance, and greed" [of shack landlords] (INR 1997, 5). The report did not mention it, but a distinctive aspect of this crisis was that women constituted a large majority of heads of households in the most vulnerable settlements (Goebel 2015, 77 and passim).

A second round of voting took place in 1996, this time for democratically constituted local authorities. Against a background of continuing violence between supporters of the ANC and the IFP, voters in the newly formed Pietermaritzburg-Msunduzi Transitional Council chose a clear majority of ANC councillors with an ANC-dominated executive. In line with the national government's Reconstruction and Development Programme (RDP), the new council announced its commitment to radically expand low-cost housing and to provide a healthy environment for all citizens regardless of race, gender, and other historical markers of discrimination. The RDP represented a balancing act between socialist ideals of redistribution of wealth and neoliberal formulae for dealing with a terrible economic crisis. Indeed, while the period 1990–94 is often celebrated for the "miracle" of political transition, it was also a time of wrenching economic recession, including massive disinvestment and capital flight and inflation as well as redundancy for hundreds of thousands of workers, particularly migrant mine workers. The Government of National Unity also faced a ballooning foreign debt when it came to power in 1994. The RDP as policy aimed in the short run to mitigate the suffering the country had inherited and to provide the long-term basis for building a more equitable and just society. Among other things, it obliged the state to provide free housing with a minimum level of services to those who could not afford them. This resulted in the construction of millions of new "RDP houses" nationally, with many millions of people receiving electricity and water for the first time, along with improved sanitation. The target in Pietermaritzburg was 5,000 new units per year, a scale surpassing even the most frenetic years of apartheid population movements.

The adoption of the national constitution in 1996 consolidated the high ideals of the liberation movement in balance with the often competing individual property rights, minority rights, and levels of government (or "spheres," so designated to remove the hint of hierarchy). The constitution smoothed over potentially antithetical green/brown perspectives on the environment with some judiciously vague language. Section 24 of the Bill of Rights bears full reiteration, both as one of the strongest such statements in the world to that point and as an eloquent equivocation. It promises everyone in South Africa the right:

> (a) to an environment that is not harmful to their health or wellbeing; and
> (b) to have the environment protected, for the benefit of present and future generations, through *reasonable* legislative and other measures that
> i. prevent pollution and ecological degradation;
> ii. promote conservation; and
> iii. secure ecologically *sustainable* development and use of natural resources while promoting *justifiable* economic and social development.[12]

The transitional council further committed itself to the global vision of sustainable development articulated through the United Nations after the 1992 Earth Summit. Known as Local Agenda 21, its many provisions included ones to strengthen local government and civil society, to empower marginalized and indigenous populations through participatory consultation and decision making, to change consumer patterns, to combat poverty, to manage demographic growth, and to establish explicit definitions and targets for state intervention. The latter included reduction of habitat loss and increased controls for erosion and pollution (greenhouse gas emissions were not yet on the political radar), plus some non-conventional understandings of environment such as noise/vibration abatement and the protection of cultural heritage (SRK Consulting 2010). The legislative framework to steer the country – including Pietermaritzburg – towards those goals followed shortly after. The National Environmental Management Act and the National Water Act, both passed in 1998, set out clear decision-making principles, conflict-management processes, environmental-assessment requirements, and enforcement provisions. They confirmed in law the polluter pays principle: "Those who are responsible for environmental damage must pay all the repair costs. These include costs to the environment, costs to human health and well-being, and the costs of reducing or preventing any further damage" (cited in Pole 2002, 31). The Department of Environment, Agriculture and Tourism in 2001 announced the national target of "ZERO WASTE" by the year 2022 (emphasis in the original, cited in Hallowes and Munnik 2008, 52).

National legislation mandated each transitional council to create an Integrated Development Plan (IDP), a Spatial Framework Plan, and a Catchment Management Agency, all with an eye to coordinating the integration of constitutional provisions into local decision-making practices. The city's first IDP placed "environment" as one of its seven strategic issues and "sustainability" as the first of its guiding principles (Msunduzi 2002). One of Janet Roukema's key informants in the municipal bureaucracy praised these gestures in the following terms: "I think we have in South Africa, some super environment legislation, it really is top notch stuff" (2006, 137). These initiatives have been supported by the progressive provision of welfare on a scale that few countries in the Global South remotely match and that is highly skewed to enable women to live independently of men (Goebel 2015).

The Greater Edendale Development Initiative (GEDI) was a significant component of council's IDP and the municipal vision for environmental justice, at least on paper. Its origins go back to 2000, when Edendale landowners first approached the newly constituted municipal government for help in addressing one specific aspect of the post-liberation crisis – ongoing land invasions onto private property and environmentally sensitive lands by desperate migrants from the rural areas. An initial grant of R20 million from the province began the process of untangling byzantine, conflicting, and poorly (if at all) documented land claims. This was to be the first step towards creating an ordered space for infrastructural development that would attract new private investment to create new jobs, empower women, and so on. Institutionalized as a discrete branch of the Msunduzi municipal bureaucracy in 2006, GEDI produced its own integrated planning document in 2009 that called for nearly R1 billion in public spending, an amount nearly equivalent to the annual capital budget for the entire municipality. Together with language that explicitly recognized the valley as an integral part of a cohesive municipality, its commitment to human rights and gender equity sharply distinguished GEDI from LHC/*uPoyinandi* or from apartheid-era plans for "separate development." Indeed, according to municipality propaganda, GEDI's "revolutionary transformative approach" would turn an impoverished hodgepodge of settlements into "a fully-fledged suburb … [a] vibrant urban district … [with] a recognisable townscape character and identity" that would attract tourists, property buyers, and shoppers of all backgrounds (Mzunduzi Municipality 2009, passim).[13] Among GEDI's specific goals, now embedded within a broader GEVDI vision that theoretically includes the former tribal area of Vulindlela, are:

- improvements to core infrastructure (roads, sewers, electrification);
- the creation of five formal urban "nodes," with proper services (shopping centres, farmers' markets, neat taxi ranks, administrative buildings, swimming pools, and other sports facilities);
- densification with quality housing for a range of incomes;
- poverty-alleviation projects through, for example, micro-finance;
- parklands and urban agriculture (including a centre for research into organic, sustainable farming and historic breeds of Nguni cattle);
- beautification and public art;
- erosion control, and the revitalization of the river and remaining forests and wetlands;
- fostering black-owned business enterprise.

GEDI/GEVDI has little original to say specifically on women and gender, although, given the demography of the city, women's empowerment is implicit in the general discourse. It does commit the city to actively promote community participation in local governance. A "community based environmental program," notably, would focus primarily on raising awareness about environmental issues and engaging people in recycling, waste management, the eradication of alien species, and small-scale regeneration projects. All of this would not only be in keeping with the ideals and obligations set out by the national constitution, including the right to "adequate" shelter, food, and water, and the protection of human dignity, but also with reference to UN-Habitat Poverty Reduction Strategy principles, the Millennium Development Goals, the Kyoto Protocol on climate change, and Local Agenda 21 (BESG 2004; Msunduzi Municipality 2009). At the time I began this research, GEDI claimed to be "the largest town planning and urban renewal project currently under way in South Africa" (Msunduzi Municipality 2006, 36), which was probably true considering the vast area covered and the size of the target population (which GEDI then and GEVDI now still place at 300,000).

Few doubted that the challenges would be immense. For now, let me simply focus on one aspect –children. The schooling system was shattered, with high levels of unresolved psychological trauma (Goedeke 2000; Denis 2005) and many children at near-starvation levels. Statistics can hardly capture the depth of suffering; an anecdote barely hints at it: as the headmistress of the best, and probably only, private school in the valley recalled, "We arrived at school and there was a little dead duckling and then I said put it in the bin please, later that day during break suddenly in the bush next to school I see that fire … It was our children who have taken … the duck out of the bin and were roasting it" (SC, Carleen Richardson, interviewed by Philippe Denis, Pietermaritzburg, 15 August 2010). This deprivation occurred in the context of an exploding HIV/AIDS epidemic that was starting to bring "humiliation and desperation" on a scale that few other cities in the world have to face (Msunduzi 2012, 366). Here again, statistics can be numbing. We can get a sense of the devastation to families among Edendale's middle class, however, by listening to eighty-year-old Clothilda Hlophe speaking about her children and grandchildren with Sinomlando interviewer Mabongi Mtshali:

CH: [My firstborn] passed away and left his widow. His wife also passed away.
MM: Sad! So she also passed away?
CH: Yes.
MM: Very sad indeed …
CH: [My second son] died during the time when councillors were installed. He had already been installed as one. He died in the way he died …
CH: [My third son] passed on in 2006.
MM: Sad! And it's quite recent. Who comes after him? (silence). Who comes after T?

CH: B comes after T. This is the one I said is uMuntakabongi. (Laughs)

MM: O yes "a person isn't grateful."

CH: B went to Varsity. He studied B.Com. He finished the Course. His wife also studied Law. The wife passed on. By the way when did she pass on? It was in 2001. They had three boys and the last born was a girl. The eldest boy passed Grade 12 and passed on ...

MM: Is [fourth son] in business?

CH: He used to be unsuccessful.

MM: What is he doing? Is he unemployed?

CH: He's home not doing anything.

MM: That's sad. Who comes after N?

CH: K. K has passed on.

MM: The last born. When was he born?

CH: K was born in 1962, May 13.

MM: When did he pass on?

CH: He was deceased in 1998 ...

[MM: equires about adopted children]

CH: S for instance ... Her mother gave her to us and said she was ours even when she got married. It was as if she knew because the mother died soon thereafter. (SC, Clotilda Hlophe, interviewed by Mabongi Mtshali, KwaMachibisa, 2 November 2009)[14]

To all the urban problems inherited from apartheid has now been added the profound and pervasive stigma attached to this illness. HIV/AIDS has also greatly increased the phenomenon of orphans and child-headed households. Hence, when asked to define the word "environment," one municipal official put it simply: "Look, the real primary issue here is HIV/AIDS, it really is. I don't have any statistics given to me but I am conscious that there are numerous orphans and child-headed households and single parents. That to me is a frightening concept because who is worried about them learning about the life, learning about the way they interact in this or that situation" (Roukema 2006, 94).

From the GEDI/GEVDI perspective, addressing HIV/AIDS and other pathologies will be done in conjunction with its broader development initiatives such as building up the infrastructure. Given the frustrations and setbacks on the health and other "social capital" fronts, it is easy to understand why officials often quickly shift focus to the material, where the municipality can point to some undeniable achievements. They include:

- vast, formal, electrified housing developments in places like France;
- in situ upgrades made to one of the densest and most entrenched informal settlements (Peace Valley);
- a new public library opened in Georgetown;
- with private investment, the construction of two quality shopping centres and service stations that have brought basic consumer facilities (and job opportunities) to the corridor, hence reducing the need to travel into the city centre to buy food, petrol, school supplies, and such; and
- improved sewage.

Some other development achievements have been less tangible. A Land Summit in 2006 convened government, community, and business leaders to begin discussions on how to unravel the many overlapping and poorly documented property claims in Edendale, and the city expropriated land to provide for rights of way for proper roads and sewers. The Local 21 Committee established a consultative process of "stakeholders," including civic associations like BESG, that aimed both to gather inhabitants' views and to raise the awareness of city technocrats about poverty and social issues. Small-scale interventions directed at winning people's hearts and minds to the concept of environmental responsibility have also unfolded through a combination of initiatives from government, non-governmental organization, and other donors. BESG, notably, ran a project that employed forty-four local residents to help clean up the environment:

> Blockages in sewers, unsafe open areas with long grass, they just found the body of a person there, so this program since it has started has helped a lot, they take care of those things. People were also just dumping in the river, the project is working hard to minimize those types of things, to keep the environment clean in these areas.
> The project is also planting trees because in this area, the contractors come if they are going to put houses, level the land like that.
> We are planting grass to prevent erosion, trees; we are also planting groundcovers to keep it. (Roukema 2006, 117)

Across the city, several other projects have aimed to reverse the blight that had set in since the late apartheid era and to change public attitudes inherited from the past about the environment. These projects include local "adopt-a-spot" or "Masakhane" initiatives focused on anti-litter blitzes, the removal of exotic flora, and wetland restoration, in many cases taken on by schools as a long-term educational activity. BESG helped negotiate the relocation or upgrading of informal settlements at Peace River and "North East Sector 2" (mostly Nthutukoville) and has launched a climate change awareness and preparation campaign. A non-profit "Section 21 company" devoted to cleaning up the rivers in line with Local Agenda 21 principles was created in 2006. The Duzi-uMgeni Conservation Trust (DUCT) now trains and employs dozens of "Enviro Champs" and "River Care" teams (Ward 2013). Since 2005, Wildlands Conservation Trust has adopted a "green economy" or "green-preneurship" approach to create jobs for poor women through such activities as waste collection (mainly bottles for sale for recycling), afforestation (piece work), market gardening, and other remunerated environmental clean-up or restorative projects, with reported modest successes (Hlahla, Goebel, and Hill 2016). Revitalization of the downtown core has also been supported through a successful anti-crime initiative and the formalization of "car guards" into uniformed parking attendants. The Bessie Head public library and refurbished Freedom Square projects have turned a dowdy building and a scrappy park into an accessible and attractive public space in the historic heart of the city.

One other "mega-project" that could potentially have an important positive impact has been approved, though not yet realized. With both green and brown goals in mind, the Camp's Drift waterfront proposal would fill in one of the ugly buffer areas between Greater Edendale and Maritzburg (Sutton 2008). For an estimated billion rands, private investors relying on public service provision would see an enclave of high-end residential buildings, hotels, sports venues (including for canoeing and golf), shops, and other recreational facilities rise over the current sludge ponds. Carefully architectured wetlands and lagoons would "naturally" filter the river water. All told, the project would generate new rates, attract tourists, create jobs, and stand like a shiny new portal beckoning visitors to Pietermaritzburg to venture out from the centre and northern suburbs towards its southwestern reaches (to the Edendale mall, perhaps, or to historic sites being developed farther up the valley).

These initiatives have unfolded against the backdrop of uneven but relatively steady progress towards resolution of the political conflict that had earlier riven the region. It took two more elections after 1994 but the ANC eventually won its coveted majority of votes and formed the provincial government. This brought an immediate boost to the Msunduzi economy. Ulundi was removed as co-capital, and all provincial government functions moved to headquarters in Msunduzi. Property values soared, and talk turned to "strategic location," "boom," "unprecedented growth," and "economic powerhouse" (Haswell 2006; Msunduzi Municipality 2006). One expects optimism from city officials and business promoters, but the spirit of progress infected the work of important civil society organizations as well: among the many testimonials to DUCT's work over its first eight years, one finds assessments such as "amazing," "clearly beneficial," "much cleaner," and "inspirational and life-changing."[15]

Meanwhile, progress on the health front has been fitful but nonetheless significant. Early responses at the national level to the catastrophies in this sector were scandalously weak (Nattrass 2007) but, according to Dyer (2012, 3[16]), Pietermaritzburg was "ahead of its time for the country" in developing cross-sectoral protocols and preventative interventions. Eight new clinics were established in the first decade of democracy across the historically underserviced parts of the city. Together with improved immunization and the expansion of welfare provisions and housing, this action contributed to steady drops in mortality and morbidity from most diseases except HIV/AIDS and the co-epidemic of tuberculosis. Even on that daunting file, a comprehensive, holistic strategy was developed by 2004 that won praise from the United Nations (at a time when the national government was drawing massive condemnation). This was followed soon after by the roll-out of antiretroviral medicines, which has stabilized the pandemic, albeit still at one of the highest levels of prevalence in the country (Dyer 2012). The roll-out of a program for medical male circumcision, in collaboration with the Zulu king and other traditional authorities and healers, gives a further degree of hope that the disease can be contained if not eradicated, notwithstanding continued pervasive gender-based violence and poverty (KZN 2015).

## A DEGROWTH PERSPECTIVE

However much one admires the boldness of the vision, and can appreciate the multiple improvements to people's daily lives since 1994, it is legitimate to ponder whether such achievements and the proposals made to date are sustainable in light of the gathering critique of conventional econometrics and the environmental limits of growth.[16] Certainly the post-2004 economic boom quickly petered out and was soon followed by the rippling global effects of the Great Recession and a period of de facto municipal bankruptcies. The most obvious physical changes to the city over the two decades of freedom meanwhile directly recall some of the anxieties around modern urbanism expressed by city planners in the 1970s or earlier: low-rise residential sprawl over greenfields, new malls with huge parking lots and highway interchanges, the extension of the central business district with "Motor City" (more shops and car dealerships), and new accretions of informal housing in sometimes highly visible places. The Camp's Drift proposal also raises the question of the sustainability of costly residences, hotels, and recreational facilities that depend so heavily on the state to manage the pollution and siltation of the main attraction (the river), a stone's throw from some of the city's most economically marginalized and unhealthy population. From a degrowth perspective – or indeed many other critical green perspectives – there is a disturbing continuity in thinking between this project and that which created high levels of pollution, erosion/siltation, and poverty in the first place.

To begin the degrowth critique of the present development strategy, we need to step back to the RDP, which unfolded under intense scrutiny and pressure from donors, international advisers, and big business. Put simply, the RDP required the state to provide expanded services within the constraints of fiscal discipline as defined by the "Washington consensus" on the imperative of structural adjustment. At the national level, this meant aggressively paying down the external debt accumulated by the apartheid regime, and adhering to neoliberal prescriptions for growth, including recommended taxation levels, ability to expatriate funds, currency convertibility, and central bank independence from elected officials. At the local level, the need for fiscal discipline was signalled with a grant from the central government that capped expenditure on each new RDP house at the rough equivalent of US$500 (including the costs of land, construction materials, labour, and related services). A guideline for the environmental impact assessment for each was set at roughly US$6.50 (Roukema 2006, 129–31).[17]

For left-wing critics of the government, these amounts were sufficient to dull people's desire for real transformation but inadequate to ensure human dignity and equitable economic development. Local South African Communist Party leader Themba Harry Gwala mockingly termed the RDP the "Revolution Delaying Programme" (Haswell 1995, 88). For city officials, these expenditures allowed for quantity and a rapid pace. Funding was admittedly insufficient to meet modern planning objectives, and perhaps even code specifications of earlier times such as the 1920s. The hope, however, was that the new housing in such quantity would give the urban poor a leg up to get themselves established with restored confidence and security, and that over time people would invest in home improvements, extensions, and gardens. This has certainly happened in many cases. The more common trend, however, was entrenched in the logic of the market. Cheap land meant land long distances from the city centre. Short cuts, hasty construction, scrimping on building materials and labour, a soulless aesthetic, and small size were the norm, particularly since the actual construction was tendered out to private contractors, who took a cut of the tender as profit. One sees the result in places like France: RDP housing that is degradingly small (about a third less in area than the pre-apartheid homes in Ashdown), with faulty sanitation, leaky roofs, and crumbling walls, far-removed from centres of meaningful economic activity, and often contributing to erosion and waste pollution. The proliferation of such tracts on far-flung fields and hillsides, disproportionately comprising female-headed households, seems almost designed to increase the sense of alienation from place. Not surprisingly, a recent survey found that residents in France and other RDP areas have almost the lowest levels of sense of place or belonging in the whole municipality, only marginally better than the city's most notorious informal housing settlement, Jika Joe (Goebel 2007).

One unanticipated consequence of RDP sprawl was in fact more informal settlement close to the city, in areas like Jika Joe. Inner-city informal housing had historically taken the form primarily of DIY extensions to existing homes or backyard shacks in Sobantu and peripheral districts like Camp's Drift. After 1990, however, the lifting of restrictions on African mobility led to the proliferation of informal structures on vacant lots within the city proper. Those lots had usually been left empty due to severe environmental constraints. Ash Road settlement, as Jika Joe is formally called, lies especially close to town, right beside the N3 highway and mostly below the fifty-year flood line. While such informal densification may eventually result in relatively stable and upgraded communities, the present trend has been for unsanitary hovels to be thrown up in a noisy, polluted, and flood-prone location, with the added mortal danger of illegal connections to the electricity grid and frequent fires. Housing conditions in Jika Joe were thus easily as bad as could be found in the worst sections of Edendale or Macibise in the 1940s. Yet people prefer to live there when the alternative is a bleak field of boxes

7.2 Walking to the taxi rank on the southwestern fringes of Greater Edendale. It takes three rides to make it to town – an hour if the commute goes well

an hour or more commute away. Those who qualify for an RDP home on the outskirts often take possession only to rent it out while keeping their wattle and daub, tin, or scrap homes closer to town.[18]

Contradictions in the economic growth model also became increasingly manifest under the post-1996 Growth, Employment and Redistribution policy (GEAR) that superseded the RDP. GEAR, even more than the RDP, hitched South Africa's macro-economic policy to the Washington consensus, including through the rapid dismantling of international trade barriers that had supported much of South Africa's industrialization during apartheid. Pietermaritzburg and the Midlands were especially hard hit by the closure of factories that could not compete with the sudden flood of cheap imports from abroad. High-employment (which is to say, "inefficient") clothing and textiles, leather, and shoe industries were most vulnerable to foreign competition. The Sutherlands tannery, the country's oldest (founded in 1862) and one of the main economic anchors of Edendale and Imbali, closed in 1995 (Harrison, Futter, and Meth 1997). The Edendale tannery shuttered soon after,

7.3 The Edendale tannery, structurally adjusted

leaving its toxic sludge ponds for someone else to attend to (Zuma 2002). Province-wide, but most heavily concentrated in and around Maritzburg, employment in the footwear industry declined by over 50 per cent from 1993 to 2002. For those who managed to keep their jobs, average remuneration over the RDP/GEAR periods declined by 10 per cent (Mosoetsa 2011, 11).

Older statistics illustrate the decline even more dramatically. In 1950, 706 residents of Edendale found employment in industry in Edendale itself, about one quarter of the formally working population. Moreover, "some Natives appeared to hold very responsible positions, and there was little European supervision" (University of Natal 1951, 81). A further 600–700 were formally employed at the Sutherlands tannery or elsewhere in neighbouring Plessislaer. Today, manufacturing in Greater Edendale is virtually non-existent. GEDI's original proposals for the future of that sector were restricted to "small scale and service industrial activity" (Msunduzi Municipality 2006, 36).[19]

The closing of dirty industries is, ultimately, a positive development from a green environmental perspective. The civil society organization GREEN claimed the closing of the Edendale tannery as one its first "success stories," while former shop steward Thembinkosi Mkhandawiri similarly blamed the bankruptcy, at least in part, on tougher environmental inspection by the parastatal Umgeni Water (Zuma 2002, 48). Yet, the environmental benefits have been mixed, in the absence of state capacity to clean up the mess left behind. The cost of investigation and lawsuits, let alone physical remediation, is simply beyond the ability of the local authority to address. As one resident in the neighbourhood expressed it:

> I feel very relieved [that the tannery closed], you couldn't eat when this smell was there and there were times when it felt like they are stirring it up. During heavy rains the smell comes back again but it's very mild now. But my main worry is the children who play in the ponds and I'm also concerned about older women who go and pick "*imifino*" [greens] near the ponds. I wish the government could put some electric fence around those ponds to stop people from going there. That place is dangerous, a cow got stuck in the ponds, when people removed it, its skin was wearing off but they slaughtered and ate it. There's poison in those ponds. (Nomvula Msimang, quoted in Zuma 2002, 46)

The environmental impact of long-term, massive structural unemployment plus the hidden carbon and other pollution costs of foreign imports should rightly be considered in this "efficiency" trade-off, and further research in this area is needed. For industries that survived trade liberalization, however, unhealthy tendencies of intensified global competition are fairly obvious. The raft of post-1998 legislation and the potential of onerously expensive retrofits, waste disposal, worker protection, and clean-ups, in the context of the cut-throat economic environment created by GEAR, gave companies and individuals positive incentives for illegal dumping, night-time burning, and hazardous labour practices. The gamble was that they could get away with such cost-savings activities or bluster their way past enforcement of the law. By the admission of government officials charged with monitoring and prosecuting environmental crimes, the chances are good that they will and are probably getting better. Indeed, the raft of new legislation and bureaucracies empowered to intervene can itself be an impediment to effective intervention. As the city's legal adviser Johan van der Merwe explained to Adrian Pole, a UKZN student researching the efficacy of the enforcement of the polluter pays principle: "Ah, there's been so many changes in legislation that's brought so many different duties on officials and Council that I think … people don't really know where to start" (Pole 2002, 52).

Contradictions and ambiguities in many of the key documents that are supposed to frame local authority responses to environmental issues exacerbate the problem. The constitution, for example, articulates both green and brown perspectives with only the vaguest suggestions on how to reconcile their frequently antithetical requirements. As an example of how this can affect local practice, invasive species like wattle and gum have long been known to have quite negative impacts on soil and water management. The green agenda would be to take them out, with the added benefit of providing a significant number of steady jobs. Yet people still prefer the invasives for their short-term economic values. The brown agenda thus suggests planting even more. In other cases, people have resisted the re-introduction of indigenous vegetation on barren or degraded hillsides on the grounds that bush provides cover for criminals. Security in the form of clear-cut fields trumps the green argument.

To the chief environmental officer, Hilton Ryder, the proliferation of rival arms of different spheres of government compounds the disarray:

> What happens is that sometimes we find that when we go to industry, we're told that somebody from environmental affairs is there, someone from ah water affairs is there, someone from Umgeni Water … Now these investigations should be carried out by local authority, who are conversant with the local conditions, who know the local leaders … and all those issues … They're close to the ground, and local authorities work with the ward councillors etc. Now to have provincial people appointed to … to actually go and carry out … physical inspections and tracing of infectious diseases, is not my understanding of how it should be done. (Pole 2002, 8–9)

Tellingly, Ryder defines how it should be done with reference to the history of how not to do it:

> You know we've had a long history of fragmentation, I mean we really had thousands of them [agencies, bureaucracies, regulatory bodies] … I can remember ten, fifteen years ago when a restaurant owner would say to me "you know there's twelve inspectors that

are able to come to these premises." There's the guy from the citrus board, there's the guy from the potato board, there's the environmental health officer, and there's this person and that, and the licensing officer. You know, that's the sort of fragmentation, it was horrific. And now you find yourself going back into that fragmentation. (8)

The question of where to target interventions is also fraught. Big companies with smokestacks and affluent shareholders attract the most obvious attention for their environmental crimes. However, the temptation to take environmental short-cuts is at least as strong in the informal economy, where unemployed people struggle to get by using whatever means they can. The stereotype of an informal economy worker (not counting sex work or drug trafficking) is a poor but smiling woman selling things on the roadside or sewing things in her home. The most serious environmental consequence of her work is litter. Yet much of the legitimate informal economy actually consists of small motor repair, cleaning engines, picking through and recycling often hazardous waste, and cobbling together cheaply manufactured items like tourist bric-à-brac that can be sold cheaply precisely because the producers avoid environmental, health, and labour regulations. Individually, the scope of harm may be small but, cumulatively, with over 50 per cent unemployment and minimal ability or political will to enforce regulations, the impacts add up. Ryder again describes the challenge, with the story of an informal operation making highly toxic lead sinkers for fishing: "It was actually quite a big operation. This guy was supplying … the whole of Natal. He had 8 workers working on it continuously night and day … a sort of group of mini-factories, where you had an engineering works and at the back of this engineering works this lead smelting thing … It was just an individual that was trying to fly a kite. I mean he was just trying to make some money" (ibid., 17).

According to Pole's informants, the majority of the large, formal industrial concerns generally cooperated with the city's requests to reduce effluent and to pay for clean technologies, clean-up of spills, and other environmental costs. A small number, however, including key hold-overs from the "demented" period of industrialization, have actively conspired to thwart the law. Secrecy and obstruction are preferred tactics. According to Mike Greatwood, then deputy city engineer, in one case of investigation "we literally had to hold their security guards at gunpoint to open the gate" (41). Another tactic is to threaten liquidation. This is powerfully persuasive to city councillors who fear the political fallout of adding to the already dire unemployment situation. At least one lengthy prosecution for multiple oil spills was abandoned out of concern that the company would simply declare bankruptcy (46), and city reluctance to push too hard to enforce its own by-laws is a constant complaint by many of Pole's informants.

As with much else in the South African context, this issue is often framed in racialized terms. Reputedly, the most notorious offenders are companies owned predominantly by Muslims of Indian descent. Some of the emotive comments by frustrated officials about the owners' perceived culture of indifference to the environment, and to the African people who live downstream of the offence, do not bear repeating. However, Ebrahim Seedat, technical manager at Capital Products (an Indian-owned company), tacitly accepted at least part of the critique:

> Margins are under pressure. What they call refining margins are under pressure, they call it profitability … Per ton is under pressure … We have a number of companies that have fallen by the wayside. Most of these companies that have fallen by the wayside are, if you'll excuse me, white companies. They have now been replaced in part by so-called Indian companies. Sadly the culture with Indian companies is more of maximizing profit and less of cleaning up the environment. (Pole 2002, 103)

The economic squeeze is by no means the only resonance with Pietermaritzburg's earlier struggles to industrialize. As in earlier periods, albeit for different reasons, a growing crisis of legitimacy in local governance undermines the stated objectives of social and environmental justice. Exactly when and how the kind of cynicism and corruption set in that led the provincial government to impose an emergency administration over the municipality in 2010 is beyond the scope of this research. BESG, which enthusiastically welcomed collaboration with the city after 1994 and has generously acknowledged key contributions by the city to providing housing and "empowering" local communities, by 2007 viewed "the efforts of the municipality as being inadequate, at best, and downright incompetent at worst" (Magwaza n.d.), with declines in service delivery and an increase in punitive actions against the poor. Many of Roukema's and Sutton's interviews also make it clear that lack of confidence in the municipal government was widespread and entrenched by the middle of the first decade of this century. Informants uninhibitedly shared their view that government paid only lip service to the environment, to public consultation, and to its own stated principles on a wide range of issues. We can discern an obvious sense of frustration even earlier in the remarkably frank interviews collected by Pole (2002, passim) with some of the key officials in charge of maintaining service and infrastructure ("haven't got the manpower," "lack of urgency," "negligence," "cutting back," "acrimony," "briberies," "very scared to do anything," "death threats," "impunity," "I think they just pay off everybody, quite honestly," "no monitoring, poor enforcement, the criminal justice system … takes three years to get anything going," and much more).

The details of the erosion of good governance in Msunduzi will emerge in the course of time through the courts and investigative journalism. For now, I want to highlight some environmental consequences of corruption. Corruption in the tendering process, whereby private companies with fraudulent or scurrilous bids win contracts to perform services at inflated rates paid for by the municipality also typically in pursuit of Black Economic Empowerment (BEE) targets, is probably the most directly harmful. Among numerous allegations of corruption was the contract to dredge the canal section of the Msunduze River (that is, Camp's Drift). It was awarded in 2008 to a company owned by a disgraced former politician, and remained unfinished for over three years after an expenditure of nearly R13 million in state funds, or three times the cost per quantity of sludge removed that was subsequently charged by the provincially funded company hired to finish the job ("Mjwara: Parties Call for Action on Tender," *Witness*, 9 June 2008, *Witness*, "Camp's Drift Canal Desilted," 23 December 2011). Another example suggests a more subtle form of collusion between management and unionized workers to trim off commercially valuable scrap from the municipal dump at the expense of informal waste pickers or the city's own potential revenue generation (Hallowes and Munnik 2008, 157).

Such corruption not only drains the municipal budget. It also indirectly undermines the ability of the municipality to achieve its goals by justifying popular resistance to paying rates and fees for services. Non-payment for municipal services was a relatively effective tactic in the anti-apartheid struggle, but it surged to roughly half of all customers in the year prior to the municipality's bankruptcy in 2010.[20] Lost revenue and expensive repairs also occur through vandalism of city property such as water meters. People then commonly justify illegal connections to water and electricity as a legitimate response (protest) to an incompetent, unresponsive, or abusive state. Cable theft for backyard smelters and home electricity hook-ups has also emerged as a significant form of "protest" and an element of the informal economy that passes huge costs onto the formal one (repairs, brown-outs, blow-ups, fires, and enhanced security, notably). People may acquire "free" electricity in this way but at additional un-priced costs to their surrounding environment and to human lives. As one investigation described it, "stolen Telkom strands used for pirate electricity connections have turned fields around informal settlements into no-go areas for sport or children's play. With at least a dozen fatally electrocuted last year, residents are now living the bizarre 'island' lifestyles of Mozambican villages that were once surrounded by suspected minefields" ("Copper Heists Change How We Live," *Witness*, 30 April 2014, 3; "Angry Jika Joe Residents Threaten to Disrupt Elections," *Witness*, 7 May 2014, 3).

Lack of capacity, in contrast to corruption, refers to honest limitations of money and personnel needed to guard against environmental abuses. New demands since 1994 have only widened the gap between means and aspirations. According to one municipal official, not a single environmental impact assessment was carried out for all the low-income housing developed from 1994 to 2006 (Roukema 2006, 144). Even where funds are available, personnel shortages make delivery difficult. Again, this has been a long-standing problem for Pietermaritzburg, but it has been greatly exacerbated by hiring freezes, redeployments, and the flight from the city of white and Indian professionals since the late 1990s. The consultants for the city's strategic environmental impact assessment described the phenomenon drily: "The municipality has … lost key staff with extensive experience; and, if and when replaced, new staff often lack the requisite experience. This has resulted in an overall and significant decrease in the municipality's capacity for environmental governance" (SRK Consulting 2010, iii).

Structural issues can make it extremely difficult for the local authority to resolve such problems. Local perspectives are simply not enabled in parliament by the national system of election by proportional representation. South Africa's post-apartheid constitution sought to guard against the kinds of regional or ethnic voting blocks that wrought havoc in many other newly independent African nations. Hence, people do not vote for a local constituent representative but for a political party, which presents its own list of candidates in order of priority. This means that voters in Msunduzi have no way to send a candidate to Pretoria who directly voices their local concerns in

7.4 A bad day at the dump

the development of policy. The housing subsidy is a prime example of how this system affects policy, as it seems to reflect the theoretical cost of greenfields development in flat parts of Gauteng rather than the topographical complexities of Msunduzi. And in what way can voters in Edendale express their opposition to national "free trade" policies that may create high tech opportunities in Gauteng or Cape Town but that further decimate the local formal economy?

For many, the suspicion is that decision makers in the ANC are actively hostile to environmental initiatives, and that this position reflects, in part, bitterness against whites' callousness towards black lives in contrast to their concern for certain majestic or cuddly animals. Lack of interest (rather than hostility) could easily have the same effect and may be explained by simple electoral calculation. The national survey by Roberts, wa Kivulu, and Davids (2010, 215) found that people in KwaZulu-Natal as a whole expressed the second lowest level of environmental concern by province in the country (narrowly behind Mpumalanga). Hlahla, Goebel, and Hill (2016) underscore a long-term implication of such negative attitudes, finding that "a 'stigma attached to working with the Earth'" among young black women hinders their recruitment even to relatively well-remunerated green economy activities. From a narrow partisan perspective, there is little political gain for ruling party candidates or local councillors to press the issue. As a result, even the equivocal language in the constitution and law ("reasonable," "justifiable") may be insupportable. This shows at the national level in a shift from equivocation to outright prevarication. South Africa's Intended Nationally Determined Contribution to climate change action is an important such instance that reflects the seriousness of the ruling party's commitment to the international community. While it unambiguously endorses the science that demands economic transition, the document also disingenuously states that the country is dependent upon a "fleet of old and inefficient coal-fired plants that are nearing, but not yet at, the end of their design lifecycles" (South Africa 2015, 2). This translates to mean that the continent's most profligate greenhouse gas producer cannot afford such a transition just yet. In truth, South Africa is currently building what will become the largest coal-fired power plant in the world, at phenomenal cost and heavily supported by a World Bank loan and the ANC's own investment arm, for intended service into the 2050s (Phaahla 2015).

For many people in the most stressed communities, the sense of alienation from the environmental and other justice ideals expressed in the constitution and law is compounded by the evident gap between rhetoric and reality in other spheres. Black Economic Empowerment, notably, was originally conceived as a means to achieve social justice by correcting the historical dominance of non-Africans over the nation's economy. Yet it has patently served the interests of a narrow, politically well-connected circle while preserving the historical contours of severe exploitation of labour. Among aspects of capitalism that BEE exonerates are inherited real estate market values, which allow the state to buy up black-owned lands at low cost while land in former white group areas is priced out of reach. A rousing speech by Phumani Zondi of the Edendale Landowners' Association at the historic 2006 Land Summit points to a deep sense of grievance at the municipality for obeying the market in the interests of the new BEE elite:

> What makes this more disappointing is that it happens when the Government of the day is talking about Black Economic Empowerment. What is Black Economic Empowerment if the very people who must be empowered are dispossessed of the means

> of production and those means of production are handed over to others who have been privileged over the years? Why is it so easy to expropriate land which is in the hands of black people when the very Government is failing to expropriate land that is in the hands of white control? People with only 13% of the land even to this day and age are the victims of expropriation in the name of development but they are left underdeveloped. Our grazing land was expropriated by this government and the compensation we received is a disgrace and an insult to our dignity. This phenomenon continues unabated. Who is prepared to listen to us? Who will understand our pain? (Zondi 2006)

One further unresolved tension lies in the role of traditional chiefs, who remain in place over nearly half the land area and in a position of authority over more than a third of the population of the municipality, including much of its poorest. Vulindlela was predominantly an IFP-controlled area and hence there may be lingering political issues between the chiefs there and the ANC-dominated city council. Alexius Amtaika also found deep distrust between the unelected, often undereducated, and sometimes corrupt chiefs on the one hand, and the educated, sometimes also corrupt, and frequently patronizing municipal officials and business people on the other. Among the latter, many believe that the chiefs' guardianship over traditional communal land tenure creates a "tragedy of the commons" effect that nullifies development initiatives – if no one owns the property, who will invest to protect or improve it? To Amtaika, the problem lies in the mode of governance: "The institution of traditional authorities actually blocks the implementation of development projects due to its authoritarian nature" (2001, 231). One suspects that, in many cases of ill-planned projects such obstruction could be a good thing. Try as I might, however, I could not find this aspect of transformation discussed in key pertinent municipal documents beyond the following, a passing statement in the latest available IDP: "There have not been, and are currently no plans to regulate land use within the Vulindlela area, which falls under a tribal authority" (Msunduzi 2015, 75).

The rights to vote and for the public to voice anger are widely and emotionally appreciated in South Africa, and the party seen as mostly responsible for attaining those rights remains overwhelmingly the voters' preference, more so in KwaZulu-Natal than most of South Africa. As Sarah

Mosoetsa (2011) poignantly attests, however, structural unemployment at very high levels and the perception of an uncaring elite have exacerbated many profound social ills, including gender-based violence, alcoholism, and gang criminality. One needs to be careful when older South Africans tell white researchers that things have gotten worse since the advent of black majority rule. When older black women even in measurably "improving households" repeatedly say as much to young black researchers, however, we need to pay attention. Let us listen to four of Mosoetsa's female informants in Africa's richest, most developed, best-constitutioned country:

> Things were much better during apartheid; there were jobs and factories did not close down. Mandela came in and there was unemployment and now that it is Mbeki, things are even worse. (MaMnisi, 31)

> Before Mandela and, even worse, Mbeki, came to power, we had jobs but now there are none. All our politicians do is take care of themselves and they remember us when there are elections. They have left the community and they now drive expensive cars. We have been left behind to fend for ourselves. Things are worse than they were when white people ruled this county. (Thulisile, 49)

> I do more work now in the house than I ever did when I was working, everyone in the house was working, and the children were going to school. It is different now that there are no jobs. I am not paid for making sure that all ten mouths are fed and that everyone wears clean clothes and lives in a clean house. It is not fair to me but what can I say; I am a woman. (Kethiwe, 41)

> Gone are the days where instead of growing my own vegetables, I used to buy them. There was money then. I was employed then and so were my brothers and sister. Now everyone is unemployed and no-one has money to spend on luxury items such as tomatoes. (MaNtobi, 41)

For tomatoes to be a luxury in a place with a growing season of eight to ten months, plentiful water supplies, and ample labour suggests something seriously amiss. Yet alongside food insecurity in that basic sense we now also see a rapid rise in coronary heart disease associated with caloric-rich, nutrition-poor, atherogenic diets. A rapid increase in the consumption of sugar and fat, and to a lesser extent red meat, is also linked to a rise in chronic "self-induced ailments" such as diabetes Type 2, gout, hypertension, and obesity (Dyer 2012, 276). Jonathan Crush and Mary Caesar (2014) raise the alarm, showing how far Msunduzi has progressed along the development path that, in North America, is marked with grave health consequences: the collapse of locally produced, informal sources of food and the rise of corporate, industrialized, globalized food supply chains that culminate in supermarkets whose principal obligation is to reward their shareholders. "In this respect, Msunduzi offers other African cities a picture of their own future" (174).

For some people, this type of ill health is quite acceptable – fat, to many, is culturally a sign of prosperity and spiritual well-being. For others, however, the new diseases, including HIV/AIDS, are part and parcel of a development paradigm that is profoundly alienating. Many residents of Edendale and environs turn to faith for answers, consolation, and hope, and Denis (2004a and 2004b) among others has shown how women in particular benefit from "religious health assets" (in current jargon, referring to community building and psycho-social support, as, for example, demonstrated by the *manyanos/kopanos*, the established churches' women's associations). But faith is also sometimes expressed in a desperate need for scapegoats. The rise of xenophobia is perhaps one of the most dispiriting signs of the latter. Indeed, as harsh as GEAR was for South African workers, structural adjustment in much of the rest of Africa has been even harsher over a longer period of time. If we count state reactions against structural adjustment, such as Zimbabwe's accelerated land reform and dismantling of its commercial agricultural sector, the crisis of neoliberalism elsewhere on the continent has generated a flood of migrants to South Africa seeking work. Many South Africans blame their economic or other misfortunes on these migrants, and Pietermaritzburg has not escaped the scourge of mob violence against them.[21]

Others turn to alternative paradigms rooted in pre-*abelungu* African traditions (*uBuntu*), the sacred texts of universalist religions, or, perhaps most commonly, a judicious mix of all in order to help them spiritually and emotionally through the turmoil. To the extent that they are organized, many of the new faith groups profess strongly against the fetishization of growth or the glamorization of consumer products and all that that implies. Yet, as Garner (1998, 2004) shows, the fastest-growing churches in Edendale do not present a coherent critique of materialism. Beyond the biblical injunction for Man to exercise domination over Nature, they appear to be largely oblivious to the environment. Garner takes a guardedly optimistic view that Apostolic and Zionist churches do encourage elements of social trust, respect, and self-discipline that could be "fruitful and sustainable as a context for shaping of an African brand of capitalism" (2004, 102). But he also notes profoundly patriarchal attitudes, the "reinforcement of highly consumptive aspects of Zulu culture," "strong legitimation" of the market economy, and rejection of state intervention to redistribute wealth. The turn to faith sometimes also brings self-harming emotional responses and neo-traditional occult practices.[22] In the context of the global trends already noted, these religious responses are hardly promising when it comes to sustainability. On the contrary, the cultural space available to introduce and propagate science-based environmental management tools has in many ways been narrowed in recent years. To put this less drily, who is going to want to venture down into the spruit to clean up the litter if they might encounter the corpse of a victim of *muthi* murder, as Kwanzi Mngadi of the Sobantu Environmental Desk wondered in an interview with Adrian Pole (2002, 64)?

≈

Over two decades have now passed since the heady early years of transition to democracy, and much has unquestionably been achieved towards the stated goals. A single municipal government extends from the kraals of Elandskop to the gated communities of Town Hill, from the tiny spaza shops of Macibise to the sprawling Liberty Midlands Mall, with concrete targets established to meet the high ideals of social justice and environmental sustainability. The Msunduze River is, reportedly, significantly cleaner in most sections (Ward 2013), while electrification has cleared the local air of at least the most visible concentrations of smog that used to hang over town in winter. As Dyer (2012, 300) points out, overall levels of smoke pollution recorded in 1998 had fallen to just 16 per cent of those during the peak years of the mid-1970s.

Yet, an air of gloom nonetheless hangs over the city, as it does over much of "the new South Africa." It still comes literally from frequent fires in the municipal dump, but also metaphorically in the form of strikes, service delivery protests, horrific crime, and the daily grind of suffering that unfolds on the pages of the mainstream press or, almost worse, that passes as normal and invisible in the form of preventable, chronic disease. Many of those stories hinge on the stark contradiction between statistical "facts" about growth and people's lived realities. There is the fact of a massive roll-out of free houses, for example, yet people are still living in conditions that would have shamed *uPoyinandi* seventy years before. There is the right to an environment that "is not harmful to health," yet an aggressive expansion of coal-burning thermal power stations, granting of licences for hydraulic fracturing (fracking) in the upper catchment area, and the public celebration of patently unsustainable consumption habits. There is a vastly expanded social welfare system, yet the most inequitable and highly racialized distribution of wealth with the worst "misery index" in the world, and rapidly worsening perceptions of corruption. How can the economy have grown more than 300 per cent in twenty years yet the number of unemployed people has risen by 20 per cent, with many working in virtual slavery conditions? And how can a city where 95 per cent of the population were reported to have access to clean piped drinking water have, in 2001, experienced a cholera outbreak affecting hundreds of people (Dyer 2012, 122, 130)?[23]

Some of these problems, obviously and infuriatingly, have been visited upon South Africa by its international friends and are beyond the reasonable ability of the country to resist. Among other things, there will need to be global frameworks to phase out tax incentives and subsidies for greenhouse gas–emitting production, the elimination of tax havens or other ways that the extremely rich hide their assets, and large-scale financial transfers from developed to developing countries to offset costs. Many critics, however, fault the national government for colluding in some of the worst practices, and argue that the rhetoric of radical change since 1990 belies profound continuities with the apartheid past or, worse, has actually intensified structural injustices that bode ill for future environmental health. Alexius Amtaika's assessment of the state of local government in South Africa under a constitution that enshrines rights to health and a healthy environment is sobering. He writes: "Regarding pollution and environmental degradation, most municipalities fail to foster constructive attitudes among citizens … They have made little effort to influence public behavior, except in a coercive way" (2013, 310). While scholars such as Jacklyn Cock (2012) see hope in the emergence of a labour-linked "eco-socialist" movement, and Robins (2008) and Kolbe (2014) in "citizen science" by which concerned, empowered citizens will challenge complacency and venality in government and the corporate world, Amtaika notes a very deep and likely contagious cynicism in the way the public is currently engaged. "Consultation is mainly a political ritual, which brings legitimacy, but does not produce qualitatively different decisions" (209). As the proponents of separate development learned, this is a recipe for disaster.

What, then, would a degrowth (truly sustainable, democratic) alternative look like? Historians are probably not the best positioned to set an agenda for the future, and I am not going to pronounce one. But, by way of conclusion, I do believe this book demonstrates two basic truths that could carry the conversation forward constructively: 1) the large majority of the people of Edendale and environs absolutely need greatly improved services and better quality of life, both for their own social justice merits and as a necessary condition for establishing a sustainable economy; and 2) growth, as historically and presently still understood, is unlikely to achieve that objective. Indeed, the growth model today promoted by international advisers, pursued by all spheres of government, and endorsed by all the major political parties has not served Edendale well over the course of its history, and is currently producing the sharp, unsustainable contradictions discussed in this book and widely acknowleged by South Africans of all backgrounds. Even when qualified by reference to green economy, green capitalism, and concrete targets for peak, plateau, and decline of greenhouse gas emissions, growth is fundamentally incompatible with the imperative to mitigate and then abate global climate change, food insecurity, emergent diseases, and widely anticipated (in fact, already observable) conflict over resources like clean water. New technologies will undoubtedly provide some answers to some of these problems. From the historical record, however, we may also expect that new technologies could easily introduce new toxins, new environmental stressors, and new social inequalities. We will need to think more creatively about alternatives to the growth model if the imminent risks are to be averted in a non-authoritarian way.

A wealth of ideas is coming out of the degrowth and other radical environmentalist movements. At the national level and below, these include the redistribution of wealth through a guaranteed basic and a regulated maximum income; incentives for urban organic farming and other local, convivial production; local community currencies and barter systems of exchange; cooperatives, ecocommunities, or "transition towns" with shared housing, efficient public transportation, and disincentives to private automobiles; digital commons (that is, free and accessible knowledge); bans on public advertisements; incentives for voluntary restraint on consumption; and much more. These may sound utopian or contrary to human nature, but they are in fact already happening at different scales in cities around the world. It should also be recalled that many of these ideas are not only implicit in the traditional ethic of *uBuntu* (and perhaps, as well, Arya Samaj, liberation theology, and other minority, eco-centric expressions of non-African religious practices in Natal). They have also been realized to a noteworthy degree in the past in public policy, even in a place like Edendale and environs: local gardening, job sharing and communal labour, war on waste, subsidized public transportation, social medicine, and so on. Degrowth initiatives should thus be able to look both to current successful international experiences and to local history for inspiration. Small cities in Africa with vestiges of convivial culture, a relatively undeveloped carbon-dependent infrastructure, and a low political profile nationally may actually be in a better position than the big cities to leapfrog to a transition economy as the urgency to do so becomes ever clearer.

# Notes

## Chapter One

1 See Msunduzi Municipality (2002, 2006, 2009) and the website http://www.msunduzi.gov.za/ for the municipality's official perception of itself, its version of its history, its vision for development, and the divergent population estimates cited here. Counting all the formerly separate jurisdictions, Msunduzi's population makes it somewhere between the eighth and twelfth largest municipality in the country, with an official census count of just over 600,000 in 2011.

2 Johannesburg's Alexandra, for example, was founded in 1912 and encompassed 800 hectares. Durban's main freehold township, Clermont, was founded in 1931 on 647 hectares. To make another size comparison readers may find easier to visualize, the original Edendale farm is just less than half the area of Manhattan (population 1.6 million). Greater Edendale by today's definition is roughly equivalent to that island in size.

3 These connections, as with those in the following paragraph, have never been systematically explored, and undoubtedly many others could be mentioned. I have simply cobbled these few together from diverse passing references in Limb (2010), Brown (1982), Buthelezi (1987), Swanson (1990), Verwey and Sonderling (1995), Couzens (1985), van Diemel (2002, also available through South African History Online, at http://www.sahistory.org.za/, accessed 2 May 2013), Denis and Duncan (2011), and Mkhize (2015).

4 Pietermaritzburg Archive Repository (PAR), SNA 1/1/40 410/1880, letter from Acting Secretary for Native Affairs J.W. Shepstone, July 22, 1880.

5 See also Dr G. Park Ross of the Union Health Office, who reported in the same paper that he knew of "no other place in the Union where conditions

were so bad just outside a large town" (*Natal Witness* [hereafter NW], 2 March 1933).

6 See "Msunduzi Crisis," *Witness*, 4 March 2010. Comparisons with other cities hint at the extent of the crisis. Estimated per capita income for the whole of Msunduzi is less than half of Durban's and a third of Johannesburg's while, according to one report, "severe" household food insecurity is now greater in Msunduzi than it is in much poorer Maputo or Blantyre (Caesar, Crush, and Hill 2013, 14). See also http://sacities.net/workwith/msunduzi and Iyer Urban Design Studio (2014). See Greatwood's interview, 2 October 2001 in Pole (2002, 48).

7 Msunduzi Museum, personal observation and the Msunduzi Municipality website, http://www.msunduzi.gov.za, as observed over multiple visits, 2006–15. The photo referred to is of the Edendale hospital, which is neither in historic Edendale nor of particular architectural merit (other than exemplifying apartheid-era contempt for local scale or existing urban character). That noted, there are two rooms devoted to remembering Edendale and other predominantly black neighbourhoods in the city's other major museum (the Natal Museum) plus a collection of colonial-era bottles collected from an Edendale dump, not currently on display (Rodhén 2008, 169, and personal observation).

8 The public history proposal is laid out in Msunduzi Municipality (2009). I would also like to point out another, more grass-roots but apparently now equally moribund, public history initiative linked to community development, described and astutely critiqued by Whelan (2012).

9 A small sample of this historiography must include Parnell (1988, 1991, 1995, 2002), Maylam and Edwards (1996), Nauright (2000), Beall, Crankshaw and Parnell (2002), Freund and Padayachee (2002), Beavon (2004), Goodhew (2004), Bonner (2005), Murray (2008), Nuttall and Mbembe (2008), McDonald (2008), Bonner and Nieftagodien (2008), Bonner and Nieftagodien with Mathabatha (2012), and Field (2001, 2012). Important exceptions to this point are Mabin's (1986) study of segregation in Kimberly, Baines's (1994) and Kirk's (1998) of Port Elizabeth, and Minkley's (1998) and Bank's (2011) of East London.

10 See the website of the Centre for Civil Society at the University of KwaZulu-Natal, which, among other things, monitors social protests around the country (http://ccs.ukzn.ac.za/).

11 The failure of neoliberal nostrums is the subject of a large and often combative literature: Desai (2002), Ballard, Habib, and Valodia (2006),

Bond (2000, 2003), McDonald and Ruiters (2005), Dawson and Sinwell (2012), Marais (2011), Ferguson (2011), and Goebel (2011, 2015), for example. Global constraints are frankly acknowledged in UN-Habitat (2014, 2015), while a surge of negativity about the present state of the economy dominates the business press. At the time of writing, the IMF (2016, 6) has downgraded growth prospects to anemic levels for the coming two years. This may be optimistic, given the political turmoil surrounding allegations of corruption and "state capture" of the highest level of government by private, transnational interests ("Zuma's Day: Criminal Charges, Accusations of Betrayal and Calls for Impeachment," *Mail and Guardian* 17 March 2016).

12 http://esa.un.org/unpd/wpp/unpp/panel_population.htm, accessed 12 October 12, 2013. See also Sai (2010) for an African perspective on the demographic threats to development aspirations, also discussed in cautious terms in UN-Habitat (2014). The African Union monitors these issues, albeit with scant reference to politically sensitive demographic change, through its Climate and Development in Africa initiative (ClimDev-Africa): see http://www.climdev-africa.org/.

13 These and the following place names are described in Msunduzi's Integrated Development Plans (e.g., Msunduzi Municipality 2002 and 2011b). I will also note another zone within Maritzburg's economic and social orbit, Table Mountain, the much smaller former location (native reserve) on the eastern side of the city just beyond Bishopstowe. Table Mountain is close enough to be a prominent feature of the urban "bowl," but because it does not fall within the current municipality boundaries and is only rarely featured in municipal debates about health or the "black belt," I did not include it in the principal research. Yet as Goebel (2015) relates, events there in the 1980s and 1990s in particular have had a big impact on life in present-day Edendale and environs.

14 The "Johannes" cited here was almost certainly the same person as John Kumalo cited above.

15 Notably, Welsh (1973), Swanson (1976, 1977), Etherington (1978), Rich (1979), Slater (1980), Bradford (1987), Marks (1986), Duminy and Guest (1989), Guest and Sellers (1994), Lambert and Morrell (1996), and Lambert (1996).

16 Such are among the terms used on the municipality's website, http://www.msunduzi.gov.za/site/gevdi. The notion of defining hierarchies of urban space as a development planning imperative is also, I note, embedded in the South African Cities Network call for research on secondary cities

(John 2012, 10–15 and passim). The exercise of ranking cities, or parts of cities, is conceived in this case primarily as a tool to help assess budget priorities. Regarding North American urbanism as an ideal, an eye-opener for me came during a public forum on urban planning strategies in which a well-known local businessman advocated no planning whatsoever, citing Dallas as an example for the city to emulate. The free market would obviously guide development in the most efficient manner.

17 Urban histories from elsewhere on the continent also largely conform to the pattern. See, for example, Fetter (1976), Brennan, Burton, and Lawi (2007), and Rodriguez-Torres (2006), with Bissell (2000) offering an especially pointed critique of continuities between colonial and post-colonial privileging of elite interests in a capricious and obfuscating planning process (in Zanzibar).

18 I suspect this topic, and the rise of demagogic leaders such as Julius Malema, will be the subject of abundant careful study in the near future. For now, let me just cite Mngxitama and Nkopo (2013) as one of innumerable examples of essentialist, accusatory rhetoric taken from the liberal press; Roper (2013) for a sample reaction to the National Press Club's naming of the rhinoceros as newsmaker of 2012; and Maré (2014) for a cogent argument against the many absurdities and inhumanities of post-1994 racialism.

19 See also Giblin (2005) on a similar concept of "prioritizing history" and Dlamini (2005) on the role of history in the politics of identity among isiZulu-speaking youth. South African historians do, I emphasize, worry about this very much and often address it very thoughtfully. Here let me just cite important reflections on post-apartheid public history in Stolten (2007), Wells (2008), Hamilton (2009), Mangcu (2011), and Meskell (2012).

20 I should add that Amtaika is generally optimistic that the problems he identifies will be rectified given the other strengths of South African democracy. While I prefer to share that optimism, I do note how South Africa's rank in Transparency International's "perceptions of corruption index" has dropped significantly even in the short time since Amtaika wrote (see note 23 in chapter 7, below).

21 See, for example Coleman (2013) and Bundy (2014) devoted to the twenty-years theme. I will also note Freund (2012, 2014) as an example of scholarship that pushes back against the revisionism – a caution to those who might understate the structural and cultural factors working against the liberal moment in the 1940s – as well as Caesar (2015) as a modest

pushback to the pushback in her assessment of the LHC experiment.

22 I should emphasize that such reactivity is by no means confined to South Africa. I once tried to explain the concept of degrowth to a Nigerian colleague who intimated it was racist of me to think it should apply in Africa. I come back to this discussion in chapter 7.

23 http://www.msunduzi.gov.za/site/gevdi, accessed 1 December 2015. I say apparently token, as most of the text remains entirely unchanged from the GEDI era, and there are only passing, scant references to Vulindlela in the most recent planning document (Iyer Urban Design Studio, 2014).

## Chapter Two

1 See Ramphele (1991) for an early alert by an African intellectual to the dangers of the racialized nature of this debate, and Harber's (2011, 7) wry observation of the rise of environmentalism around the notorious Diepsloot informal settlement: "the closer blacks get, the greener whites get." A shocking eruption of this phenomenon caused a national uproar in early 2016, to which DA (2016) refers.

2 See Msunduzi's various Intergrated Development Plans (Msunduzi Municipality 2002, 2012 and 2015a) for articulations of the "balanced" perspective.

3 See the website http://abahlali.org/. Not to suggest a direct connection to Abahlali, but the late Wangari Maathai's ruminations also reflect that kind of optimism in grass-roots environmentalism (Maathai 2009). Other radical environmental scholars are discussed later in this chapter as well as in chapter 7.

4 Bond (2002), McDonald and Ruiters (2005), and Marais (2011) are among the many who make this point at the national level.

5 The figures are based on a conservative estimate of 18 per cent total infection rate and a low estimate of the city population. Working from provincial estimates and a more realistic population count, it is probable that there are closer to 150,000–200,000 people with HIV/AIDS in Msunduzi today, living preponderantly on the Edendale/Vulindlela side.

6 GilFillan (1920) is an example of this in global terms.

7 See, for example, Graham (2012), particularly her discussion of the works and speeches of J.M. Coetzee.

8 On the so-called black peril, see chapter 5, as well as Martens (2002, 2003, 2009) and Epprecht (2008). Bryant ([1909] 1966), Currey (1968), and the

novel *Bayete!* by Natalian segregationist George Heaton Nicholls (1923) are local examplars of this type of "analysis," and Blackburn ([1908] 1991) a biting satire of it (and of many other colonial and missionary prejudices).

9 See also Clark (1972, 49) and Currey (1968) with similar images of treelessness and erosion scars from the 1850s to 1865. Bews's map (1913) depicts six discrete pockets of forest on the mountainside visible from the city and reaching down into the valleys of Edendale, with wide swathes of open space above and around them.

10 This would seem to be an interpretation of the story that Shaka had a pregnant woman cut open so he could examine the foetus.

11 The progenitor of this thesis was actually an anthropologist, Max Gluckman (1960); it has been greatly enriched since by the scholarship of archaeologists such as Martin Hall (1976) and historian Jeff Guy (1980).

12 Key works in this historiography that have informed my own research include Beinart (1984, 1989, 2003), Anderson (2002), Moore and Vaughan (1994), Leach and Green (1997), Leach and Mearns (1996), Mackenzie (1995), McCann (1999), Alexander (2000), Dovers, Edgecombe, and Guest (2002), Neuman (2002), Showers (2005), Cock and Koch (2006), Tropp (2006), Hodge (2007), Sunseri (2009), and Meskell (2012), among many others. Early overviews of the emerging field can be found in Beinart (2000) and Carruthers (1990, 2006), for example. There has meanwhile been some pushback against overly broad critiques of colonial science. Bennett (2011), notably, found that South African foresters "actively rejected" inappropriate scientific knowledge from Europe and successfully experimented on naturalizing Australian trees to the local environment, with important economic benefits to the country.

13 For concise overviews of the main issues in the historiography of health in South Africa, see Digby (2008), Phillips (2012), and Parle and Noble (2014).

14 See Swanson (1977), which I discuss with reference to subsequent elaborations of the concept in chapter 4.

15 For critical perspectives on feminist approaches to environmental studies, see Merchant (1989), Leach and Green (1997), Dodson (2002), and Goebel (2002), for example. Feminist analyses of urbanization also often engage this literature, and I take great inspiration both from pioneers (Hellmann 1948) and works of critical scholarship by, among many, White (1990), Lee (2009), Bank (2011), and Goebel (2015).

16 An introduction to this burgeoning literature can be accessed through D'Alisa, Demaris, and Kallis (2015), as well as Schneider, Kallis, and

Martinez-Alier (2010), LaTouche (2010), Assadourian (2011), Cattaneo et al. (2012), and http://www.degrowth.org/. Taylor (2015) extends the critique to climate change adaptation in the Global South, while Jerven (2015) provides a devastating deconstruction of conventional econometrics/cliometrics (ways of measuring and promoting economic growth, including through naive use of historical statistics). The concept has scarcely been applied in African contexts (Ramose 2015 is a rare exception) but is implicit in much of the political ecology scholarship and radical environmental activism to be discussed below.

## Chapter Three

1 Colonial sources generally anglicize the term (lobolo), while isiZulu speakers differentiate between the verb and noun forms (*ukulobola*, meaning to offer cattle for, and *ilobolo*, the actual cattle for the marriage). *Ukulobola* could stretch over many years. Two important Zulu ethnographers interpreting this and other aspects of Zulu culture, including stresses experienced in an urban, modernizing context, are Vilakazi (1962) and Ngubane (1977).

2 See Ellis (1988, 1998). This point is also attested to in artwork from the 1840s to the 1890s (Wedderburn 1991). Natal's early economic development, including struggles over control of African labour, is the subject of considerable scholarship, which I will acknowledge below.

3 Van Sittert (1998) discusses this issue with primary focus on the eastern Cape. I have not found any research specifically on Natal, but we can surmise that a place like Edendale (progressive-minded but poor) would likely have conformed to the pattern van Stittert found – that is, farmers as very aggressive poisoners and employees as casual handlers of poison. I note as well Robinson (1900, 87), who attributes the defeat of a massive locust invasion in 1894 to "arsenic dousing," presumably of the crops intended for human consumption.

4 Pietermaritzburg Archive Repository (PAR), memo on Edendale by Professor F.B. Burchell (trustee), September 1937, CNC 37a PMB-CNC 23/1, and CNC 175 973/1914.

5 PAR, LHC vol. 14 B, Annexure, memo dated 9 January 1956. This particular memo did not mention eucalyptus, but its ubiquity elsewhere in Natal was by then a well-documented concern. For a contemporary overview of the erosion crisis, see Adler (1957).

6 PAR, 3/PMB 4/3/47-1077, Rev. A. Nicholls to Town Clerk, 11 November 1929, meaning. The term "irregular use" referred, I believe, to defecation and the disposal of animal cadavers.

7 B.A. Dormer, "Tuberculosis Survey: Edendale," 16 June 1938, PAR, 2/PMB 3/1/1/2/5.

8 PAR, CSO 748 1320/1880, letter to J.W. Shepstone, Acting Secretary of Native Affairs, from Frederick Mason, Pietermaritzburg, 25 March 1880. See also Shepstone's promotion of the principle to the Cape Commission on Native Laws and Customs in 1881 (Cape Colony 1883, vol. ii 66–8) and PAR, SNA 1/1/40 410/1880, to which I will return to in chapter 6.

9 PAR, SNA 1/1/296 1429/1902, petition complaining of the overbearing conduct of Stephen Mini, 1 May 1902.

10 At least according to Mini's later claims, as stated in his opposition to incorporation into the city (*NW*, 11 April 1933) – a story I will pick up in chapter 5.

11 PAR, PWD/2/150 2637/1905; PAR, SNA 1/1/388, 3857/1907.

12 PAR, CNC 38 1632/1911.

13 See South Africa (1932), also Brookes and Hurwitz (1957) and Tropp (2006) on neighbouring eastern Cape. Similar impacts observed elsewhere in colonial Africa are discussed, for example, by Alexander (2000) in Southern Rhodesia, Anderson (2002) in Kenya, and Moore and Vaughan (1994) in Northern Rhodesia.

14 The squatter camp of whites at Oribi consisted of twelve huts that were finally demolished in 1952 after the inhabitants had been moved to a new subsidized housing scheme known as Riverbend (Pietermaritzburg [1904–90] 1952, 67).

## Chapter Four

1 Pietermaritzburg Archive Repository (PAR), LHC vol. 9/1a 29 Apr. 1950.

2 Heather Peel (1987, 1988) was the first historian of Pietermaritzburg to apply the concept of the sanitation syndrome to Maritzburg, but it has since been invoked in important monographs by Christopher Merrett (2009) and Julie Dyer (2012). The most recent iteration at the time of my writing is by Jeff Guy (2013, 374), who stretches the concept yet further back in time to the early 1870s. I want to stress that my present querying of Swanson does not mean that I reject the critique of colonial science, nor in any way exculpate settler racism. On the contrary, I have made similar

points elsewhere on how unsuspected tropes conveying racial prejudice or cultural arrogance percolated through the early scientific literature about Africans, African culture, and "African sexuality" to sometimes invidious ends and (critically) to be internalized by African elites themselves (Epprecht 2008).

3 Cell (1982, 1986), Goerg (1998), Stock (1988), Packard (1989), Freund (2007), Harrison, Todes, and Watson (2008), and Bank (2011) are important studies that extend the concept more broadly geographically and to a range of different disease "pretexts," as does Nightingale (2012) indirectly through the work of Echenberg (2007) on responses to the plague globally. I acknowledge that several key contributors to the history of urban segregation pointedly do not refer to Swanson, notably, Parnell (1991, 1993), and that I am not the first to offer a critique – Rich (1979) and Maylam (1995), for example, do so with aplomb from a materialist perspective, while McNeely (2008) tentatively tests the Maritzburg evidence. See also Wright (2006–07) for a subtle rebuke. Interestingly, Maylam (2001, 163) backs off from his intial critique in a later book, claiming that "plenty of evidence" supports Swanson (but not citing any). It is also noteworthy that the *Journal of African History* promoted the article as among fifty over five decades of publishing that are noteworthy for having "shaped the development of the field." As of 20 May 2013, it remained the journal's single most frequently cited article in all those decades; see http://journals.cambridge.org/action/displayJournal?jid=AFH&tab=mostcited#tab.

4 Reported in the *Natal Witness* under the sensational title "Living Conditions of Middle Ages: City Menaced by 'Black Belt,'" 12 September 1941.

5 I do acknowledge that isiZulu-language newspapers remain an untapped source for now – *mea culpa* – and these may tell another story. Judging from opinion in the nationalist *Ilanga lase Natal* in later years (1950s), however, my prediction is that views in that paper on the subject of a native location would have likely conformed to those that African leaders expressed in the *Natal Witness* and their various petitions to government discussed in this chapter.

6 For example, "Alpha," writing to oppose a native village proposal (*NW*, 6 April 1855). Other commonly used terms included *interloper, refugee, squatter*, and *intruder*, the latter being the term the first lieutenant governor disingenuously used to define the majority African population against the tiny minority of ostensibly "aboriginal natives" (see Bryant

[1929] 1965, 239). Many of the so-called intruders were in fact people returning to ancestral lands after the displacements of the 1820s.

7 National Archive Repository (NAR), NTS 9433 6/388 vol. 2.

8 PAR, 3/PMB [Town Clerk correspondence], 4/3/341 TC 80a/1939.

9 PAR, LHC vol. 11 B (Annexures), memo in response to Natal Indian Organization, 19 November 1952, citing Pietermaritzburg valuations of property by racial group.

10 PAR, CSO 44 part 1, 20 January 1848 letter from T. Shepstone, C.L. Gibb, and D. Lindley to the colonial secretary.

11 Town council minutes, draft regulations by Mr Smarfit, as published in *NW*, 5 October 1855.

12 The wisdom of that tardy by-law can be gauged by a survey the following year that found that no fewer than two-thirds of two hundred newly erected houses for African workers within the city and white suburbs still did not provide toilets (Pietermaritzburg [1904–90] 1923, 41; 1925, 8).

13 To wit, city council did not consider public latrines connected to sewers for Africans' use to be a justifiable expense for another six decades – Pietermaritzburg [1904–90] 1919).

14 This idea was interpreted by Guy (2013, 374) as early proof of the sanitation syndrome.

15 The occasion of this somewhat out-of-character position is discussed in the following chapter under "black peril."

16 The chiefs' petition was submitted to the magistrate in Pinetown, suggesting that at least some of the authors were from the nearby Table Mountain reserve and that both Durban and Maritzburg were on their minds.

17 Dyer (2012, 306) cites a source from 1890 putting the number of brothels in the city at thirty, of which ten were staffed by white women.

18 Maritzburgers seems to have been typical in this regard. See Swanson (1983) on Durban or Rodriguez-Torres (2006) on Nairobi for comparison. Further to Maritzburg, Atkins (1993, 136) notes an element of nostalgia among whites for the relatively clean African *amaWasha* compared to the Indian launderers who partially displaced them in the hand-laundry business in the 1880s. Wright (2006–07) places Indians at the centre of colonial debates over public health in late colonial Natal.

19 PAR, 3/PMB 394/1902, 4 April 1902.

20 According to H.S. Msimang's notes for a 1975 speech included with his unpublished manuscript, "The History of Edendale," Alan Paton Centre and Struggle Archives (AP), PC 11/1/6/2/1-8.

21 Msimang papers, autobiographical writings, AP PC 14/1/1/2.

22 These events lie at the heart of the early radical historiography and are analyzed, *inter alia,* in Cell (1982), Bozzoli (1979), Marks and Rathbone (1982), and Bradford (1987).

23 As the city engineer put it in his report in 1920 (Pietermaritzburg [1904–90] 1920, 76).

24 NAR, NTS 6018 319/307, 6 November 1924. Council rejected out of hand that it was "in any way responsible," while suggesting that Sutherlands was in fact a NAD creation that "seriously prejudiced" its initiative to find an appropriate site for a proper native village (2 December 1924).

25 PAR, 2/PMB 3/1/1/2/5, B.A. Dormer, "Tuberculosis Survey: Edendale," 16 June 1938.

26 Cited by C. Merrett (2009, 110 and 129n53), in correspondence with the town treasurer and the sport associations.

27 NAR, NTS 6018/133/313n, 24 July 1925, petition to the minister of NAD. See also Peel (1987) for a discussion of this debate.

28 That is, 1926, as originally proposed in Pietermaritzburg ([1904–90] 1923), 15. See also NAR, NTS 6018 133/313, correspondence with Pietermaritzburg Municipality: Establishment of Locations, in particular the town clerk's assurances to the NAD, 24 November 1924, and the latter's stipulation of removal as a prerequisite for approval, 3 July 1925.

29 NAR, NTS 6018/133/313n, Secretary for Native Affairs to Chief Native Commissioner, Natal, 17 August 1925; see also Pietermaritzburg ([1904–90] 1928).

30 NAR, NTS 6018 ref 133.313n, petition to Secretary of NAD, 29 December 1925.

31 See also the letter by a European resident of Camp's Drift who claimed to be a fluent isiZulu-speaker and to have talked to "hundreds" of African residents about the issue, *NW*, 13 March 1925.

32 See, for example, the letters from "White in Willowfontein," *NW*, 17 March 1925; "Camp's Drift," *NW*, 13 March 1925; and H. Mason, *NW*, 24 March 1925.

33 As denounced by the Edendale advisory board (LHC 13/1b, 30 August 1954) and reported sarcastically in an *Ilanga lase Natal* column by "Rolling Stone" (H.E.I. Dhlomo, 18 August 1954), 17.

34 Pietermaritzburg ([1904–90] 1961, 126, and 1981, 97), the latter claiming 17,000.

## Chapter Five

1 Historical assessments of the colonial roots of the "crisis of masculinity," with important qualifications and nuance, include Moodie with Ndatshe (1994), Morrell (1998, 2001), Jochelson (2001), Epprecht (2008), and Graham (2012), while Fassin (2007), Hunter (2010), and Timberg and Halperin (2012), for example, directly relate it to HIV/AIDS. I hasten to add that most of this scholarship is grounded in pioneering feminist studies, upon which I also rely, notably Gaitskell (1982), Walker (1990), Bozzoli and Nkotsoe (1991), Bradford (1992), and Gasa (2007). See also one of the very rare instances of African women themselves remembering the segregation era, Reyner ([1949] 1998).

2 The classic psychological study that anticipates this analysis is Sachs (1937), which was originally entitled *Black Hamlet* but in its later American edition became *Black Anger.*

3 By coincidence (although not really, given Zuma's high-profile polygynous marriages, allegations of extramarital affairs and paternity, assertions of traditional masculine values, including disdain for homosexuals, and stated belief that a shower compensates for not using a condom during sex with an HIV-positive woman), two Cape Town artists independently produced paintings that depicted Zuma with his penis prominently displayed. One (a white man) became the target of furious attacks by government officials and letter writers, while the other (black) was basically ignored (Human Sciences Research Council 2012). See also Ndlovu-Gatsheni (2013, 136), who takes the first as an exemplar of white South Africans' unconscious arrogance, apparently unaware of the second.

4 Accumulated deaths for KwaZulu-Natal are not well publicized, but one actuarial report from 2008 placed the number then at 849,000 (Nicolay 2008, 4), with an estimated 115,000 new deaths per year, and 1,561,000 -people living with HIV (see also Dyer 2012, 319, for what appear to be conservative numbers). It should be emphasized that the disease overwhelmingly and disproportionately affects the black population.

5 Dubow and Jeeves (2005, 36) coined the term *liberal interregnum*, but anticipating it see de Beer (1986), Jeeves (2000), and the Karks' personal retrospective reflections on the experiment (Kark and Kark 2001).

6 Respectively, "The light of Natal" (1889–96) edited by Edendale's Solomon Kumalo, and described by Peter Limb (2010, 93) as the "organ of the

*kholwa*," and "Paper of the nation" (1898–1904). See also Meintjes (1986) and Kumalo (2009). "Exempted" in this context meant from the Natal Native Code – that is, "traditional" law as first codified by Shepstone. Africans could apply to the colonial government for an exemption based on their assets, taxes, monogamous status, and level of literacy; if exempted, English law would apply to them. Healy-Clancy (2013, 59) figures that scarcely more than 5,000 Africans in the whole colony (only 5 per cent of all *amakholwa*) held such status when the right was rescinded prior to Union.

7 For example, Ndlovu (2009) and Atkins (1993, 31–51 and passim); also see Ngubane's (1977) discussion of wives and sorcery.

8 The term *soft patriarchy* has been applied to southern Africa with reference to modern struggles around "biblical masculinity" (van Klinken 2013). See Meintjes (1990), Badassy (2011), and Healy-Clancy (2012) for a fuller discussion of nineteenth- and early-twentieth-century *amakholwa* gender ideology.

9 Recall from chapter 4 that this was precisely the reason the *Natal Witness* later ridiculed the Durban argument in favour of a native village in 1903.

10 Meintjes (1988b, 457–60). By this time, Europeans, Indians, and coloureds had begun to buy up property. The largest female landowner in Edendale was Cornelia Kothe, the German widow of the community's leading merchant (with 101 acres, the second-biggest property in the district). Ten of the listed female owners had African surnames; in other cases, the sex of the owner cannot be easily identified from the given name; hence there may have been more female landowners.

11 See, for example, Pietermaritzburg Archives Repository (PAR), LHC vol. 9/1b, memos of 4 August and 22 November 1950, in which the commission follows up on a complaint against her monopoly practices.

12 PAR, SNA 1/1/81 185/1885.

13 National Archives Respository (NAR), NTS 6018 133/313, 24 July 1925 petition to minister of NAD.

14 PAR, LHC vol. 11/1, 6 August 1952 letter from LHC Secretary Donaldson re: appeal by Mrs Ndhlovu against the demolition of her slum dwelling.

15 PAR, CNC 37a PMB - CNC 23/1, G.A. Park Ross (Union Health Department to Chief Native Commissioner, 28 October 1936.

16 Alan Paton Centre and Struggle Archives (AP), PAR, 3 PMB 4/3/113 274/1933, Msimang "Distressful Decline," chap. 6, PC 11/1/6/2/1-8, n.d.

17 Pietermaritzburg ([1904–90] 1936, 109); PAR, 3/PMB 4/3/205 832, police memo on "Natives at Edendale"; CNC-Natal (Lugg) to the Secretary

for Native Affairs (in Pretoria), 25 July 1936, confidentially estimated the Sutherlands refugees to be around 1,800. PAR, CNC 37a PMB-CNC 23/1.

18 PAR, 3/PMB 4/3/318 TC 955/1938 CNC, Re: Control of Edendale, "Proposed Regulations drafted by Acting Chief Native Commissioner – Natal" (30 October 1934). To be fair, probably no one really believed the job could be done on the shoestring. To put the subsidy in perspective, the upkeep for Maritzburg's main sporting ground cost an average of roughly £3,000 per year throughout the interwar period. Asphalting the cycle track in 1938 cost almost the equivalent of the proposed Edendale town board budget (C. Merrett 2009, 57, 91). The first serious estimate on the costs of putting Edendale "in order," by LHC Engineer W.M. Chrystall, put the figure at one million pounds. PAR, LHC, vol. 3/2 "Development of Edendale," 8 February 1945.

19 PAR, SNA 1/1/296 1429/1902, Petition complaining of the overbearing conduct of Stephen Mini, 1 May 1902.

20 The letter is from Kothe himself, denouncing Mini's racialism.

21 PAR, CNC 37a PMB-CNC 23/1, Mini to CNC, 14 July 1937 and 13 September 1937. When police got wind of this private taxation scheme, they immediately ordered him to desist. As for the request for expansion at province's expense, this was so far out of the realm of political feasibility that it probably caused more mirth than concern.

22 PAR, CNC 37a PMB - CNC 23/1.

23 I examine the social medicine experiment in the next chapter, but will note here the Karks' (2001) memoir of the Pholela experiment, and the effusive praise offered by his contemporary and backer, Edgar Brookes (1954).

24 PAR, LHC vol. 1, 31 March 1943, passim, including Proclamation 413, 23 Ocotober 1941, appointments, procedures, and first meetings.

25 AP, PC 14/1/1/1, Msimang, Autobiographical Writings, n.d.

26 LHC 3/1 Special Meeting of Advisory Board, 8 February 1944. Let us be sceptical of these promises, but Wadley did not only express them to African leaders (viz., "ultimately passing to such [directly elected] representatives of the entire responsibility" for the commission's mandate, LHC3/1, Meeting with A.W.G. Champion, Chief Mini [etc.], 18 July 1944). He was also explicit about his vision in confidential memos to his superiors in government. As he explained to the Union secretary of health (NAB, GES 963 898/13C, 26 September 1946), "Local government and self-government are almost inseparable conceptions," and when the latter was achieved, "the undemocratic instrument of the Local Health Commission will have served its purpose, and the need for its existence will have disappeared."

27 I have not been able to find what happened to him after that stint in Parliament but surmise (from his later politics) that he may have been among those South Africa Party members alienated by its union with the National Party in 1934. They broke away to form the pro-English Dominion Party, which met electoral disaster in the 1938 elections. As Durban's mayor in the mid-1920s, Wadley would have been party to discussions there on how to respond to the NUAA, which response was slow in the Durban case compared to Pietermaritzburg (Torr 1996, 248–9).

28 NAR, NTS 9433 6/388, vol. 2, 26 November 1942 memo from Wadley to Secretary of Native Affairs D. Smit, and Wadley to Financial Secretary Natal, copied to Smit, 23 June 1943.

29 This makes a noteworthy contrast with Durban, which never did ante up for the Clermont experiment in commission rule (Swanson 1996, 294).

30 PAR, LHC vol. 2, meeting of advisory board, 22 September 1943.

31 Notably, *Natal Witness*, 9, 11, and 17 March 1944. Dr Landau was also on public record estimating that no fewer than 2,000 people in Edendale had syphilis (Nuttall 1984, 50). That is one in six of the population and about double the rate noted at the Pholela clinic (Kark and Cassel 1952, 103).

32 I base this observation on Lugg's own memoir (1970, 96), where he expresses mild regret at having "had the distinction" of burning Chief Bambatha's hut during the notoriously brutal repression of that rebellion, and on passing references to some quite reactionary interventions over the following decades (see, e.g., Dubow 1989, 115, 118). Lugg only rarely appears as present and active in LHC documents other than this one.

33 Minutes of meeting, 18 July 1944, PAR, LHC vol. 3/1. Selby Msimang was also not named as present, although it would be surprising if he had not been at the meeting. He was by then secretary of the EDPHA advisory board, and a long-standing political adversary of Mini and Champion.

34 Bonner and Nieftagodien (2008, 50) note that Mtimkulu was among several Africans and coloured employees dismissed from their positions in 1935 ostensibly for incompetence, and that he took his case to the Native Representative Council with angry accusations that whites sought "to tramp the Native down."

35 For example, NAR, NTS 9433 6/388, vol. 2, report on meeting of CNC, LHC, and ELHA representatives (including Nxumalo and Mrs J. Caluza), 11 March 1943; report on Lot-Holders' complaint, CNC-Natal to SNA, 4 March 1944.

36 PAR, LHC vol. 10/1a, Msimang to NC-PMB, 23 April 1951.

37 NAR, NTS 9433 6/388, vol. 2, letter from Mtimkulu to the minister of NAD, 11 February 1942.

38 NAR, NTS 9433 6/388, vol. 2, see Lot-Holders' complaints to the Secretary of Native Affairs, 1 March 1944; minutes of meeting between Col. B.W. Martin, acting CNC, and V. Addison, NC-PMB, with Zachariah Miya, chairman of ELHA, G. Mtimkulu, and other members, 4 March 1944.

39 NAR, NTS 9433 6/388, vol. 2, G. Mtimkulu to the governor general, 24 February 1946.

40 NAR, NTS 9433 6/388, vol. 2, 31 December 1943. By comparison to a more normal gender ratio in a peri-urban space, Wilson and Mafeje (1963, 1) pegged Langa (Cape Town) at 316 males to every 100 females. Marks (1991, 220) notes a Natal survey reporting roughly 90,000 African men to 37,000 African women in all of urban Natal.

41 NAR, NTS 9433 6/388, vol. 2, memo from G. Mtimkulu to minister of NAD, 23 February 1943.

42 Ibid.

43 H.S. Msimang papers, AP, PC 4/1/1/6. See Bradford (1987), Marks (1986), Limb (2010), Soske (2012), and Mkhize (2015) for more on Champion's mixed legacy in the nationalist movement.

44 PAR, LHC vol. 3/1, 18 July 1944, 388.

45 Ibid., 383.

46 Brookes (1954, 5) and Heard and Ncobeni (1986, 284) reported an over 50 per cent reduction in infant mortality in Pholela's first decade, achieving a lower rate than obtained in parts of peri-urban Maritzburg and Durban even three decades later.

47 Msimang papers (memoir), AP, PC/14/1/1, pp. 4–5.

48 PAR, LHC vol. 3/1, 18 July 1944, 382.

49 Laundau's ultimate bureaucratic boss in the Union Department of Public Health was Dr George Gale, who had been the medical officer of health (hereafter MOH) for Maritzburg in the mid-1930s. In that role, he had been a pioneer of progressive and highly successful approaches to treating VD (Gale 1939).

50 As Marks (2000) reminds us, for example, George Gale came from a missionary background and, in his own words, was not above advocating "a smart slap" to African men to teach them manners (Gale 1939, 266). See also Phillips (2005) for white doctors' tendency to blame coloureds for their own ill health at an urban health centre. The paternalism of social medicine zealots is much discussed by Wylie (2001), Caesar (2008), and throughout exchanges between Marks (2014 and 2015), Digby (2012, 2014, 2015) and Freund (2012, 2014), while Marks (1994, 209) makes the important point that African

nurses "imbibed the commonsense racist discourse of their white mentors," with long-lasting harmful impacts on the health care system.

51 PAR, LHC vol. 3/1, 18 July 1944, 382.

52 Ibid., 381.

53 PAR, LHC vol. 8/1a, report, 14 April 1949. See also NAR, NTS 9433 6/388, vol. 2, for a Mrs G. Mhlongo's complaint against the tribal police, Minutes of meeting between Col. B.W. Martin and ELHA complainants, 4 March 1944.

54 PAR, LHC vol. 4/1, memo from assistant MOH, March 1945.

55 PAR, LHC vol. 8/1a, F.W.P. Culver, chief health officer to MOH Seymour, 11 May 1949.

56 PAR, LHC vol. 15/2b, Advisory board minutes, 26 September 1956; LHC vol. 15/2a, Annexures to minutes, Report on Control of Commonage (Donaldson, 21 November 1956).

57 PAR, LHC vol. 9b, Annexures to minutes, memo on Georgetown market, 7 December 1950.

58 PAR, LHC vol. 13/1b, F.J. Mazibuko, protest by Ashdown advisory board against the LHC survey on income, 4 June 1954, for which the LHC expressed regret.

59 PAR, LHC vol. 12b, confidential memo from MOH, 25 August 1953.

60 PAR, LHC vol. 9/1a, letter from B.P.T. Nyembezi et al., 29 April 1950.

61 PAR, LHC vol. 12/2a, 24 December 1953.

62 We should be careful about extrapolating from Shula Marks's study of nurses nationally, but her observation that the apartheid state viewed black nurses as "ideal collaborators" (1994, 210) likely holds true in the Edendale case, with their training in a highly regimented, authoritarian, patriarchal, and racialized professional culture, and their social status strongly linked to consumerism.

## Chapter Six

1 Pietermaritzburg Archive Repository (PAR), CSO 748 1320/1880.

2 PAR, 3/PMB 4/3/596/1922; 3/PMB 4/3/828/1926, 22 October 1926.

3 Sometimes also referred to as *uPoyinandi*. Mkhize (2015, 157) has established the etymology of this word, which has evolved in contemporary usage to mean "pensions," or "social welfare" in general. Following a common practice in isiZulu, it derives from the personal name of an important figure in the history – Mr Paliniyandi, the Indian landlord of the Plessislaer building that housed the original LHC offices.

4 PAR, LHC vol. 13/2a, minutes of the advisory board meeting, 22 November 1954; Caesar (2015, 114).

5 Other LHC officials had their doubts about the validity of the census with regards to Africans. Secretary Donaldson put their population at 40,000 to 50,000 in that year (memo 7 June 1957, PAR, LHC, vol. 16/1b), which if accurate would have made non-Africans about 10 per cent of the total population.

6 PAR, 3/PMB 4/5/363, City Engineer V. Harris to Town Clerk on negotiations with Zwartkop authorities, 25 October 1972. Recall how adamantly the city had refused to accept any responsibility for Edendale or even Sutherlands, let alone Zwartkop, in the 1930s. This dramatic offer, which included building a thousand new homes per year, was predicated on the central government's paying for most of it and assuming responsibility for administration once built, which Pretoria rejected.

7 PAR, 3/PMB 4/5/363, Confidential memo from City Engineer V. Harris to Town Clerk, 9 July 1974 re: 8 Jan 1973 comprehensive agreement between City and the Dept. of Bantu Administration and Development. Re Edendale/Imbali complex; and Msimang (1974) on how the promise of self-government for Edendale "was flagrantly and unashamedly dishonoured … without notice to its inhabitants or consultation."

8 The numbers referred to students pursuing daytime studies only, not those in residence.

9 Quotations from the Sinomlando GEDI interviews are from draft transcripts in my personal archives, provided courtesy of Radikobo Ntsimane. Polished versions are published on the Sinomlando website (http://sinomlando.ukzn.ac.za/), which, unfortunately, is not active at the time of proofing this manuscript.

10 This topic is explored in a very rich historiography to which I am indebted. See, in particular, Marks (1986, 1991), many of the essays in Carton, Laband, and Sithole (2009), and Mokoena (2011), as well as Maré (2014) on inherited racialist thinking in today's South Africa. For debates from the *kwaPoyinandi* era that reveal the views of Msimang and other Edendale intellectuals, see Soske (2009, 2012) and Mkhize (2015). I tap into the Sinomlando interviews not to promote the isolated recollections of a small number of people as definitive or representative, but to illustrate people's memories of fluidity or complexity in contrast to the dominant narratives. I also note the identity issues discussed by Whelan (2012), whereby the strength and continuity of *amakholwa* "cultural nobility" over at least seven generations emerged as a significant inhibiting factor to

the success of the historical development project she oversaw in the early 2000s.

11 PAR, SNA 1/1/296 1429/1902, petition complaining of the overbearing conduct of Stephen Mini, 1 May 1902.

12 Testimony of Chief Stephen Mini, J.T. Gumede, and the Rev. Abner Mtimkulu of the Natal Native Congress, before the Select Committee on Native Affairs, 15 and 18 June 1917, http://www.sahistory.org.za/archive/testimony-chief-stephen-mini-j-t-gumede-and-rev-abner-mtimkulu-natal-native-congress-select. Accessed 4 September 2014.

13 PAR, LHC vol. 16/2a, Rev. Gcabashe, in joint advisory board minutes 1 October 1957.

14 Eric Matshatsha, interviewed by R. Lundie 26 July 1995, AP, 95/APB/8. Information on the Basotho of Matatiele is hard to come by but Bardsley (1982) discusses the formative years of colonial settlement and richly problematizes the racial/tribal identities so evidently in flux at the time.

15 PAR, LHC vol. 18/1b, 11 August 1959. I want to be careful not to promote an ethnic stereotype, but the issue of migrants from Basutoland finding it difficult (or undesirable) to adjust to the conditions of township life (or indeed, to the nascent urban areas within Basutoland) is a theme that comes up in several other studies from the period of the 1920s–1950s (see Bonner 1990; Epprecht 2000; Goodhew 2004; Kynoch 2005, for example).

16 PAR, LHC vol. 7/1a, memo on the Chief Native Commissioner's decision not to recommend the abolition of the Kolwa chieftainship, 16 June 1948.

17 The likely peak population was pegged at 891, "The Commission's Population," 19 May 1957, PAR, LHC vol. 18/1a. As noted, there is no in-depth history of the various coloured communities in either Edendale or Maritzburg – they did not even warrant any entries in Laband and Haswell (1988). The following is taken primarily from the excellent display devoted to them in the Msunduzi-Voortrekker Museum.

18 See, for example, Raymond Rudman letter to General Kemp, MLA, on behalf of "White Afrikanerdom," 16 October 1938, National Archives Respository (NAR), NTS 9433 6/388, vol. 2.

19 PAR, 3/PMB 4/3/427, 828 and 596, notably, which include petitions and letters by whites in support of the status quo from 1922 to 1936.

20 Uncle Remus, "A Week-end in City of Flowers," *Ilanga lase Natal,* 15 September 1956. I say "somewhat advanced" since integration was not complete: African men, the reporter included, were allowed in the building but were not permitted to dance with Indian women.

21 PAR, LHC vol. 9/2, 8 April and 22 November 1950.

22 The decision to opt for buckets was initially justified as an interim measure in response to the high cost of installing sewers, given the topography. In the face of the uproar, however, the LHC promised to put in proper toilets. Yet the technical (hence, financial) issues remained, to such an extent that the LHC reported that the project was "still stalled" ten years later, providing another lightning rod for criticism: LHC, Natal annual report of the secretary for the year ended 30th June 1954, and abstract of accounts. See PAR, TC 314/3 for the R. Rudman cyclostyle, 25 November 1946 and letter to LHC, 11 April, 1947; NAR, NTS 9433 6/388, vol. 2, 1 March 1944 for a sample of the many lot-holders' complaints put to the Secretary of Native Affairs. For the NAD's unhelpful (and poorly explained) intervention, see PAR, LHC vol. 3/1, special meeting of advisory board, 2 August 1944.

23 PAR, LHC vol. 18/2a, Edendale advisory board minutes, 27 October 1959.

24 NAR, NTS 9433 6/388, vol. 2, "Future Administration of Edendale," 19 January 1943.

25 PAR, LHC vol. 1, 31 March 1943.

26 PAR, LHC vol. 3/b, 2 August 1944.

27 PAR, LHC vol. 3/1, 22 March 1944.

28 For example, PAR, LHC vol. 12/2b, 25 September 1953, for the second joint advisory board meeting, in which Wadley vetoed another request by the advisory boards on moving towards autonomy: "As things are at present, Africans were not competent in law to elect their own self-governing bodies."

29 PAR, LHC, vol. 4/1, memo 48/3. The minutes of that 25 April 1945 meeting indicate that discussion of the report was deferred until MOH Landau could comment (which, if it happened, does not appear in subsequent minutes).

30 According to Msimang's handwritten account of the first two years of advisory board affairs, AP, PC 14/1/4/4, p. 12. There is no record of whether the LHC ever took this motion to the province; in any case, the appointments never happened.

31 PAR, LHC vol. 4/1, advisory board minutes, 27 June 1945.

32 PAR, LHC vol. 7/1b, advisory board annual report, 12 July 1948.

33 Msimang papers, AP, PC 14/booklet 2, p. 8.

34 PAR, LHC vol. 4/1. Msimang to LHC, 13 March 1946.

35 Maritzburg and district elected United Party candidates in that and seven successive elections (Thompson 1988, 199).

36 PAR, LHC vol. 11/1, first joint meeting of those boards, 6 August 1952.

37 PAR, LHC vol. 16/1b, "Informal Discussion with Commissioners," 7 June 1957.

38 PAR, LHC vol. 15/2b, Provisional Draft Town Planning Scheme No. 1.

39 At least according to a complaint filed by Edgar Mini to the secretary of native affairs against Lugg, 8 January 1943, NAR, NTS 9433 6/388 vol. 2.

40 PAR, LHC vol. 7/1a, "Chieftainship at Edendale," 2 April 1948, and advisory board minutes, LHC vol. 9/1b, 31 August 1950.

41 PAR, LHC, vol. 2 April 1943–March 1944, minutes of meeting between LHC and PMB representatives, 6 August 1943.

42 This sorry episode did not garner mention in the city's annual report, nor could I find any reference in the Town Clerk correspondence. It did briefly make the local news, however (*NW*, 10 January 1946), with, as far as I can see, only a single response from an enraged correspondent, "Atlantic Charter," *NW*, 18 January 1946.

43 See PAR, LHC vol. 15/2 a-b, water supply for peri-urban Pietermaritzburg, 1956, and PAR, LHC vol.16/2a, 11 December 1957, memo from J.B. Watkins-Baker and S. Newmark (pp. 2351 and 1779).

44 PAR, LHC vol. 9/2b, 12 October 1950, minutes of the meeting of the LHC, Edendale advisory board, and R. Ashton, NC-PMB.

45 Msimang papers, AP, PC/14/1/1/0 1969 (handwritten diary).

46 PAR, LHC vol. 9/2b, 10 February 1951.

47 PAR, LHC vol. 8/2a, advisory board minutes 24 November 1949.

48 The city had to defend this decision right up to the Supreme Court, a case it finally won in 1959. I note, simply to give a feel for the times, the headline announcing this: "City Wins Its Case against Apartheid Buses," *NW*, 9 December 1959.

49 Msimang papers, AP, PC 14/1/1/2; PAR, LHC vol. 3/2, memo from Msimang on development of industries, October 1944.

50 PAR, LHC vol. 5/2a, 9 December 1946; vol. 7/2a Georgetown market, 21 September 1948; and vol. 7/2b, 27 January 1949.

51 Secretary D.R. Donaldson acknowledged this setback in his memo to commissioners, 5 December 1950, PAR, LHC vol. 9/2a.

52 PAR, LHC vol. 15/2a annexures, 5 November 1956.

53 To cite but two, Msimang wrote to the city at the time of the attempted extension of pass laws by Maritzburg in 1950 to praise the "spirit of goodwill they [the commissioners] have shown unsparingly and consistently … [their] complete absence of harassing the people or an

exhibition of the most irritating officialdom. The Commission has given the general impression that it lays emphasis on the need to carry the people along with them at every stage and make them feel that none-the-less they are one of and with them": AP, PC 11/1/6/2/1-8, "The History of Edendale" (unpublished ms), Msimang letter to Ashton, October 1950, chapter B. The second example is from a November 1979 speech, in which Msimang retrospectively described the LHC as having achieved "monumental work": AP, PC 11/1/6/3/13, "Social Economic and Political Development of Edendale: Current Problems."

54 PAR, LHC vol. 3/1, departmental committee meeting recommendations on proposed Hollingwood PHA, 19 June 1944.

55 PAR, LHC vol. 9b, Annexures to minutes, 17 January 1951 report on the first twelve-month "experiment" with *Ikhwezi*, LHC (1952), and, on its closure, vol. 14/2 Minutes, 30 January 1956.

56 AP, 96 APB 7, interview with Bunty Biggs, 25 August 1996.

57 PAR, LHC vol. 12/1b, 24 July 1953.

58 Education was not an LHC responsibility and there is no paper trail in its archival records on this issue. One of the Sinomlando informants, however, recalled it as a key factor in his political mobilization.

VICTOR MADISHA: Do you remember when the people were protesting against the use of Afrikaans at schools? Before it was imported to Johannesburg it was started here by us.

NICHOLAS ZIQUBU: In 1976?

VM: It was 1957. I was still at Ashdown High school. It was very difficult for us those days because every subject was to be taught with Afrikaans. And it was a time when the ANC was beginning to split. Now all those who were leaders in the organisation came to canvass to the youth. They got us in 1959 which means that we were conscripted to the organisation you see. We join the Youth Brigade. (SC, Victor Madisha, interviewed by Nicholas Ziqubu, Edendale, 19 October 2009).

59 PAR, LHC vol. 15/2B, 26 September 1956.

60 PAR, LHC vol. 15/2a, for statistics on rates payments, and vols. 16/2b and 18/2b on provisional estimates.

61 *NW*, 24 December 1954; and on the gathering opposition to Pretoria's ruling, *NW*, 14, 16, and 22 May 1956; also *Ilanga lase Natal*, 2 June and 6 October 1956; see Peel (1988) for a brief discussion of this controversy.

62 PAR, LHC vol. 18/1a, "The Commission's Population," 19 May 1959.

63 For example, "PMB Lags Behind in Industrial Expansion," *NW*, 18 February 1957; "Race for New Industries: City Lags Behind," ibid., 22 February 1957; "Local Health Commission," ibid., 11 June 1957; "Mountain Rise Could Be a Great Failure: 'Industrialists Must Be Attracted,'" ibid., 12 February 1958.

64 PAR, LHC vol. 16/1b, memo by Engineer E.L. Bennett on group areas, June 1957. The commission at that time still wanted its border areas to be available for private development on a "specified" (non-racial) basis, and regarded the placement of a huge African residential area and buffer zone on those lands as a threat.

65 PAR, LHC vol. 18/2b, 3 March 1960, p. 2800.

66 C.B. Downes, G.J.R. Bulman, R.H. Rodseth, and J.M. Pugin right through to 1975.

67 Dr M.D.C. de Wit, quoted in Pietermaritzburg ([1904–90] 1964, 23).

68 No study was ever conducted on the fate of these people; however, a close study of an analogous "black spot" removal near Ladysmith suggests horrendous, long-term impacts on health and morale. In that case, no less that 11 per cent of the children under five who were removed in 1967 were dead after ten years. Only 2 per cent of the community participated in the formal political process that had been established supposedly to govern it (Desmond 1978, 7, 15).

69 Wills (1994, 300) gives a total of 800 Africans removed from the city.

70 AP, PC 14/5/4/16, extracts from the LHC annual report, 1964.

71 PAR, 3/PMB 4/3/318 TC 955/1938 CNC, "Re: Control of Edendale."

72 Msimang, autobiographical writings, AP, PC 14/1/1/1; Brown (n.d.).

73 PAR, LHC vols. 16/2a and 16/2b, Ashdown advisory board, 18 November 1957 and 4 February 1958.

74 AP, Sinomlando Mpophomeni interviews, Michael Xaba (no. 19, by Joe Xaba, 3 September 2010) and Victoria Shibase (no. 13, by Radikobo Ntsimane, 2 December 2010).

75 Msimang, "Edendale's Dilemma," AP, C 14/1/4/4/3.

76 Hlope is the son of one of Edendale's most prominent landowners and businessmen, Gideon Hlope.

77 A good history of Imbali has yet to be written, particularly of the period of direct administration by the city up to 1973. Tellingly, for example, Imbali did not get a chapter in Laband and Haswell (1988) although at the time of that book's publication Imbali was likely the most populous section of the "outer city." I have culled the above from Butler, and Harley with Aitchison (1993) and Denis and Duncan (2011).

78 Much has been written on this provincially and nationally, including from the "crisis of masculinity" angle (Campbell 1992) and the culture of transactional sex (Hunter 2010). Locally, the best accounts are BESG (2003a and b) and Dyer (2012).

## Chapter Seven

1 Among many outstanding contributions to local history for this narrative, see Laband and Haswell (1988), Kentridge (1990), C. Merrett (2009), Mkhize (2012), Denis, Ntsimane, and Cannell (2010), and Denis and Duncan (2011). Important "official" versions include South Africa (1999), Houston et al. (2013), Sithole (2014), and those parts of the GEDI proposals that touch upon public history (Msunduzi Municipality 2006, 2009).

2 Chairperson of Edenvision Medical Holdings Limited, Dr Pubie Padayachee, quoted in Chris Ndaliso, "New Hospital for Edendale" 14 June 2010. http://www.skyscrapercity.com/showthread.php?t=1296081, accessed 1 July 2013.

3 Interview by R. Lundie, 26 August 1995, Alan Paton Centre and Struggle Archives (AP, 95/APB/8).

4 The *Natal Witness* (NW) reported on the department's work as a regular feature throughout the war years. The department was closed down a year after the war's end: NW, 26 July 1946.

5 These are ongoing themes in the press through Russell's years in office and beyond, and there was scarcely a week without some jeremiad against or grandiose scheme for industrial growth. See, for example, NW, 15 May 1951, 18 August 1951, and 17 November 1953. P. Merrett (1988) provides some biographical background on this remarkable woman.

6 Pietermaritzburg Archive Bureau (PAR), LHC vol. 15/2 a-b, Water supply for peri-urban Pietermaritzburg, n.d. [1956].

7 PAR, 3/PMB 4/3/318 TC 955/1938 CNC, Re: Control of Edendale, confidential memo from City Engineer Harris to town clerk, 9 July 1974.

8 Ibid.

9 NW, 23 June, 1989. The next day he claimed that the river had already cleaned itself and therefore no further action was required. NW, 24 June 1989.

10 One of the ironies of the times was that the liberation of Zimbabwe from white minority rule in 1980 actually mitigated Maritzburg's staffing crisis by sparking an exodus of white professionals to the city from that country (Pietermaritzburg [1904–90] 1981, 97).

11 The mayor, Pamela Reid, was referring to the Willowton decision discussed below but might also have had the free waste disposal idea in mind.

12 http://www.gov.za/documents/constitution/chapter-2-bill-rights#24, accessed 1 March 2016, my emphasis.

13 The billion rands (then around US$150 million) were to be garnered from a wide range of sources, including the city itself, the Development Bank of Southern Africa, and the Departments of Housing and Environment. The time frame for spending was unspecified and would, presumably, have to be matched or exceeded by private investment.

14 The interview has been edited down in order to focus on the theme in an otherwise sprawling discussion.

15 http://www.duct.org.za/; see also http://verdantlifekzn.com/, accessed 10 September 2014.

16 Notable South African critics include Le Quesne (2000), Bond (2000, 2002), and Cock (2006, 2012). Karumbidza (2014) provides a broad reflection on the limitations of liberal green thinking under existing capitalist conditions for the whole of the continent, described simply as "untenable" by UN-Habitat (2015, 11).

17 Amtaika's figures (2001, 392) suggest significantly higher amounts per unit. He reports that, up to the year 2000, the transitional council delivered or planned nearly 28,000 new low-cost homes for a total cost of R344 million. This translates into only slightly shy of the target quantity but significantly over cost, at roughly R12,000 or (at the then current exchange rate) US$2,000 per unit. That figure seems to include the cost of supporting infrastructure, like roads but, even if not, it is still remarkably low when set beside the sale price of apartheid-style matchbox homes. These can now be found on the real estate market at upward of half a million rands (US$30,000, at 2016 exchange rates).

18 As a sampling of press coverage testifies, including S. Peters,"Terrified Foreigners Return to Jika Joe Fearing Attacks," *Daily News*, 21 January 2014, http://www.iol.co.za/dailynews/news/terrified-foreigners-return-to-jika-joe-fearing-attacks-1.1634226#.U5m1lvldX6M, accessed 12 June 2014; "Easter Fire Spreads through Jika Joe Settlement," http://publiceyemaritzburg.co.za/10439/easter-fire-spreads-jika-joe-settlement/; and "Mayor Calls for Calm at Jika Joe," http://maritzburgsun.co.za/8419/mayor-calls-calm-jika-joe/, accessed 10 August 2014. Some of the worst shacks have been demolished since the big fire and, as of May 2015, many people have been housed in metal, barracks-like structures.

19 The most recent plan, I should note, is far more ambitious, seeking to generate tens of thousands of new manufacturing jobs mainly in the Imbali/Mason's Mill area (Iyer Urban Design Studio, 2014, 29). It concedes that this will remain "a challenge," dependent on state incentives to private enterprise and the construction of new highways (31).

20 The city reported that it has since improved, in part through an aggressive disciplinary process rather ominously called "Operation Pitbull" (Msunduzi 2011).

21 S. Peters, "Terrified Foreigners Return to Jika Joe Fearing Attacks," *Daily News*, 21 January 2014, http://www.iol.co.za/dailynews/news/terrified-foreigners-return-to-jika-joe-fearing-attacks-1.1634226#.U5m1lvldX6M, accessed 12 June 2014.

22 Garner (1998, 2004) alludes to some of this in his comparative study of new and old religions in Edendale, but one may infer similar practices in Greater Edendale as Ashforth (2005) and Niehaus (2013) found in similarly stressed communities elsewhere in the country. I will note that even my history tour guide, completely unbidden, brought up the topic of *muthi* (medicine) murders, while sorcery is a recurrent topic in the press.

23 Dyer (2012, 122, 130). Statistics are contentious of course. I draw the above mostly from a pro-ANC document using the most conventional metrics derived mostly from the IMF (Coleman 2013). Concerns about fracking are reported, for example, by the KwaZulu-Natal Agricultural Union and Midlands Conservancies, http://www.kwanalu.co.za/default.asp?action=news&Id=1039 and http://midlandsconservanciesforum.wordpress.com/2014/05/28/fracking-issues-right-here/, accessed 10 September 2014. On corruption, Amtaika (2013) cites Transparency International's 2009 ranking of South Africa at the fifty-fifth worst country in the world, which by 2013 had dropped to seventy-seventh, well below several of its immediate neighbours (Botswana, Namibia, and Lesotho) and below or tied with other African countries (e.g., Ghana, Senegal, and Tunisia).

# A Note on Sources

This book is based primarily on archival research of English-language documents. These included government documents and private correspondence held at the South African National Archive Repository (NAR) in Pretoria and the Pietermaritzburg Archives Repository (PAR). The Local Health Commission files are all in the latter, comprising bound, numbered volumes that gather together the commission's documents in more or less chronological order, mostly unpaginated and non-indexed. In addition, I consulted popular press and literature at the Bessie Head Library in Maritzburg (some of which is also available online), and private memoirs, correspondence, and oral history transcripts at the Alan Paton Centre and Struggle Archives (AP) at the University of KwaZulu-Natal (UKZN). Among journalistic sources, I draw primarily on the *Natal Witness* (after 1996, the *Witness*), the oldest newspaper in the colony and, for most of its run, a powerful platform for the expression of liberal conscience. While trumpeting itself as a voice for the whole colony, and then province, it also contained significant local Pietermaritzburg content, including town council debates published verbatim and a vibrant letters to the editor section.

The shortcomings of archival sources in colonial Africa and apartheid Natal specifically have been much discussed (see, for example, McClendon 2002). Many key decisions affecting the history of the city took place behind closed doors and were minuted indifferently, if at all. Many documents have been lost or destroyed either accidently by fire (as when the repository at city hall burned down in 1891) or deliberately (as in the dying days of the apartheid regime). The "colonial nature" of the archives is a further impediment to knowledge about marginalized communities in the sense of the powerfully eurocentric, androcentric, and class-biased provenance of archived documents. As a stark example, the KwaZulu-Natal archives' collection, which comprises thousands of photographs, has exactly ten catalogue entries for Edendale, most of which depict white visitors or scenic waterfalls. More recent sources are generally more balanced, although hangovers from the old days persist in surprisingly durable ways. I also wish to insert a note of caution here about the durability and accessibility of new documentary material. I gathered considerable "gray literature" in the form of brochures, reports, and pamphlets in both hard and electronic copy. These may or may not survive the formal archiving process but they do still exist in my personal archive.

Given these limitations, and notwithstanding the insights I have gained about reading against the colonial grain from scholars such as Harris (2007), Hamilton (2009), and Mangcu (2011), a project of this nature really calls out for oral history gathered from among the majority population and/or key informants. Bozzoli with Nkotsoe (1991), Jacobs (2003), Goebel (2004), and Lee (2009), among many of the scholars cited in the text, exemplify both the need and the value of that method, with due caution as Field (2012) and Wells (2008), among others, articulate. Oral history is something I have very much enjoyed gathering and have benefited from in my earlier work. The unusual path I took to writing this book, however, meant that I did not plan for oral history work myself. For example, in the early days of the project I accompanied ecohealth team members on numerous fact-gathering meetings in which key informants, often in the municipal government or parastatals, shared their full and frank thoughts on the state of health care, sanitation, and other service provision. Since I did not have their approval or consent, I could not directly use my notes from these sessions, although they certainly shaped the ways I asked questions subsequently (the same applies to photographs taken). For oral evidence, I relied principally upon that collected through the Sinomlando Centre for Oral History and Memory Work (http://sinomlando.ukzn.ac.za/), using the original, unofficial transcripts, which in some cases have been lightly edited for publication on the website. A number of other oral histories are available at the Alan Paton Centre and Struggle Archive while still further recollections can be found in graduate student dissertations at the UKZN. In one particularly helpful instance, Adrian Pole (2002) includes a very rich "archive" of all his interview transcripts as an appendix to his dissertation.

In addition, over the years 2005–12, I was a participant observer at numerous municipal meetings, including the Greater Edendale Development Initiative and the historic 2006 Land Summit. Just watching, listening, and sometimes debating people opened my eyes to undercurrents of tension or presumptions that are not generally apparent in the public record. I meanwhile avidly consumed the public history on display or on file at the Msunduzi/Voortrekker, Natal, and other local museums, and as offered in heritage tours of Greater Edendale and other historical sites throughout South Africa. Aerial photography that visually documented the changes to physical form over the key decades from the 1930s to 1970s rounded out my documentary sources. Such photographs alone cry out for a meticulous analysis that could, for example, chart the emergence of erosion gulleys, new settlements, and changing vegetation and could be a wonderful tool to spark oral history recollections in a future project.

# References

Academy of Science of South Africa (ASSAF). 2015. *Diversity in Human Sexuality: Implications for Policy in Africa*. Pretoria: Academy of Science of South Africa.

Adler, E.D. 1957. "Soil Conservation." In *Natal Regional Survey*. Vol. 13. *Agriculture in Natal: Recent Developments – A Symposium*, edited by H.R. Burrows, 123–34. Cape Town: Oxford University Press and University of Natal.

Alexander, Jocelyn. 2000. "Technical Development and the Human Factor: Sciences of Development in Rhodesia's Native Affairs Department." In *Science and Society in Southern Africa*, edited by Saul Dubow, 212–37. Manchester: Manchester University Press.

Amtaika, Alexius. 2001. "The Role of the Local State in Social and Economic Development: A Case Study of the Greater Pietermaritzburg Local Council in KwaZulu-Natal." PhD diss., University of the Witwatersrand.

– 2013. *Local Government in South Africa since 1994: Leadership, Democracy, Development and Service Delivery in a Post-Apartheid South Africa*. Durham, NC: Carolina Academic Press.

Anderson, David. 2002. *Eroding the Commons: The Politics of Ecology in Baringo, Kenya, 1890–1963*. Oxford: James Currey.

Ashforth, Adam. 1990. *The Politics of Official Discourse in Twentieth-Century South Africa*. Oxford: Clarendon Press.

– 2005. *Witchcraft, Violence, and Democracy in South Africa*. Chicago: University of Chicago Press.

Assadourian, Erik. 2011. "The Path to Degrowth in Overdeveloped Economies." In *State of the World, 2012: Moving toward Sustainable Prosperity*, edited by Worldwatch Institute, 22–37. Washington: Covelo; London: Island Press.

Atkins, Keletso. 1993. *"The Moon Is Dead! Give Us Our Money!" The Cultural Origins of an African Work Ethic, Natal, South Africa, 1843–1900*. Portsmouth, NH: Heinemann; London: James Currey.

Badassy, Prinisha. 2011. "A Severed Umbilicus: Infanticide and the Concealment of Birth in Natal, 1860–1935." PhD diss., University of KwaZulu-Natal.

Baines, Gary. 1994. "The Contradictions of Community Politics: The African Petty Bourgeoisie and the New Brighton Advisory Board, c. 1937–1952." *Journal of African History* 35 (1): 79–97.

Ballantine, Christopher. 1993. *Marabi Nights: Jazz, "Race" and Society in Early Apartheid South Africa*. Johannesburg: Ravan Press.

Ballard, Richard, Adam Habib, and Imraan Valodia, eds. 2006. *Voices of Protest: Social Movements in Post-Apartheid South Africa*. Scottsville: University of KwaZulu-Natal Press.

Bank, Leslie J. 2011. *Home Spaces, Street Styles: Contesting Power and Identity in a South African City*. New York: Pluto Press; Johannesburg: Wits University Press.

Bardsley, Gillian. 1982. "Politics and Land in Matatiele, 1844–1900: A Report from the Archives." *Social Dynamics: A Journal of African Studies* 8 (2): 46–82.

Barker, Lady (Mary Anne). 1877. *A Year's Housekeeping in South Africa*. London: Macmillan.

Beall, Jo, Owen Crankshaw, and Susan Parnell. 2002. *Uniting a Divided City: Governance and Social Exclusion in Johannesburg*. London: Earthscan.

Beavon, Keith. 2004. *Johannesburg: The Making and Shaping of the City*. Pretoria: University of South Africa Press.

Beinart, William. 1984. "Soil Erosion, Conservationism and Ideas about Development: A Southern African Exploration, 1900–1960." *Journal of Southern African Studies* 11 (1): 52–83.

– 1989. "Introduction: The Politics of Colonial Conservation." *Journal of Southern African Studies* 15 (2): 143–62.

– 2000. "African History and Environmental History." *African Affairs* 99 (395): 269–302.

– 2003. "South African Environmental History in the African Context." In *South Africa's Environmental History: Cases and Comparisons*, edited by Stephen Dovers, Ruth Edgecombe, and Bill Guest, 215–26. Cape Town: David Philip.

– 2012. "Beyond 'Homelands': Some Ideas about the History of African Rural Areas in South Africa." *South African Historical Journal* 64 (1): 5–21.

Beningfield, Jennifer. 2006. *Frightened Land: Land, Landscapes and Politics in South Africa in the Twentieth Century*. London: Routledge.

Bennett, Brett M. 2011. "Naturalising Australian Trees in South Africa: Climate,

Exotics, and Experimentation." *Journal of Southern African Studies* 37 (2): 265–80.
BESG (Built Environment Support Group). 2003a. *The Msunduzi HIV/AIDS Strategy: A Partnership Response to HIV/AIDS at Local Government Level.* Pietermaritzburg: Msunduzi Municipality and BESG.
– 2003b. *The Coping and Adaptive Strategies of Low-Income Households Affected by HIV/AIDS.* Pietermaritzburg: Urban Sector Network and BESG.
– 2004. *Evaluation of the Greater Edendale Development Initiative's Poverty Reduction Strategy.* Pietermaritzburg: BESG.
– 2009. *BESG: A Short History.* Pietermaritzburg: BESG.
Bews, John William. 1913. "An Ecological Survey of the Midlands of Natal, with Special Reference to the Pietermaritzburg District." *Annals of the Natal Museum* 2(4): 485–545.
Bhana, Surendra. 1997. "Indianness Reconfigured, 1944–1960: The Natal Indian Congress in South Africa." *Comparative Studies of South Asia, Africa and the Middle East* 17 (2): 100–7.
Bird, John. 1888. *The Annals of Natal, 1495 to 1845.* Vol. 1. Pietermaritzburg: P. Davis and Sons.
Bissell, William Cunningham. 2000. "Conservation and the Colonial Past: Urban Planning, Space and Power in Zanzibar." In *Africa's Urban Past,* edited by David M. Anderson and Richard Rathbone, 246–61. Oxford: James Currey; Portsmouth, NH: Heinemann.
Bizley, William. 1984. "John William Bews: A Commemorative Note." *Natalia* 14: 17–21.
Blackburn, Douglas. [1908] 1991. *Leaven: A Black and White Story.* Pietermaritzburg: University of Natal Press.
Bond, Patrick. 2000. *Elite Transition: From Apartheid to Neoliberalism in South Africa.* Pietermaritzburg: University of Natal Press.
– 2002. *Unsustainable South Africa: Environment, Development and Social Protest.* London: Merlin Press; Pietermaritzburg: University of Natal Press.
– 2003. "The Degeneration of Urban Policy after Apartheid." In *Confronting Fragmentation: Housing and Development in a Democratizing Society*, edited by Philip Harrison, Marie Huchzermeyer, and Mzwanele Mayekiso, 40–56. Cape Town: University of Cape Town Press.
Bonner, Philip. 1990. "'Desirable or Undesirable Basotho Women?' Liquor, Prostitution and the Migration of Basotho Women to the Rand, 1920–1945." In *Women and Gender in Southern Africa to 1945*, edited by Cherryl Walker, 221–50. Cape Town: D. Phillip.

– 2005. "Eluding Capture: African Grass-roots Struggles in 1940s Benoni." In *South Africa's 1940s: Worlds of Possibilities,* edited by Saul Dubow and Alan Jeeves, 170–91. Cape Town: Double Storey.

Bonner, Philip, and Noor Nieftagodien. 2008. *Alexandra: A History.* Johannesburg: Wits University Press.

Bonner, Philip, and Noor Nieftagodien with Sello Mathabatha. 2012. *Ekurhuleni: The Making of an Urban Region.* Johannesburg: Wits University Press.

Bozzoli, Belinda, ed. 1979. *Labour, Townships, and Protest: Studies in the Social History of the Witwatersrand.* Johannesburg: Ravan Press.

– 2004. *Theatre of Struggle and the End of Apartheid.* Athens: Ohio University Press.

Bozzoli, Belinda, with Mmantho Nkotsoe. 1991. *Women of Phokeng: Consciousness, Life Strategy and Migrancy in South Africa, 1900–1983.* Portsmouth, NH: Heinemann.

Bradford, Helen. 1987. *A Taste of Freedom: The ICU in Rural South Africa, 1924–1930.* New Haven: Yale University Press.

– 1992. "'We Women Will Show Them': Beer Protests in the Natal Countryside, 1929." In *Liquor and Labor in Southern Africa,* edited by Charles Ambler and Jonathan Crush, 208–34. Athens: Ohio University Press; Pietermaritzburg: University of Natal Press.

Brain, Joy B. 1985. "Health and Disease in White Settlers in Colonial Natal." *Natalia* 15: 64–77.

– 1988. "Infectious Diseases." In *Pietermaritzburg, 1838–1988: A New Portrait of an African City,* edited by John Laband and Robert Haswell, 193–5. Pietermaritzburg: University of Natal Press and Shuter and Shooter.

Brennan, James R., Andrew Burton, and Yusuf Lawi, eds. 2007. *Dar es Salaam: Histories from an Emerging African Metropolis.* Dar es Salaam: Mkuki na Nyota Publishers.

Brett, Michael. 1988. "The Living Environment." In *Pietermaritzburg, 1838–1988: A New Portrait of an African City,* edited by John Laband and Robert Haswell, 2–5. Pietermaritzburg: University of Natal Press and Shuter and Shooter.

Brookes, Edgar H. 1954. "The Health of the Bantu." *Race Relations Journal* 21 (2): 1–15.

Brookes, Edgar H., and N. Hurwitz. 1957. *The Native Reserves of Natal.* Natal Regional Survey Vol. 7. Cape Town: Oxford University Press.

Brooks, Shirley. 2009. "Royal Precedents and Landscape Midwives: Claiming the

Zululand Wilderness." In *Zulu Identities: Being Zulu, Past and Present*, edited by Benedict Carton, John Laband, and Jabulani Sithole, 293–303. Scottsville: University of KwaZulu-Natal Press.

Brown, Peter. 1982. "Henry Selby Msimang." *Natalia* 12: 71–3.

– n.d. "A History of the Liberal Party of South Africa, 1953–1966." Selby Msimang Papers, PC 2/1/4/2. Alan Paton Centre and Struggle Archives, Pietermaritzburg.

Brown, Peter, and John Aitchison. 1988. "Opposing Apartheid in the Pietermaritzburg Region." In *Pietermaritzburg, 1838–1988: A New Portrait of an African City*, edited by John Laband and Robert Haswell, 209–11. Pietermaritzburg: University of Natal Press and Shuter and Shooter.

Bryant, A.T. (1929) 1965. *Olden Times in Zululand and Natal: Containing Earlier Political History of the Eastern-Nguni Clans*. Cape Town: C. Struik.

– (1909) 1966. *Zulu Medicine and Medicine-Men*. Cape Town: C. Struik.

Bullard, Robert, and Beverly Wright, eds. 2009. *Race, Place, and Environmental Justice after Hurricane Katrina*. Boulder, CO: Westview.

Bulpin, Thomas. 1966. *Natal and the Zulu Country*. Cape Town: Books of Africa

Bundy, Colin. 2014. *Short-Changed? South Africa Since Apartheid*. Cape Town: Jacana.

Burrows, H. Raymond, ed. 1951. *Archeology and Natural Resources of Natal*. Natal Regional Survey. Vol. 1. Cape Town: Oxford University Press.

Buthelezi, Mangosuthu G. 1987. "Unveiling of the Tombstone of the Late H. Selby Msimang (Unkonkana WeFusi), Founder-Member of the Banned African National Congress, and Later Central Committee Member of Inkatha, and of Mrs Mirriam Noluthando Msimang: A Tribute." Ulundi: Office of the Chief Minister, KwaZulu.

Butler, Mark, and Anne Harley with John Aitchison. 1993. *Imbali: A Centre for Adult Education Study*. Pietermaritzburg: University of Natal, Centre for Adult Education.

Caesar, Mary. 2008. "Gender, Politics and Social Medicine in South Africa, 1940–1959." Master's thesis, Queen's University.

– 2015. "A 'New Experiment in Local Government': The Local Health Commission: A Study of Public Health and Local Governance in Black Urban Areas in Natal, South Africa, 1930–1959." PhD diss., Queen's University.

Caesar, Mary, Jonathan Crush, and Trevor Hill. 2013. *The State of Food Insecurity in Msunduzi Municipality, South Africa*. Cape Town and Kingston: African Food Security Urban Network.

Campbell, Catherine. 1992. "Learning to Kill? Masculinity, the Family, and Violence in Natal." *Journal of Southern African Studies* 18 (3): 614–28.
Cape Colony. 1883. *Report and Proceedings with Appendices of the Government Commission on Native Laws and Customs.* Cape Town: W.A. Richardson.
Carruthers, Jane. 1990. "Toward an Environmental History of Southern Africa: Some Perspectives." *South African Historical Journal* 23 (1): 184–95.
– 2006. "Tracking in Game Trails: Looking Afresh at the Politics of Environmental History in South Africa." *Environmental History* 11 (4): 804–29.
– 2014. "All for One, One for All? Leveraging National Interests with Regional Visions in Southern Africa." *South African Historical Journal* 66 (2): 207–16.
Carton, Benedict. 2000. *Blood from Your Children: The Colonial Origins of Generational Conflict in South Africa.* Scottsville: University of KwaZulu-Natal Press.
Carton, Benedict, John Laband, and Jabulani Sithole, eds. 2009. *Zulu Identities: Being Zulu, Past and Present.* Scottsville: University of KwaZulu-Natal Press.
Cattaneo, Claudio, G. D'Alisa, G. Kallis, and C. Zografos, eds. 2012. "Degrowth Futures and Democracy." *Futures* 44 (6): 515–23.
Cell, John. 1982. *The Highest Stage of White Supremacy: The Origins of Segregation in South Africa and the American South.* Cambridge: Cambridge University Press.
– 1986. "Anglo-Indian Medical Theory and the Origins of Segregation in West Africa." *American Historical Review* 91 (2): 307–35.
Charlton, Sarah. 2006. "Learning from the Local: Experiences of Informal Settlement Upgrading in KwaZulu-Natal." *South African Review of Sociology* 37 (1): 48–64.
Clark, Grant. 2004. "'Lost Tribe' Gets Land Back." *Mail and Guardian*, 1 October. Accessed 23 May 2013. http://mg.co.za/article/2004-10-01-lost-tribe-gets-land-back.
Clark, John. 1972. *Natal Settler-Agent: The Career of John Moreland, Agent for the Byrne Emigration-Scheme of 1849–51.* Cape Town: A.A. Balkema.
Cock, Jacklyn. 2006. "Connecting the Red, Brown and Green: The Environmental Justice Movement in South Africa." In *Voices of Protest: Social Movements in Post-Apartheid South Africa*, edited by Richard Ballard, Adam Habib, and Imraan Valodia, 203–24. Scottsville: University of KwaZulu-Natal Press.
– 2012. "The Challenge of Ecological Transformation in Post-Apartheid South Africa: The Re-Emergence of an Environmental Justice Movement." In *Contesting Transformation: Popular Resistance in Twenty-First-Century*

*South Africa*, edited by Marcelle C. Dawson and Luke Sinwell, 166–82. New York: Pluto Press.
Cock, Jacklyn, and Eddie Koch. 2006. *Going Green: People, Politics, and the Environment in South Africa*. Cape Town: Oxford University Press.
Coka, Gilbert. 1939. "The Story of Gilbert Coka, of the Zulu Tribe of Natal, South Africa, Written by Himself." In *Ten Africans*, edited by Margery Perham, 273–322. London: Faber and Faber.
Coleman, Colin. 2013. *Two Decades of Freedom: A 20-Year Review of South Africa*. Johannesburg: Goldman-Sachs International.
Colenso, J.W. 1855. *Ten Weeks in Natal*. London: Macmillan.
Cope, Nicholas. 1990. "The Zulu Petit Bourgeoisie and Zulu Nationalism in the 1920s: Origins of Inkatha." *Journal of Southern African Studies* 16 (3): 431–51.
Cosnett, J. E. 1975. "Edendale Hospital, Pietermaritzburg: The First Twenty-One Years." *South African Medical Journal* 49: 1486–91.
Couzens, Tim. 1985. *The New African: A Study of the Life and Work of H.I.E. Dhlomo*. Johannesburg: Ravan.
Crush, Jonathan, ed. 1995. *Power of Development*. London and New York: Routledge.
Crush, Jonathan, and Mary Caesar. 2014. "City without a Choice: Urban Food Insecurity in Msunduzi." *Urban Forum* 25: 165–75.
Currey, R.N., ed. 1968. *Letters and Other Wrtitings of a Natal Sheriff: Thomas Phipson, 1815–76*. Cape Town: Oxford University Press.
DA (Democratic Alliance). 2016. "DA Lays Criminal Charges against Penny Sparrow." 4 January. Accessed 20 January 2016. https://www.da.org.za/2016/01/da-lays-criminal-charges-against-penny-sparrow/.
D'Alisa, Giacomo, Federico Demaria, and Giorgos Kallis, eds. 2015. *Degrowth: A Vocabulary of a New Era*. New York and London: Routledge.
Dawson, Marcelle C., and Luke Sinwell, eds. 2012. *Contesting Transformation: Popular Resistance in Twenty-First-Century South Africa*. New York: Pluto Press.
de Beer, Cedric. 1986. *The South African Disease: Apartheid Health and Health Services*. London: Catholic Institute for International Relations.
de Waal, Alex, and Alan Whiteside. 2003. "'New Variant Famine': AIDS and Food Crisis in Southern Africa." *Lancet* 362 (9391): 1234–7.
Deane, Dee Shirley. 1978. *Black South Africans: A Who's Who. 57 Profiles of Natal's Leading Blacks*. Cape Town: Oxford University Press.
Denis, Philippe. 2004a. "'We Also Had to Live with Apartheid in Our Homes.' Stories of Women in Sobantu, South Africa." *Studia Historiae Ecclesiasticae* 30 (1): 151–67.

– 2004b. “African Traditional Religion and Christian Identity in a Group of Manyano Leaders.” *Missionalia* 32 (2): 177–89.
– ed. 2005. *Never Too Small to Remember: Memory Work and Resilience in Times of AIDS*. Sinomlando Centre for Oral History and Memory Work. Pietermaritzburg: Cluster Publications.
Denis, Philippe, and Graham Duncan. 2011. *The Native School That Caused All the Trouble: A History of the Federal Theological Seminary of Southern Africa*. Dorpspruit [Pietermaritzburg]: Cluster Publications.
Denis, Philippe, Radikobo Ntsimane, and Thomas Cannell. 2010. *Indians vs Russians: An Oral History of Political Violence in Nxamalala (1987–1993)*. Dorpspruit [Pietermaritzburg]: Cluster Publications.
Dent, G.R., and C.L. Sibusiso Nyambezi. 1969. *Scholar's Zulu Dictionary: English-Zulu, Zulu-English*. Pietermaritzburg: Shuter and Shooter.
Desai, Ashwin. 2002. *We Are the Poors: Community Struggles in Post-Apartheid South Africa*. New York: Monthly Review Press.
Desmond, Cosmas. 1978. *Limehill Revisited: A Case Study of the Longer-Term Effects of African Resettlement*. Pietermaritzburg: Development Studies Research Group.
Dhlomo, H.I.E. 1954. “Shaka: His Character, Philosophy, Policy and Achievements.” *Ilanga lase Natal*, 2 October 1954.
Dhlomo, R.R.R. 1928. *An African Tragedy: A Novel in English by a Zulu Writer*. Lovedale: Lovedale Institution Press.
Diamond, Jared. 1997. *Guns, Germs and Steel: The Fates of Human Societies*. New York: W.W. Norton.
Diesel, Alleyn, comp. 2007. *Shakti: Stories of Indian Women in South Africa*. Johannesburg: Witwatersrand University Press.
Digby, Anne. 2008. “The Medical History of South Africa: An Overview.” *History Compass* 6 (5): 1194–210.
– 2012. “Evidence, Encounters and Effects of South Africa's Reforming Gluckman National Health Services Commission, 1942–1944.” *South African Historical Journal* 64 (2): 187–205.
– 2014. “Debating the Gluckman Commission: A Rejoinder.” *South African Historical Journal* 66 (3): 588–92.
– 2015. “Debating the Gluckman Commission: A Final Rejoinder.” *South African Historical Journal* 67 (1): 91–94.
Dlamini, Jacob. 2009. *Native Nostalgia*. Auckland Park: Jacana.
Dlamini, Sibusisiwe Nombusa. 2005. *Youth and Identity Politics in South Africa, 1990–1994*. Toronto: University of Toronto Press.
Dodson, Belinda. 2002. “Searching for a Common Agenda: Ecofeminism and

Environmental Justice." In *Environmental Justice in South Africa*, edited by David A. McDonald, 81–108. Athens: Ohio University Press; Cape Town: University of Cape Town Press.

Dovers, Stephen, Ruth Edgecombe and Bill Guest, eds. 2002. *South Africa's Environmental History: Cases and Comparisons*. Cape Town: David Philip.

Dubow, Saul. 1989. *Racial Segregation and the Origins of Apartheid in South Africa, 1919–1936*. Basingstoke: Macmillan.

– ed. 2000. *Science and Society in Southern Africa*. Manchester: Manchester University Press.

Dubow, Saul, and Alan Jeeves, eds. 2005. *South Africa's 1940s: Worlds of Possibilities*. Cape Town: Double Storey Books.

Duminy, Andrew, and Bill Guest, eds. 1989. *Natal and Zululand from Earliest Times to 1910: A New History*. Pietermaritzburg: University of Natal Press.

Du Plooy, Paula. 1988. "African Reaction to the Beer Halls." In *Pietermaritzburg, 1838–1988: A New Portrait of an African City*, edited by John Laband and Robert Haswell, 142–3. Pietermaritzburg: University of Natal Press and Shuter and Shooter.

Dyer, Julie. 2012. *Health in Pietermaritzburg (1838–2008): A History of Urbanisation and Disease in an African City*. Pietermaritzburg: Natal Society Foundation.

Echenberg, Myron. 2007. *Plague Ports: The Global Urban Impact of Bubonic Plague, 1894–1901*. New York and London: New York University Press.

Edwards, Iain. 1996. "Cato Manor, 1959: Men, Women, Crowds, Violence, Politics and History." In *The People's City: African Life in Twentieth Century Durban*, edited by Paul Maylam and Iain Edwards, 102–42. Portsmouth, NH: Heinemann; Pietermaritzburg: University of Natal Press.

Elder, Glen S. 2003. *Hostels, Sexuality, and the Apartheid Legacy: Malevolent Geographies*. Athens: Ohio University Press.

Ellis, Beverly. 1988. "Pietermaritzburg and Its Environs." In *Pietermaritzburg, 1838–1988: A New Portrait of an African City*, edited by John Laband and Robert Haswell, 29–32. Pietermaritzburg: University of Natal Press and Shuter and Shooter.

– 1998. "The Impact of the White Settlers on the Natural Environment of Natal, 1845–1870." Master's thesis, University of Natal-Pietermaritzburg.

Emery, Frank. 1985. "Early Colonists' Perceptions of Natal." *Journal of Natal and Zulu History* 8: 1–9.

Epprecht, Marc. 1993. "Domesticity and Piety in Colonial Lesotho: The Private Politics of Basotho Women's Pious Associations," *Journal of Southern African Studies* 19 (2): 202–21.

– 2000. *"This Matter of Women Is Getting Very Bad": Gender, Development and Politics in Colonial Lesotho, 1870–1965*. Pietermaritzburg: University of Natal Press.
– 2008. *Heterosexual Africa? The History of an Idea from the Age of Exploration to the Age of AIDS*. Athens: Ohio University Press; Scottsville: University of KwaZulu-Natal Press.
– 2013. *Sexuality and Social Justice in Africa*. London: Zed Books.
Essa, Azad. 2015. "South Africa in Midst of 'Epic Drought.'" *Al Jazeera News*, 4 November. Accessed 14 December 2015. http://www.aljazeera.com/news/2015/11/south-africa-midst-epic-drought-151104070934236.html.
Etherington, Norman. 1978. *Preachers, Peasants and Politics in Southeast Africa, 1835–1880: African Christian Communities in Natal, Pondoland and Zululand*. London: Royal Historical Society.
Fassin, Didier. 2007. *When Bodies Remember: Experience and Politics of AIDS in South Africa*. Berkeley: University of California Press.
Ferguson, James. 1990. *The Anti-Politics Machine: "Development," Depoliticization, and Bureaucratic Power in Lesotho*. Cambridge: Cambridge University Press.
– 2011. "Rethinking Poverty and Assistance in Southern Africa." In *Zuma's Own Goal: Losing South Africa's "War on Poverty,"* edited by B. Maharaj, A. Desai, and P. Bond, 399–416. Trenton, NJ: Africa World Press.
Fetter, Bruce. 1976. *The Creation of Elisabethville, 1910–1940*. Stanford, CA: Hoover Institution Press.
Field, Sean, ed. 2001. *Lost Communities, Living Memories: Remembering Forced Removals in South Africa*. Cape Town: David Philip.
– 2012. *Oral History, Community, and Displacement: Imagining Memories in Post-Apartheid South Africa*. New York: Palgrave Macmillan.
Flint, Karen. 2008. *Healing Traditions: African Medicine, Cultural Exchange and Competition in South Africa, 1820–1948*. Athens: Ohio University Press.
Forget, Gilles, and Jean Lebel. 2001. "An Ecosystem Approach to Human Health." *International Journal of Environmental Health* 7 (2): S3–S37.
Freund, Bill. 2007. *The African City: A History*. Cambridge: Cambridge University Press.
– 2012. "The South African Development State and the First Attempt to Create a National Health System: Another Look at the Gluckman Commission of 1942–44." *South African Historical Journal* 64 (2): 170–86.
– 2014. "Reflections on the 1944 National Health Services Commission: A Response to Shula Marks." *South African Historical Journal* 66 (3): 593–6.

Freund, Bill, and Vishnu Padayachee. 2002. *(D)Urban Vortex: South African City in Transition*. Scottsville: University of Kwa-Zulu Natal Press.
Gaitskell, Deborah. 1982. "'Wailing for Purity': Prayer Unions, African Mothers and Adolescent Daughters, 1912–40." In *Industrialisation and Social Change in South Africa*, edited by Shula Marks and Richard Rathbone, 338–57. London: Longman.
Gale, George W. 1939. "The Incidence and Control of Venereal Diseases among the Natives of an Urban Area." *South African Medical Journal* 13: 265–70.
– 1954. *John William Bews: A Memoir*. Pietermaritzburg: University of Natal Press.
Garner, Robert Charles. 1998. "Religion and Economics in a South African Township." PhD diss., Cambridge University.
– 2004. "African Independent Churches and Economic Development in Edendale." In *Engaging Modernity: Methods and Cases Studying African Independent Churches*, edited by Dawid Ventner, 81–103. Westport, CT and London: Praeger.
Gasa, Nomboniso, ed. 2007. *Women in South African History: Basus'iimbokodo bawel'imilambo / They Remove Boulders and Cross Rivers*. Cape Town: Human Sciences and Research Centre Press.
Gennrich, Daniela, John Inglis, and Hilary Kromberg, eds. 2009. *Journeying for Justice: Stories of an Ongoing Faith-Based Struggle*. Pietermaritzburg: Cluster Publications and Jive Media.
Giblin, James. L. 2005. *A History of the Excluded: Making Family a Refuge from State in Twentieth-Century Tanzania*. Athens: Ohio University Press.
GilFillan, S.C. 1920. "The Coldward Course of Progress." *Political Science Quarterly* 35 (3): 393–410.
Gluckman, Max. 1960. "The Rise of the Zulu Empire." *Scientific American* 202 (April): 157–68.
Goebel, Allison. 2002. "Gender, Environment and Development in Southern Africa." *Canadian Journal of Development Studies* 23 (2): 293–316.
– 2004. *Gender and Land Reform: The Zimbabwe Experience*. Montreal: McGill-Queen's University Press.
– 2007. "Sustainable Urban Development? Low-cost Housing Challenges in South Africa." *Habitat International* 31 (3–4): 291–302.
– 2011. "'Our Struggle Is for the Full Loaf': Protests, Social Welfare and Gendered Citizenship in South Africa." *Journal of Southern African Studies* 37 (2): 369–88.
– 2015. *On Their Own: Women and the Right to the City in South Africa*. Montreal: McGill-Queen's University Press.

Goebel, Allison, and Belinda Dodson. 2011. "Housing and Marginality for Female-Headed Households: Observations from Msunduzi Municipality (Pietermaritzburg, South Africa)." *Canadian Journal of African Studies* 45 (2): 240–72.
Goebel, Allison, Belinda Dodson, and Trevor Hill. 2010. "Urban Advantage or Urban Penalty? A Case Study of Female-Headed Households in a South African City." *Health and Place* 16 (3): 573–80.
Goebel, Allison, Trevor Hill, Rob Fincham, and Mary Lawhon. 2010. "Transdisciplinarity in Urban South Africa." *Futures: The Journal of Policy, Planning and Futures* 42 (5): 475–83.
Goedeke, Sonya. 2000. "Imbali Guidance Project: Facilitating Development among Guidance Educators in Kwa Zulu Natal, South Africa." *Journal of Social Development in Africa* 15 (1): 27–48.
Goerg, Odile. 1998. "From Hill Station (Freetown) to Downtown Conakry (First Ward): Comparing French and British Approaches to Segregation in Colonial Cities at the Beginning of the Twentieth Century." *Canadian Journal of African Studies* 32 (1): 1–31.
Goodhew, David. 2004. *Respectability and Resistance: A History of Sophiatown.* Westport, CT: Praeger.
Gordon, Ruth. 1981. *The Place of the Elephant: A History of Pietermaritzburg.* Pietermaritzburg: Shuter and Shooter.
Graham, Lucy Valerie. 2012. *State of Peril: Race and Rape in South African Literature.* Oxford: Oxford University Press.
Greater Edendale Environmental Network (GREEN). 2002. Untitled draft report. Accessed 20 January, 2016. http://www.greennetwork.org.za/docs/ffloodsressearchpaper.pdf.
Grove, Richard. 1989. "Scottish Missionaries, Evangelical Discourses and the Origins of Conservation Thinking in Southern Africa, 1820–1900." *Journal of Southern African Studies* 15 (2): 163–87.
Guest, Bill. 1988. "Economic Development of the Capital City, 1838–1910." In *Pietermaritzburg, 1838–1988: A New Portrait of an African City*, edited by John Laband and Robert Haswell, 120–28. Pietermaritzburg: University of Natal Press and Shuter and Shooter.
Guest, Bill, and John M. Sellers, eds. 1994. *Receded Tides of Empire: Aspects of the Economic and Social History of Natal and Zululand since 1910.* Pietermaritzburg: University of Natal Press.
Gussler, Judith. 1973. "Social Change, Ecology and Spirit Possession among the South African Nguni." In *Religion, Altered States of Consciousness,*

*and Social Change*, edited by E. Bourguignon, 88–126. Columbus: Ohio University Press.

Guy, Jeff. (1979) 1994. *The Destruction of the Zulu Kingdom*. Pietermaritzburg: University of Kwa-Zulu Natal Press.

– 1980. "Ecological Factors in the Rise of Shaka and the Zulu Kingdom." In *Economy and Society in Pre-Industrial South Africa*, edited by Shula Marks and Anthony Atmore, 102–19. London: Longman.

– 1990. "Gender Oppression in Southern Africa's Precapitalist Societies." In *Women and Gender in Southern Africa to 1945*, edited by Cherryl Walker, 33–47. Cape Town: David Philip; London: James Currey.

– 2013. *Theophilus Shepstone and the Forging of Natal: African Autonomy and Settler Colonialism in the Making of Traditional Authority*. Scottsville: University of Kwa-Zulu Natal Press.

Hall, Martin. 1976. "Dendroclimatology, rainfall and human adaptability in the later Iron Age of Natal and Zululand." *Annals of the Natal Museum* 22 (3): 693-703

Hallowes, David, and Victor Munnik. 2008. *Wasting the Nation: Making Trash of People and Places*. Pietermaritzburg: Groundwork.

Halperin, Daniel, and Helen Epstein. 2007. "Why Is HIV So Severe in Southern Africa? The Role of Multiple Concurrent Partnerships and Lack of Male Circumcision: Implications for AIDS Prevention." *South African Journal of HIV Medicine* 16: 19–27.

Hamilton, Carolyn. 2009. "Uncertain Citizenship and Public Deliberation in Post-apartheid South Africa." *Social Dynamics* 35 (2): 355–74.

Harber, Anton. 2011. *Diepsloot*. Johannesburg and Cape Town: Jonathan Ball.

Harries, Patrick. 2007. *Butterflies and Barbarians: Swiss Missionaries and Systems of Knowledge in South-East Africa*. Athens: Ohio University Press; Johannesburg: Wits University Press.

Harris, Verne. 2007. *Archives and Justice: A South African Perspective*. Chicago: Society of American Archivists.

Harrison, Philip, Margaret Futter, Fiona Meth. 1997. *An Industry in Distress: Pietermaritzburg's Footwear Sector*. Durban: Centre for Social and Development Studies, University of Natal.

Harrison, Philip, Alison Todes, and Vanessa Watson. 2008. *Planning and Transformation: Learning from the Post-Apartheid Experience*. London and New York: Routledge.

Hassim, Shireen. 1993. "Family, Motherhood, and Zulu Nationalism: The Politics of the Inkatha Women's Brigade." *Feminist Review* 43 (Spring): 1–25.

Haswell, Robert. 1995. "Themba Harry Gwala (1920–1995)." *Natalia* 25: 88–91.
– 2006. "Local Economy Never More Bouyant." *Witness*. 28 April: 12.
Hattersley, Alan F. 1936. *More Annals of Natal, with Historical Introductions and Notes*. Pietermaritzburg: Shuter and Shooter.
– 1938. *Pietermaritzburg Panorama: A Survey of One Hundred Years of an African City*. Pietermaritzburg: Shuter and Shuter.
– 1950. *The British Settlement of Natal: A Study in Imperial Migration*. Cambridge: Cambridge University Press.
– 1951. *Portrait of a City*. Pietermaritzburg: Shuter and Shooter.
– 1955. *A Hospital Century: Grey's Hospital Pietermaritzburg, 1855–1955*. Cape Town and Amsterdam: A.A. Balkema.
Healy-Clancy, Meghan. 2012. "Women and the Problem of Family in Early African Nationalist History and Historiography." *South African Historical Journal* 63 (3): 450–71.
– 2013. *A World of Their Own: A History of South African Women's Education*. Scottsville: University of KwaZulu-Natal Press.
Heard, N.M., and Lafina Ncobeni. 1986. "Community-based Infant and Child Mortality Rates for Peri-urban Pietermaritzburg." *South African Medical Journal* 70 (5): 283–4.
Hellmann, Ellen. 1948. *Rooiyard: A Sociological Survey of an Urban Native Slum Yard*. Cape Town: Oxford University Press for Rhodes Livingstone Institute.
Herzfeld, Michael. 1997. *Cultural Intimacy: Social Poetics in the Nation-State*. New York and London: Routledge.
Hlahla, Sitabile, Allison Goebel, and T.N. Hill. 2016. "Green Economy: A Strategy to Alleviate Urban Poverty and Safeguard the Environment? KwaZulu-Natal, South Africa." *Urban Forum* 27 (1): 113-27.
Hodge, Joseph Morgan. 2007. *Triumph of the Expert: Agrarian Doctrines of Development and Legacies of British Colonialism*. Athens: Ohio University Press.
Holden, William C. (1855) 1963. *History of the Colony of Natal*. Cape Town: Struik.
Holliday, J.D. (1865) 1890. *Dottings on Natal: As Published in 1865 and Sundry Tit-Bits of Colonial Experience*. Pietermaritzburg: P. Davis and Sons.
Houston, Gregory, Shepi Mati, Dineo Seabe, Jeff Peires, Denver Webb, Siphesihle Dumisa, Kombi Sausi, Bernard Mbenga, Andrew Manson, and Nedson Pophiwa. 2013. *The Liberation Struggle and Liberation Heritage Sites in South Africa: Report*. Pretoria: Human Sciences Research Council.
Human Sciences Research Council. 2012. "Hail to the Thief II. The Spear:

Debating Brett Murray's Exhibition." Unpublished presentation to HSD (Pretoria) Intellectual Hour, 29 May.

Hunter, Mark. 2010. *Love in the Time of AIDS: Inequality, Gender and Rights in South Africa.* Scottsville: University of KwaZulu-Natal Press.

IMF. 2016. "World Economic Outlook Update." 19 January. Accessed 20 January 2016. https://www.imf.org/external/pubs/ft/weo/2016/update/01/pdf/0116.pdf.

Ingram, J. F. 1898. *Story of an African City.* Pietermaritzburg: C. Coester.

INR (Institute of Natural Resources). 1997. "A Short-Term Response to Flooding in the Msunduzi River Catchment." Pietermaritzburg: Institute of National Resources.

IPCC (Intergovernmental Panel on Climate Change). 2014. *Climate Change 2014: Impacts, Vulnerability, and Adaptation.* Geneva: United Nations. Accessed 31 March 2014. https://ipcc-wg2.gov/AR5/report/full-report/.

Isaacman, Allen F. 1996. *Cotton Is the Mother of Poverty: Peasants, Work, and Rural Struggle in Colonial Mozambique, 1938–1961.* Portsmouth, NH: Heinemann.

Iyer Urban Design Studio. 2014. *Edendale Urban Hub: Design Report.* Johannesburg: Iyer Urban Design Studio. Accessed 15 May 2015. http://www.msunduzi.gov.za/site/user_data/files/EDENDALE_URBAN_HUB___DESIGN_REPORT_R.pdf.

Jacobs, Nancy J. 2003. *Environment, Power, and Injustice: A South African History.* Cambridge: Cambridge University Press.

Jeater, Diana. 1995. "'The Way You Tell Them': Language, Ideology and Development Policy in Southern Rhodesia." *African Studies* 54 (2): 1–15.

– 2007. *Law, Language, and Science: The Invention of the "Native Mind" in Southern Rhodesia, 1890–1930.* Portsmouth, NH: Heinemann.

Jeeves, Alan. 2000. "Health, Surveillance and Community: South Africa's Experiment with Medical Reform in the 1940s and 1950s." *South African Historical Journal* 43 (1): 244–66.

– 2004. "South Africa in the 1940s: Post-War Reconstruction and the Onset of Apartheid." *South African Historical Journal* 50 (1): 1–11.

Jerven, Morten. 2015. *Africa: Why Economists Get It Wrong.* London: Zed Books.

Jochelson, Karen. 2001. *The Colour of Disease: Syphilis and Racism in South Africa, 1880–1950.* New York: Palgrave Macmillan.

John, Lynelle. 2012. *Secondary Cities in South Africa: The Start of a Conversation.* Braamfontein: South African Cities Network. Accessed

20 January 2016. http://sacitiesnetwork.co.za/wp-content/uploads/2014/07/secondary_cities_in_south_africa_with_more_detail.pdf.

Kark, Sydney L., and J. Cassel. 1952. "The Pholela Health Centre: A Progress Report." *South African Medical Journal* 26 (6): 101–4.

Kark, Sidney L., and Emily Kark. 2001. *Promoting Community Health: From Pholela to Jerusalem*. Johannesburg: Witswatersrand University Press. First published 1999.

– 1962. "A Practice of Social Medicine." In *A Practice of Social Medicine*, edited by Sidney Kark and Guy W. Steuart, 3–40. Edinburgh and London: E. and S. Livingstone.

Karumbidza, Blessing. 2014. "A Green Economy Based on a Greed Agenda Does Not Guarantee a Just Transition for Africa." Special issue, *Perspectives Africa* 2 (September): 13–16.

Kentridge, Matthew. 1990. *Unofficial War: Inside the Conflict in Pietermaritzburg*. Cape Town: David Philip Press.

Kirk, Joyce F. 1998. *Making a Voice: African Resistance to Segregation in South Africa*. Boulder, CO: Westview.

Kolbe, Andrea. 2014. "Citizen Science and Water Quality in the Umgeni Catchment Area, KwaZulu-Natal, South Africa." Master's thesis, Queen's University.

Kumalo, R. Simangaliso. 2009. *Methodists with White History and a Black Future: The People Called Methodists in KwaZulu-Natal*. Pietermaritzburg: Cluster.

Kynoch, Gary. 2005. *We Are Fighting the World: A History of the Marashea Gangs in South Africa, 1947–1999*. Athens: Ohio University Press.

KZN. 2015. "Health in KwaZulu-Natal, Dr SM Dhlomo: Taking Medical Male Circumcision to New Heights." Accessed 20 July 2015. http://www.kznhealth.gov.za/mmc.htm.

Laband, John. 1988. "Recasting the Portrait of Pietermaritzburg." In *Pietermaritzburg, 1838–1988: A New Portrait of an African City*, edited by John Laband and Robert Haswell, xxiii–xxxiii. Pietermaritzburg: University of Natal Press and Shuter and Shooter.

Laband, John, and Robert Haswell, eds. 1988. *Pietermaritzburg, 1838–1988: A New Portrait of an African City*. Pietermaritzburg: University of Natal Press and Shuter and Shooter.

la Hausse, Paul. 1992. "Drink and Cultural Innovation: The Origins of the Beerhall in South Africa, 1902–1916." In *Liquor and Labor in Southern Africa*, edited by Charles Ambler and Jonathan Crush, 78–114. Athens: Ohio University Press; Pietermaritzburg: University of Natal Press.

la Hausse de Lalouviere, Paul. 2000. *Restless Identities: Signatures of Nationalism, Zulu Ethnicity and History in the Lives of Petros Lamula and Lymon Maling, 1881–1948*. Scottsville: University of Natal Press.

– 2009. "'Death Is Not the End': Zulu Cosmopolitanism and the Politics of Zulu Cultural Revival." In *Zulu Identities: Being Zulu, Past and Present*, edited by Benedict Carton, John Laband, and Jabulani Sithole, 256–72. Scottsville: University of KwaZulu-Natal Press.

Lambert, John. 1988. "The Pietermaritzburg Market and the Transformation of the Natal Midlands in the Colonial Period." In *Pietermaritzburg, 1838–1988: A New Portrait of an African City*, edited by John Laband and Robert Haswell, 130–4. Pietermaritzburg: University of Natal Press and Shuter and Shooter.

– 1995. *Betrayed Trust: Africans and the State in Colonial Natal*. Scottsville: University of Natal Press.

– 1996. "The Impoverishment of the Natal Peasantry, 1893–1910." In *Enterprise and Exploitation in a Victorian Colony*, edited by Bill Guest and John M. Sellers, 287–308. Pietermaritzburg: University of KwaZulu-Natal Press.

– 2009. "'Happy Are Those Who Are Dead': Crises in African Life in Late Nineteenth-century and Early Twentieth-century Colonial Natal." In *Zulu Identities: Being Zulu, Past and Present*, edited by Benedict Carton, John Laband, and Jabulani Sithole, 214–21. Scottsville: University of KwaZulu-Natal Press.

Lambert, John, and Robert Morrell. 1996. "Domination and Subordination in Natal, 1890–1920." In *Political Economy and Identities in KwaZulu-Natal: Historical and Social Perspectives*, edited by Robert Morrell, 63–95. Durban: Indicator Press.

LaTouche, Serge. 2010. "Growing a Degrowth Movement." In *State of the World, 2010: Transforming Cultures*, edited by Linda Starke and Lisa Mastnym, 181. New York: Norton/World Watch.

Leach, Melissa, and Cathy Green. 1997. "Gender and Environmental History: From Representation of Women and Nature to Gender Analysis of Ecology and Politics." *Environment and History* 3 (3): 343–70.

Leach, Melissa, and Robin Mearns, eds. 1996. *The Lie of the Land: Challenging Received Wisdom on the African Environment*. Oxford: James Currey; Portsmouth, NH: Heinemann.

Lee, Rebecca. 2009. *African Women and Apartheid: Migration and Settlement in Urban South Africa*. London and New York: I.B. Tauris.

Le Quesne, Thomas. 2000. "The Divorce of Environmental and Economic Policy under the First ANC Government, 1994–1999." *South African Journal of Environmental Law and Policy* 7 (1): 1–20.

Lester, Alan. 1998. *From Colonization to Democracy: A New Historical Geography of South Africa*. London and New York: I.B. Tauris.
LHC (Local Health Commission). 1943–57. *Annual Report for the Year Ended 30th June*. Pietermaritzburg: Local Health Commission.
Limb, Peter. 2010. *The ANC's Early Years: Nation, Class and Place in South Africa before 1940*. Pretoria: Unisa Press.
Linscott, Graham. 2013. *Into the River of Life: A Biography of Ian Player.* Johannesburg and Cape Town: Jonathan Ball.
Low, D.A., and John Lonsdale. 1975. "Towards the New Order, 1945–1963." In *The History of East Africa*. Vol. 3, edited by D.A. Low and A. Smith, 12–16. Oxford: Oxford University Press.
Lugg, Henry C. 1970. *A Natal Family Looks Back*. Durban: T.W. Griggs.
Luig, Ute, and Achim von Oppen, eds. 1997. "The Making of African Landscapes." Special issue, *Paideuma: Mitteilungen* 43.
Luthuli, Albert. 1966. *Let My People Go: An Autobiography.* London: Collins.
Maathai, Wangari. 2009. *The Challenge for Africa*. New York: Pantheon.
Mabin, Anthony. 1986. "Labour, Capital, Class Struggle and the Origins of Residential Segregation in Kimberly, 1880–1920." *Journal of Historical Geography* 12 (1): 4–26.
MacKenzie, John M. 1988. *The Empire of Nature: Hunting, Conservation, and British Imperialism*. Manchester: Manchester University Press.
Mackenzie, Fiona. 1995. "Selective Silence: A Feminist Encounter with Environmental Discourse in Colonial Africa." In *Power of Development*, edited by J. Crush, 100–112. London and New York: Routledge.
Macleod, David. 2007. "The Dusi and the Meeting with the People of the Valley." In *Men, Rivers and Canoes*, edited by Ian Player, 167–77. Empangeni, SA: Echoing Green Press.
Macleod, Fiona. 2006. "Ministries Aim to Trash Green Laws," *Mail and Guardian*, 17 March.
Maggs, Tim. 1988. "Pietermaritzburg: The First 2,000,000 Years." In *Pietermaritzburg, 1838–1988: A New Portrait of an African City*, edited by John Laband and Robert Haswell, 14–17. Pietermaritzburg: University of Natal Press and Shuter and Shooter.
Magwaza, Mayibuye. n.d. *The Built Environment Support Group and Government: A Brief History*. Pietermaritzburg: BESG.
Maharaj, Brig, and S. Narsiah, 2002. "From Apartheid Apologism to Post-Apartheid Neo-Liberalism: Paradigm Shifts in South African Urban Geography." *South African Geographical Journal* 84 (1): 88–97.

Mander, J.J. 1995. *Greater Maritzburg Reconstruction Project. Phase I: Perspective. Environmental Component.* Scottsville: Institute of Natural Resources.
Mangcu, Xolela, ed. 2011. *Becoming Worthy Ancestors: Archive, Public Deliberations and Identity in South Africa.* Johannesburg: Wits University Press.
Mann, Robert James. 1859. *The Colony of Natal: An Account of the Characteristics and Capabilities of This British Dependency.* London: Jarrold and Sons.
Maphanga, Koekie Constance. 1997. "People's Perceptions of Flood Hazards and Their Attitudes towards Resettlement: A Case Study of the Community of Azalea in the Greater Edendale Complex, KwaZulu-Natal, South Africa." Master's thesis, University of Natal.
Marais, Hein. 2011. *South Africa Pushed to the Limit: The Political Economy of Change.* London and New York: Zed Books
Maré, Gerhard. 2014. *Declassified: Moving Beyond the Dead End of Race in South Africa.* Cape Town: Jacana.
Marks, Shula. 1986. *The Ambiguities of Dependence in South Africa.* Johannesburg: Ravan.
– 1991. "Patriotism, Patriarchy and Purity: Natal and the Politics of Zulu Ethnic Nationalism." In *The Creation of Tribalism in Southern Africa*, edited by Leroy Vail, 215–40. Berkeley and Los Angeles: University of California Press.
– 1994. *Divided Sisterhood: Race, Class and Gender in the South African Nursing Profession.* New York: St Martin's Press.
– 2000. "Doctors and the State: George Gale and South Africa's Experiment in Social Medicine." In *Science and Society in Southern Africa*, edited by Saul Dubow, 188–211. Manchester: Manchester University Press.
– 2014. "Reflections on the 1944 National Health Services Commission: A Response to Bill Freund and Anne Digby on the Gluckman Commission." *South African Historical Journal* 66 (1): 169–87.
– 2015. "Response to Anne Digby, 'Debating the Gluckman Commission': A Final Rejoinder." *South African Historical Journal* 67 (1): 85–90.
Marks, Shula, and Richard Rathbone, eds. 1982. *Industrialisation and Social Change in South Africa.* London: Longman.
Martens, Jeremy C. 2002. "'Settler Homes, Manhood, and 'Houseboys': An Analysis of Natal's Rape Scare of 1886." *Journal of Southern African Studies* 28 (2): 379–400.
– 2003. "Polygamy, Sexual Danger, and the Creation of Vagrancy Legislation in Colonial Natal." *Journal of Imperial and Commonwealth History* 31 (3): 24–45.

– 2009. "Enlightenment Theories of Civilisation and Savagery in British Colonial Natal: The Colonial Origins of the (Zulu) African Barbarism Myth." In *Zulu Identities: Being Zulu, Past and Present*, edited by Benedict Carton, John Laband, and Jabulani Sithole, 122–33. Scottsville: University of KwaZulu-Natal Press.

Mather, Charles. 1995. "Environment as Weapon: Land, Labour and African 'Squatters' in Rural South Africa." In *People and Environment in Africa*, edited by Tony Binns, 231–8. New York: John Wiley and Sons.

Maylam, Paul. 1995. "Explaining the Apartheid City: 20 Years of South African Urban Historiography." Special issue, *Journal of Southern African Studies* 21 (1): 19–38.

– 2001. *South Africa's Racial Past: The History and Historiography of Racism, Segregation, and Apartheid*. Aldershot, UK: Ashgate.

Maylam, Paul, and Iain Edwards, eds. 1996. *The People's City: African Life in Twentieth Century Durban*. Portsmouth, NH: Heinemann; Pietermaritzburg: University of Natal Press.

Mbembe, Achille, and Sarah Nuttall. 2008. "Introduction: Afropolis." In *Johannesburg: The Elusive Metropolis*, edited by S. Nuttall and A. Mbembe, 1–33. Johannesburg: Wits University Press.

Mbjanwa, Bheki. 2014. "'No Trees Cut' at Shembe Event," *Daily News*, 28 April.

McCann, James. 1999. *Green Land, Brown Land, Black Land*. Portsmouth, NH: Heinemann.

McClendon, Thomas V. 2002. *Gender and Generations Apart: Labor Tenants and Customary Law in Segregation Era South Africa, 1920s to 1940s*. Portsmouth, NH: Heinemann.

– 2010. *White Chief, Black Lords: Shepstone and the Colonial State in Natal, South Africa, 1845–1878*. Rochester, NY: University of Rochester Press.

McClintock, Anne. 1995. *Imperial Leather: Race, Gender, and Sexuality in the Colonial Contest*. New York: Routledge.

McCord, James. 1946. *My Patients Were Zulus*. London: Frederick Muller.

McCracken, D.P. 1986. "The Indigenous Forests of Colonial Natal and Zululand." *Natalia* 16: 19–38.

– 1988. "Parks and Gardens." In *Pietermaritzburg, 1838–1988: A New Portrait of an African City*, edited by John Laband and Robert Haswell, 59–62. Pietermaritzburg: University of Natal Press and Shuter and Shooter.

McDonald, David A. ed. 2002. *Environmental Justice in South Africa*. Athens: Ohio University Press; Cape Town: University of Cape Town Press.

– 2008. *World City Syndrome: Neoliberalism and Inequality in Cape Town*. New York: Routledge.
McDonald, David A., and Greg Ruiters, eds. 2005. *The Age of Commodity: Water Privatization in Southern Africa*. London: Earthscan.
McNeely, Ann. 2008. "Pietermaritzburg's 'Sanitary Conscience': Race and Municipal Water Planning from 1900 to 1968." Master's thesis, Queen's University.
Meineke, Eric, and G.M. Summers. 1983. *Municipal Engineering in Pietermaritzburg: The First Hundred Years*. Pietermaritzburg: City Engineer's Department.
Meintjes. Sheila. 1984. "Property Relations amongst the Edendale Kholwa." *Journal of Natal and Zulu History* 7: 11–29.
– 1986. "The Early African Press in Natal: *Inkanyiso yase Natal*, April 1889–1896." *Natalia* 16: 5–11.
– 1988a. "Edendale." In *Pietermaritzburg, 1838–1988: A New Portrait of an African city*, edited by John Laband and Robert Haswell, 66–70. Pietermaritzburg: University of Natal Press and Shuter and Shooter.
– 1988b. "Edendale 1850–1906: A Case Study of Rural Transformation and Class Formation in an African Mission in Natal." PhD diss., School of Oriental and African Studies, University of London
– 1990. "Family and Gender in the Christian Community at Edendale, Natal, in Colonial Times." In *Women and Gender in Southern Africa to 1945*, edited by Cherryl Walker, 125–45. Cape Town: David Philip; London: James Currey.
Merchant, Carol. 1989. *Ecological Revolutions: Nature, Gender, and Science in New England*. Chapel Hill: University of North Carolina Press.
Merrett, Christopher. 2009. *Sport, Space, and Segregation: Politics and Society in Pietermaritzburg*. Scottsville: University of KwaZulu-Natal Press.
Merrett, Pat. 1988. "Mrs E.E. Russell." In *Pietermaritzburg, 1838–1988: A New Portrait of an African City*, edited by John Laband and Robert Haswell, 213–16. Pietermaritzburg: University of Natal Press and Shuter and Shooter.
Meskell, Lynn. 2012. *The Nature of Heritage: The New South Africa*. Chichester: Wiley-Blackwell.
Methley, James Erasmus. 1850. *The New Colony of Port Natal, with Information for Immigrants*. London: Houlston and Stoneman.
Middleton, Karen. 1999. "Who Killed 'Malagasy Cactus'? Science, Environment and Colonialism in Southern Madagascar (1924–1930)." *Journal of Southern African Studies* 25 (2): 215–48.
Millward, R.H. 1909. "Natal, the Garden Colony." *National Geographic Magazine* 20 (3): 278–91.

Minkley, Gary. 1998. "'Corpses behind Screens': Native Spaces in the City." In *Blank: Architecture, Apartheid and After*, edited by Hilton Judin and Ivan Vladislavic, 202–19. Rotterdam: NAi Publishers.
Mkhize, Sibongiseni. 2012. "'A Setback to the Harmonious Race Relations in This Charming City of Scented Flowers': The August 1959 Riots in Pietermaritzburg." *Natalia* 42: 65–79.
– 2015. "Class Consciousness, Non-racialism and Political Pragmatism: A Political Biography of Henry Selby Msimang, 1886–1982." PhD diss., University of the Witwatersrand.
Mlaba, Zibusi. 2007. "Foreword to the Second Edition." In *Men, Rivers and Canoes*, edited by I. Player, xv–xvii. Empangeni: Echoing Green Press.
Mngxitama, Andile, and Athi-Nangamso Esther Nkopo. 2013. "There's No Unlearning Whiteness, Despite What Anti-racists Say." *Mail and Guardian*, 11 January, 17, 21.
Mokoena, Hlonipa. 2011. *Magema Fuze: The Making of a* Kholwa *Intellectual.* Scottsville: University of KwaZulu-Natal Press.
Moloi, Godfrey. 1991. *My Life.* Johannesburg: Jonathan Ball.
Moodie, T. Dunbar, with Vivienne Ndatshe. 1994. *Going for Gold: Men, Mines and Migration.* Berkeley: University of California Press.
Moodley, Shomenthree. 1997. "Ecological and Economic Aspects of Treating Vegetable Oil Industrial Effluent at Darvill Wastewater Works in Pietermaritzburg." Master's thesis, University of Natal.
Moore, Henrietta L., and Meghan Vaughan. 1994. *Cutting Down Trees: Gender, Nutrition, and Agricultural Change in Northern Province of Zambia, 1890–1990.* Portsmouth, NH: Heinemann; London: James Currey; Lusaka: University of Zambia Press.
Morrell, Robert. 1992. *Poor but White: Essays on the History of Poor Whites in Southern Africa, 1880–1940.* Pretoria: University of South Africa Press.
– 1998. "Of Boys and Men: Masculinity and Gender in Southern African Studies." *Journal of Southern African Studies* 24 (4): 605–30.
– 2001. *From Boys to Gentlemen: Settler Masculinity in Colonial Natal, 1880–1920.* Pretoria: University of South Africa Press.
Mosoetsa, Sarah. 2011. *Eating from One Pot: The Dynamics of Survival in Poor South African Households.* Johannesburg: Wits University Press.
Msimang, H. Selby. 1970. "An Address by Mr H. Selby Msimang." H. Selby Msimang Papers, PC 14/1/4/5/3. Alan Paton Centre and Struggle Archives, Pietermaritzburg.

– 1974. "Edendale Residents Lack Faith in Government Promises." *Natal Witness,* 12 September.
– 1975a. "Why Edendale Cannot Be Part of KwaZulu Yet." *Natal Witness,* 5 June.
– 1975b. "Edendale Fears of Slum Conditions in the Area." *Natal Witness,* 28 August.
– n.d. "The History of Edendale." Selby Msimang Papers, PC 11/1/6/2/1-8. Alan Paton Centre and Struggle Archives, Pietermaritzburg.
Msunduzi Municipality. 2002. *Integrated Development Plan*. Pietermaritzburg.
– 2003. "Greater Edendale Poverty Reduction Profile and Strategic Framework: Understanding Poverty and Vulnerability – Towards a Poverty Reduction Strategy for Edendale, Msunduzi Municipality." Pietermaritzburg: Msunduzi Municipality.
– 2006. "The Knights of GEDI." In *Pietermaritzburg: Economic Powerhouse in the Making.* Pietermaritzburg: Purple Boa Creations.
– 2009. "Greater Edendale: Urban and Township Regeneration Project: Towards the Development of a Cohesive and Sustainable Urban Place." Pietermaritzburg: Msunduzi Municipality.
– 2011. "Press Release Msunduzi Local Municipality." 18 January.
– 2012. *Integrated Development Plan, 2012/13–2016/17: Isixaxa / Pulling Together. Re-Imagine Pietermaritzburg Beyond Vision 2025.* Pietermaritzburg: Msunduzi Municipality, the Provincial Intervention Team. Accessed 28 February 2016. http://www.msunduzi.gov.za/site/user_data/files/Draft_IDP_3rd_draft_as_revised___W.pdf
– 2015. *Integrated Development Plan [IDP] Review for 2015/16.* Accessed 16 March 2016. http://www.msunduzi.gov.za/site/user_data/files/MSUNDUZI_FINAL_IDP_REVIEW_2015_2016.pdf
Murray, Martin J. 2008. *Taming the Disorderly City: The Spatial Landscape of Johannesburg after Apartheid.* Cape Town: University of Cape Town Press.
Murray, Martin J., and Garth A. Myers, eds. 2006. *Cities in Contemporary Africa.* New York: Palgrave Macmillan.
"Natalian A." 1897. *A South African Boy: Schoolboy Life in Natal.* London: Marshall Russell.
Napier, Clive, and Phil Mtimkulu. 1989. "Pietermaritzburg's Sobantu Village: A Case Study of Control and Violence in a Natal Township." Occasional Paper 1, University of South Africa.
Natal, Colony of. 1852. *Proceeding and Report the Government Commission*

*Appointed to Inquire into the Past and Present State of the Kafirs in the District of Natal*. Pietermaritzburg: J. Archbell and Son.
– 1892–1908. *Blue Book on Native Affairs*. Pietermaritzburg: P. Davis and various publishers.
– 1907. *Native Affairs Commission, 1906–7: Evidence*. Pietermaritzburg: P. Davis.
Natal, Province of. 1949. *Report of the Director of Provincial Medical and Health Services for the Year 1949*. Pietermaritzburg: Province of Natal.
Nattrass, Nicoli. 2007. *Mortal Combat: AIDS Denialism and the Struggle for Antiretroviral Treatment*. Cape Town: David Philip.
– 2012. *The AIDS Conspiracy: Science Fights Back*. New York: Columbia University Press.
Nauright, John. 2000. "'An Experiment in Native Self-Government': The Alexandra Health Committee, the State and Local Politics, 1916–1933." *South African Historical Journal* 43 (1): 223–43.
Ndinda, Catherine. 2003. "Housing Delivery in Nthutukoville, South Africa: Successes and Problems for Women." *Journal of International Women's Studies* 5 (1): 29–41.
Ndlovu, Sifiso. 2009. "A Reassessment of Women's Power in the Zulu Kingdom." In *Zulu Identities: Being Zulu, Past and Present*, edited by Benedict Carton, John Laband, and Jabulani Sithole, 111–20. Scottsville: University of KwaZulu-Natal Press.
Ndlovu-Gatsheni, Sabelo J. 2013. *Empire, Global Coloniality and African Subjectivity*. New York and Oxford: Berghahn Books.
Neuman, Roderick P. 2002. *Imposing Wilderness: Struggles over Livelihood and Nature Preservation in Africa*. Berkeley: University of California Press.
Ngubane, Harriet. 1977. *Body and Mind in Zulu Medicine*. London: Academic Press.
Nguyen, Vinh-Kim. 2010. *The Republic of Therapy: Triage and Sovereignty in West Africa's Time of AIDS*. Durham: Duke University Press.
Nicholls, George Heaton. 1923. *Bayete! "Hail to the King!"* London: George Allen & Unwin.
Nicolay, Nathea. 2008. "Summary of Provincial HIV and AIDS Statistics for South Africa." Metropolitan/Calloway Leadership Institute. Accessed 27 January 2016. http://www.callawayleadership.com/downloads/CLI_LE_episode18_summary_HIV_stats_SA.pdf.
Niehaus, Isak. 2013. *Witchcraft and a Life in the New South Africa*. Cambridge: Cambridge University Press.

Nightingale, Carl H. 2012. *Segregation: A Global History of Divided Cities.* Chicago: University of Chicago Press.

Nuttall, Sarah, and Achille Mbembe, eds. 2008. *Johannesburg: The Elusive Metropolis.* Johannesburg: Wits University Press.

Nuttall, Timothy A. 1984. "Collaboration and Conflicting Interests in a Black Community: Responses to the Introduction of Local Government in Edendale, Natal." Bachelor honors essay, University of Natal.

OECD (Organisation for Economic Co-operation and Development. 2014. "Shifting Gear: Policy Challenges for the Next 50 Years." OECD Economics Department Policy Notes, No. 24, July. Paris: OECD.

Oosthuizen, Gerhardus C. 1992. "Islam among the Zanzibaris of South Africa." *History of Religions* 31 (3): 305–20.

Packard, Randall M. 1989. *White Plague, Black Labor: Tuberculosis and the Political Economy of Health and Disease in South Africa.* Berkeley: University of California Press.

Parle, Julie. 1988. "The Impact of the Depression upon Pietermaritzburg during the 1860s." Master's thesis, University of Natal.

– 2012. "Introduction: 'The Past and Its Possibilities: Perspectives of Southern Africa.' The Southern African Historical Society's 23rd Biennial Conference, 27–29 June 2011, University of KwaZulu-Natal (Howard College Campus)." *South African Historical Journal* 64 (2): 159–69.

Parle, Julie, and Vanessa Noble. 2014. "New Directions and Challenges in Histories of Heath, Healing and Medicine in South Africa." *Medical History* 58 (2): 147–65.

Parnell, Susan. 1988. "Racial Segregation in Johannesburg: The Slums Act, 1934–39." *South African Geographical Journal* 70 (2): 112–26.

– 1991. "Sanitation, Segregation and the Natives (Urban Areas) Act: African Exclusion from the Malay Location, 1897–1925." *Journal of Historical Geography* 17 (3): 271–88.

– 1993. "Creating Racial Privilege: The Origins of South African Public Health and Town Planning Legislation." *Journal of Southern African Studies* 19 (3): 471–88.

– 1995. "Re-thinking Urban South Africa." *Journal of Southern African Studies* 21 (1): 39–61.

– 2002. "Winning the Battles but Losing the War: The Racial Segregation of Johannesburg under the Natives (Urban Areas) Act of 1923." *Journal of Historical Geography* 28 (2): 258–81.

Paton, Alan. 1980. *Towards the Mountain: An Autobiography.* New York: Charles Scribners' Sons.

Pearse, R.O. 1981. *Joseph Baynes, Pioneer.* Pietermaritzburg: Shuter and Shooter.

Peel, Heather. 1987. "Sobantu Village: An Administrative History of a Pietermaritzburg Township, 1924–1959." Bachelor of arts thesis, University of Natal.

– 1988. "Sobantu Village." In *Pietermaritzburg, 1838–1988: A New Portrait of an African City,* edited by John Laband and Robert Haswell, 82–4. Pietermaritzburg: University of Natal Press and Shuter and Shooter.

Peters, S. 2014. "Terrified Foreigners Return to Jika Joe Fearing Attacks." *Daily News,* 21 January. Accessed 12 June 2014. http://www.iol.co.za/dailynews/news/terrified-foreigners-return-to-jika-joe-fearing-attacks-1.1634226#.U5m1lvldX6M.

Phaahla, Elias. 2015. "The State of Electricity in South Africa – Part II: The Progress in Medupi." *Helen Suzman Foundation.* Accessed 27 January 2016. http://hsf.org.za/resource-centre/hsf-briefs/the-state-of-electricity-in-south-africa-part-ii-the-progress-in-medupi.

Phillips, Howard. 1987. "The Local State and Public Health Reform in South Africa: Bloemfontein and the Consequences of the Spanish Flu Epidemic of 1918." *Journal of Southern African Studies* 13 (2): 210–33.

– 2005. "The Grassy Park Health Centre: A Peri-Urban Pholela?" In *South Africa's 1940s: Worlds of Possibilities,* edited by Saul Dubow and Alan Jeeves, 108–28. Cape Town: Double Storey.

– 2012. *Plague, Pox and Pandemics: A Jacana Pocket History of Epidemics in South Africa.* Cape Town: Jacana.

Pietermaritzburg. 1904–90. *Corporation of Pietermaritzburg Year Book.* Annual volumes. Pietermaritzburg: Natal Witness.

Pietermaritzburg Chamber of Business. 2006. *Pietermaritzburg: Economic Powerhouse in the Making, 2006.* Pietermaritzburg: Purple Boa Creations.

Pietermaritzburg-Msunduzi. 2006. *Msunduzi Municipality Full Term Report, 2006.* Pietermaritzburg: City Hall.

Player, Ian, ed. (1964) 2007. *Men, Rivers and Canoes.* Empangeni: Echoing Green Press.

– 1972. "Wilderness and the Environment." *Natalia* 2: 27–9.

Pole, Adrian. 2002. "Factors Affecting the Successful Implementation of the Polluter Pays Principle: A Case Study of the Bayne's Spruit." Master's thesis, University of KwaZulu-Natal.

Posel, Deborah, and Graeme Simpson, eds. 2002. *Commissioning the Past:*

*Understanding South Africa's Truth and Reconciliation Commission.* Johannesburg: Witwatersrand University Press.
Radebe, Jeff. 2009. "Rev. Enos Sikhakhane and the Art of Positioning Edendale on the Political and Religious Map of the World." Presentation for the Department of Justice and Constitutional Development, 5 December. Accessed 12 September 2011. http://www.info.gov.za/speeches/2009/09121414251004.htm.
Ramose, Mogobe. 2015. "Ubuntu." In *Degrowth: A Vocabulary of a New Era*, edited by Giacomo D'Alia, Federico Demaria, and Giorgos Kallis, 212–14. New York and London: Routledge.
Ramphele, Mamphela. 1991. "New Day Rising." In *Restoring the Land: Environment and Change in Post-apartheid South Africa*, edited by Mamphela Ramphele and Chris McDowell, 1–12. London: Panos.
Ramutsindela, M.F. 2002. "The Philosophy of Progress and South African Geography." *South African Geographical Journal* 84 (1): 4–11.
Ranger, Terence. 1999. *Voices from the Rocks: Nature, Culture and History in the Matopos Hills of Zimbabwe.* London: James Currey; Harare: Baobab; Bloomington: Indiana University Press.
– 2004. "Nationalist Historiography, Patriotic History and the History of the Nation: The Struggle Over the Past in Zimbabwe." *Journal of Southern African Studies* 30 (2): 214–34.
– 2007. "Reconstructing Zimbabwe's Past: The Professional Historian's Return." *Safundi* 8 (2): 257–66.
Reyner, Rebecca Hourwich. (1949) 1998. *Zulu Woman: The Life Story of Christina Sibiya.* New York: Feminist Press at City University of New York.
Rich, Paul. 1979. "Ministering to the White Man's Needs: The Development of Urban Segregation in South Africa, 1913–1923." *African Studies* 37 (2): 177–92.
Robbins, David. 1988. "Edendale Hospital." In *Pietermaritzburg, 1838–1988: A New Portrait of an African City*, edited by John Laband and Robert Haswell, 189–92. Pietermaritzburg: University of Natal Press and Shuter and Shooter.
Robbins, David, and Wyndham Hartley. 1985. *Inside the Last Outpost.* Pietermaritzburg: Shuter and Shooter.
Roberts, Benjamin, Mbithi wa Kivulu, and Yul Derek Davids, eds. 2010. *South African Social Attitudes. Second Report: Reflections on the Age of Hope.* Pretoria: Human Sciences Research Council.
Robins, Steven L. 2008. *From Revolution to Rights in South Africa: Social Movements, NGOs and Popular Politics after Apartheid.* Suffolk: James Currey; Pietermaritzburg: University of KwaZulu-Natal Press.

Robinson, Jennifer. 2006. *Ordinary Cities: Between Modernity and Development*. London and New York: Routledge.
Robinson, Sir John. 1900. *A Life Time in South Africa: Being the Recollections of the First Premier of Natal*. London: Smith, Elder and Co.
Rocheleau, Diane, Barbara Thomas-Slayter, and Ester Wangari, eds. 1996. *Feminist Political Ecology: Global Issues and Local Experiences*. London and New York: Routledge.
Rodhén, Cecilia. 2008. "Lost in Transformation: A Critical Study of Two South African Museums." PhD diss., University of KwaZulu-Natal.
Rodriguez-Torres, Deyssi, ed. 2006. *Nairobi: The Paradox of a Fragmented City*. Dar es Salaam: Mkuki na Nyota Publishers.
Roper, Chris. 2013. "Save the Rhino: Castrate a White Dude." *Mail and Guardian*, 23 January. Accessed 3 December 2013. http://mg.co.za/article/2013-01-23-00-save-the-rhino-castrate-a-white-dude.
Ross, Louisa Grace. (1864–5) 1972. *Life at Natal a Hundred Years Ago, by a Lady*. Cape Town: C. Struik.
Roukema, Janet. 2006. "Urban Environmental Governance in Msunduzi Municipality, South Africa: Getting from Policy to Practice." Master's thesis, University of Western Ontario.
Russell, Robert. 1911. *Natal: The Land and Its Story*. Pietermaritzburg: P. Davies.
Sachs, Wulf. 1937. *Black Hamlet: The Mind of the Black Negro Revealed by Psychoanalysis*. London: Geoffrey Bles.
Sai, Fred T. 2010. "Population, Growth, Reproductive Health and the Future of Africa." In *A Pivotal Moment: Population, Justice and the Environmental Challenge*, edited by Laurie Mazur, 218–26. Washington and Covelo: Island Press.
Savage, Leah. 2007. "Urban Ecosystems and Human Health in South Africa: Examining the Relationships between Housing, Energy, Indoor Air Quality and Respiratory Health." MES diss. Queen's University.
Savory, Phyllis. 1965. *Gerard Bhengu: Zulu Artist*. Cape Town: Howard Timmins.
Schneider, François, Giorgos Kallis, and Joan Martinez-Alier. 2010. "Crisis or Opportunity? Economic Degrowth for Social Equity and Ecological Sustainability. Introduction to This Special Issue." *Journal of Cleaner Production* 18 (6): 511–18.
Schnurr, Matthew A. 2013. "Cotton as Calamitous Commodity: The Politics of Agricultural Failure in Natal and Zululand, 1844–1933." *Canadian Journal of African Studies* 47 (1): 115–32.

Seethal, Cecil Emmanuel. 1993. "Civic Organizations and the Local State in South Africa (1979-1993)." PhD diss., University of Iowa.
– 1996. "Reconstruction and Postdevelopmentalism in South Africa: The Case of an Informal Settlement in Pietermaritzburg-Msunduzi (1990–1995)." *South African Geographical Journal* 78 (2): 64–74.
Seymour, R.P. 1958. *A Local Health Study: Public Health Area and District.* Pietermaritzburg: Local Health Commission.
Sheridan, Michael, and Celia Nyamweru, eds. 2008. *African Sacred Groves: Ecological Dynamics and Social Change.* Oxford: James Currey; Athens: Ohio University Press; Pretoria: University of South Africa Press.
Showers, Kate. B. 2005. *Imperial Gullies: Soil Erosion and Conservation in Lesotho.* Athens: Ohio University Press.
Sicard, S. 1981. "The '*Zanzibaris*' in *Durban*, South Africa." *Journal: Institute* of Muslim Minority Affairs 3 (1): 128–37.
Sithole, Jabulani. 2014. "uMgungundlovu, a Capital of a Province Rich in Liberation Heritage." In *uMgungundlovu Liberation Heritage Route.* Msunduzi: uMgungundlovu District Municipality.
Slater, Henry. 1980. "The Changing Pattern of Economic Relationships in Rural Natal, 1838–1914." In *Economy and Society in Pre-Industrial South Africa*, edited by Shula Marks and Anthony Atmore, 148–70. London: Longman.
Soske, Jon. 2009. "'Wash Me Black Again': African Nationalism, the Indian Diaspora and KwaZulu-Natal, 1944–1960." PhD diss., University of Toronto.
– 2012. "Unravelling the 1947 'Doctors' Pact': Race, Metonymy and the Evasions of Nationalist History." In *One Hundred Years of the ANC: Debating Liberation Histories Today*, edited by Arianna Lissoni, Jon Soske, Natasha Erlank, Noor Nieftagodien, and Omar Badsha, 163–90. Johannesburg: Wits University Press.
South Africa. 1917. *First Report of the Select Committee on Native Affairs.* Cape Town: Cape Times Ltd., Government Printer.
– 1932. *Report of the Native Economic Commission, 1930–1932.* John Edward Holloway for the South Africa Native Economic Commission. 1932. Pretoria: Government Printer.
– 1940. *Administration of Areas Which Are Becoming Urbanised but Which Are Not Under Local Government Control.* Pretoria: Government Printer.
– 1960. *Records of the Natal Executive Council Part 1, 1846–1848.* Cape Town: National Comercial Printers for Office of the Chief Archivist.
– 1999. *Truth and Reconciliation Commission of South Africa Report.* Vol. 3. London: Macmillan.

– 2015. "South Africa's Intended Nationally Determined Contribution." Accessed 20 January 2016. https://www.environment.gov.za/sites/default/files/docs/sanational_determinedcontribution.pdf.
South African Native Affairs Commission (SANAC). 1905. *Report, with Annexures Nos. 1–9*. Vol. 3. Cape Town: Cape Times Ltd.
SRK Consulting. 2010. *Msunduzi Final Draft Strategic Environmental Assessment*. Pietermaritzburg: SRK Consulting.
Statham, F. Reginald. (1881) 1969. *Blacks, Boers, and British: A Three-Cornered Problem*. New York: Negro University Press.
Steady, Filomena Chioma, ed. 2009. *Environmental Justice in the New Millennium: Global Perspectives on Race, Ethnicity, and Human Rights*. New York: Palgrave Macmillan.
Stock, Robert. 1988. "Environmental Sanitation in Nigeria: Colonial and Contemporary." *Review of African Political Economy* 15 (42): 19–31.
Stolten, Hans Erik. 2007. *History Making and Present Day Politics: The Meaning of Collective Memory in South Africa*. Uppsala: Nordiska Afrikainstitutet.
Sunseri, Thaddeus. 2009. *Wielding the Ax: State Forestry and Social Conflict in Tanzania, 1820–2000*. Athens: Ohio University Press.
Surplus People Project. 1983. *Natal*. Vol. 4 of *Forced Removals in South Africa*. Cape Town and Pietermaritzburg: University of Natal.
Sutton, Colleen Marianne. 2008. "Urban Open Space: A Case Study of Msunduzi Municipality, South Africa." Master's thesis, Queen's University.
Swanson, Maynard. 1965. "The Rise of Multiracial Durban." PhD diss., Harvard University.
– 1976. "'The Durban System': Roots of Urban Apartheid in Colonial Natal." *African Studies* 35 (3–4): 159–76.
– 1977. "The Sanitation Syndrome: Bubonic Plague and Urban Native Policy in the Cape Colony, 1900–1909." *Journal of African History* 18 (3): 387–410.
– 1983. "'The Asiatic Menace': Creating Segregation in Durban, 1870–1900." *International Journal of African Historical Studies* 16 (3), 401–21.
– 1990. "Changing the Guard: Champion-Msimang-Luthuli." In *Collected Seminar Papers on the Societies of Southern Africa in the 19th and 20th Centuries*. Vol. 16: 50–6. London: Institute of Commonwealth Studies.
– 1996. "The Joys of Proximity: The Rise of Clermont." In *The People's City: African Life in Twentieth Century Durban*, edited by Paul Maylam and Iain Edwards, 274–98. Portsmouth, NH: Heinemann; Pietermaritzburg: University of Natal Press.

Taylor, Marcus. 2015. *The Political Ecology of Climate Change Adaptation: Livelihoods, Agrarian Change and the Conflicts of Development*. London and New York: Routledge.
Thompson, Paul. 1988. "The Pietermaritzburg Voter and Parliamentary Elections." In *Pietermaritzburg, 1838–1988: A New Portrait of an African City*, edited by John Laband and Robert Haswell, 198–200. Pietermaritzburg: University of Natal Press and Shuter and Shooter.
Thorrington-Smith, Eric, et al. 1973. *Pietermaritzburg: A Town Planning Report for the Borough*. Pietermaritzburg and Durban: Thorrington-Smith, Rosenberg, and McCrystal Consultants.
Timberg, Craig, and Daniel Halperin. 2012. *Tinderbox: How the West Sparked the AIDS Epidemic and How the World Can Finally Overcome It*. New York: Penguin.
Torr, Louise, 1996. "Lamontville: A History, 1930–60." In *The People's City: African Life in Twentieth Century Durban*, edited by Paul Maylam and Iain Edwards, 245–73. Portsmouth, NH: Heinemann; Pietermaritzburg: University of Natal Press.
Tropp, Jacob A. 2006. *Nature of Colonial Change: Environmental Relations in the Making of the Transkei*. Athens: Ohio University Press.
Truluck, Anne (researcher and compiler). 1990. *Pietermaritzburg 1990: The Fractured City*. Pietermaritzburg: Black Sash.
UN-Habitat. 2014. *The State of African Cities, 2014: Re-imagining sustainable Urban Transitions*. Nairobi: United Nations Human Settlements Programme.
– 2015. *Towards an African Urban Agenda*. Nairobi: United Nations Human Settlements Programme and UN Economic Commission for Africa.
University of Natal Department of Economics. 1951. *Experiment at Edendale: A Study of a Non-European Settlement with Special Reference to Food Expenditure and Nutrition*. Pietermaritzburg: University of Natal Press.
van Diemel, Raymond. 2002. *In Search of "Freedom, Fair Play and Justice": Josiah Tshangana Gumede, 1867–1947: A Biography*. Belhar, South Africa: R. van Diemel.
van Klinken, Adriaan S. 2013. *Transforming Masculinities in African Christianity: Gender Controversies in Times of AIDS*. Farnham, UK and Burlington, VT: Ashgate.
van Sittert, Lance. 1998. "Keeping the Enemy at Bay: The Extermination of Wild Carnivora in the Cape Colony, 1889–1910." *Environmental History* 3 (3): 333–56.

Verwey, Elizabeth J., and Nelly E. Sonderling, eds. 1995. *New Dictionary of South African Biography*. Vol. 1. Pretoria: Human Science Research Council.
Vilakazi, Absolom. 1962. *Zulu Transformations: A Study of the Dynamics of Social Change*. Pietermaritzburg: University of Natal Press.
Walker, Cherryl, ed. 1990. *Women and Gender in Southern Africa to 1945*. Cape Town: D. Philip.
Ward, Kim, ed. 2013. *The DUCT Story, 2005–2013*. Scottsville: Duzi-uMgeni Conservation Trust.
Watson, Vanessa. 2014. "African Urban Fantasies: Dreams or Nightmares?" *Environment and Urbanization* 26 (1): 215–31.
Webb, Colin de B., and John Wright, eds. 1986. *The James Stuart Archive of Recorded Oral Evidence Relating to the History of the Zulu and Neighbouring Peoples*. Vol. 4. Pietermaritzburg: University of Natal Press; Durban: Killie Campbell Africana Library.
Wedderburn, Jeanne Inel. 1991. "Geography and Art: The Pictorial Record of Nineteenth Century Pietermaritzburg." Master's thesis, University of Natal-Pietermaritzburg.
Wells, Julia C. 2008. "Are We Nation Building Yet? The Role of Oral Historians in Documenting the Transition Out of Apartheid." In *Oral History in a Wounded Country: Interactive Interviewing in South Africa*, edited by Phillippe Denis and Radikobo Ntsimane, 22–42. Pietermaritzburg: University of KwaZulu-Natal Press.
Welsh, David. 1973. *The Roots of Segregation. Native Policy in Colonial Natal, 1845–1910*. 2nd ed. Cape Town: Oxford University Press.
Whelan, Debbie. 2012. "Whose Colony, and Whose Legacy? Layers of Power and Hybrid Identities in Edendale, Pietermaritzburg, South Africa." In *Colonial Architecture and Urbanism in Africa: Intertwined and Contested Histories*, edited by Fassil Demissie, 107–26. Farnham, UK: Ashgate.
White, Luise. 1990. *The Comforts of Home: Prostitution in Colonial Nairobi*. Chicago: University of Chicago Press.
Whitelaw, Gavin. 2009. "A Brief Archaeology of Precolonial Farming in KwaZulu-Natal." In *Zulu Identities: Being Zulu, Past and Present*, edited by Benedict Carton, John Laband, and Jabulani Sithole, 47–61. Scottsville: University of KwaZulu-Natal Press.
Wills, Trevor M. 1988a. "The Segregated City." In *Pietermaritzburg, 1838–1988: A New Portrait of an African City*, edited by John Laband and Robert Haswell, 33–44. Pietermaritzburg: University of Natal Press and Shuter and Shooter.

– 1988b. "From Rickshaws to Minibus Taxis." In *Pietermaritzburg, 1838–1988: A New Portrait of an African City*, edited by John Laband and Robert Haswell, 138. Pietermaritzburg: University of Natal Press and Shuter and Shooter.
– 1991. "Pietermaritzburg." In *Homes Apart: South Africa's Segregated Cities*, edited by Anthony Lemon, 90–103. London: Paul Chapman; Bloomington: Indiana University Press.
– 1994. "Segregation, Separation and Desegregation: Pietermaritzburg since 1910." In *Receded Tides of Empire: Aspects of the Economic and Social History of Natal and Zululand since 1910*, edited by Bill Guest and John Sellers, 283–309. Pietermaritzburg: University of Natal Press.
Wilson, Monica, and Archie Mafeje. 1963. *Langa: A Study of Social Groups in an African Township*. Cape Town: Oxford University Press.
Witt, Harald. 2002. "The Emergence of Privately Grown Industrial Tree Plantations." In *South Africa's Environmental History: Cases and Comparisons*, edited by Stephen Dovers, Ruth Edgecombe, and Bill Guest, 90–111. Cape Town: David Philip.
– 2005. "'Clothing the Once Bare Brown Hills of Natal': The Origin and Development of Wattle Growing in Natal, 1860–1960." *South African Historical Journal* 53: 99–122.
Wolmer, William. 2007. *From Wilderness Vision to Farm Invasions: Conservation and Development in Zimbabwe's South-East Lowveld*. Oxford: James Currey; Harare: Weaver Press.
Wright, John. 1988. "Before Mgungundlovu." In *Pietermaritzburg, 1838–1988: A New Portrait of an African City*, edited by John Laband and Robert Haswell, 18–21. Pietermaritzburg: University of Natal Press and Shuter and Shooter.
Wright, Marcia. 2006–07. "Public Health among the Lineaments of the Colonial State in Natal, 1901–1910." Journal of Natal and Zulu History 24–5 (1): 135–63.
Wylie, Diane. 2001. *Starving on a Full Stomach: Hunger and the Triumph of Cultural Racism in Modern South Africa*. Charlottesville: University Press of Virginia.
Zondi, Phumani. 2006. "Greater Edendale Land Summit 2006: The Role of Private Landowners as Developers." Unpublished presentation, GEDI, 9 June.
Zuma, Nosipho M. 2002. "Implementing and Regulating Air and Water Pollution in Pietermaritzburg." Master's thesis, University of Natal.

# Index

*Page numbers in italics refer to figures*

Printed and bound by CPI Group (UK) Ltd, Croydon, CR0 4YY

15/09/2026

14962458-0001